RECENT DEVELOPMENTS IN MOBILE COMMUNICATIONS – A MULTIDISCIPLINARY APPROACH

Edited by **Juan P. Maícas**

Recent Developments in Mobile Communications – A Multidisciplinary Approach
Edited by Juan P. Maícas

Published by InTech

Janeza Trdine 9, 51000 Rijeka, Croatia

Publishing Process Manager Gorana Scerbe

Technical Editor Teodora Smiljanic

Cover Designer InTech Design Team

Image Copyright SoulCurry, 2011. Used under license from Shutterstock.com

Edition 2014

Additional hard copies can be obtained from orders@intechopen.com

Recent Developments in Mobile Communications – A Multidisciplinary Approach,
Edited by Juan P. Maícas
p. cm.
ISBN 978-953-307-910-3

For EU product safety concerns:
IN TECH d.o.o., Prolaz Marije Krucifikse Kozulić 3, 51000 Rijeka, Croatia,
info@intechopen.com or visit our website at intechopen.com.

Meet the editor

Dr. Juan P. Maícas is currently an Associate Professor of Strategy and Associate Dean for International Affairs at the Faculty of Economics and Business of the University of Zaragoza (Spain).

Professor Maícas's research investigates how firms compete in IT industries and the sources of competitive advantage. To date, his research has focused especially on competition in the mobile communications industry, e-business strategy and market entry timing and first-mover advantages. Professor Maícas's research has been published in leading outlets – e.g. Strategic Management Journal, Information Systems Research, Journal of Information Technology, Telecommunications Policy – and has been awarded several research grants and honours.

Contents

Preface

Recent Developments in Mobile Communications - A Multidisciplinary Approach aims to serve as a multidisciplinary reference for the mobile telecommunications industry. The intention of this book is to offer both comprehensive and up-to-date surveys of recent developments and the state-of-the-art of various economical and technical aspects of mobile telecommunications markets. The book will be useful for scholars, policy makers, engineers and managers within the industry.

The chapters of this book seek to reflect the principal topics in the field that have either received intensive research attention or are in need of an integrative survey. They can be read independently, though many offer a complementary view. Although there is no clear relationship between all the chapters, it seems convenient to structure the material into economic-oriented and technical-oriented.

The economic-oriented part begins with the chapter by Karamti and Kammoun, who study the effect of privatization, competition and regulation on the performance of mobile service providers. Their main findings suggest that competition is associated with increased penetration and lower prices. Privatization in itself may have little effect - it needs to be combined, for instance, with an independent regulator, to have the desired effects.

Castaldi, Addeo, Massaro and Mazzoni (Chapter 2) analyse the impact of servitization on mobile phone demand using a multidimensional segmentation framework (LAM model). Their results show that the LAM model could be fruitfully applied to market segmentation.

In Chapter 3, Zeiss, Davies and Pospischil offer an overview of Wholesale Applications Community (WAC) and compare it to existing mobile app eco-systems and technologies. They investigate the details of WAC with respect to its API definition, runtime implementation, technologies used and SDK-based development.

Martín López, Monforte Nicolás and Merino Moreno (Chapter 4) describe the current state of the telecommunications market and its shortages in terms of the generation of a future-proof business model that can give both market and financial sense to the new investments planned for the creation of the Next-Generation Networks (NGN) that will lead to 4G and beyond.

In Chapter 5, Elaluf-Calderwood, Eaton, Herzhoff, and Sorensen present a framework for analysing value networks. They provide a complementary analysis of mobile platforms as convergent systems: the composition of the ecosystem, its interactions and the relationships and conflicts that are generated by those interactions. This chapter presents a potential framework for the analysis of the composition of both ecosystems and value networks using tussles and control points.

In Chapter 6, Vilares and Coelho develop an integrated methodology for estimating the return on quality investments applied to the mobile telecommunications industry. The goal of this work is to propose a methodology that will enable the identification of profitable quality investments. This includes identifying the profitability of different levels of investments and prioritizing alternative investments, taking into account their return or profitability.

Maícas and Sesé, in Chapter 7, provide an overview of the current state of the networks effects literature, highlighting their role in the mobile communications industry in order to gain a better understanding of one of the key forces that underlie the development of these markets. They review the relevant works in the field of network effects and network markets, offer empirical support for the conceptual arguments and identify how network effects operate in mobile communications.

The technical-oriented part of the book begins with Chapter 8, where Professor Secmen describes crucial antenna parameters for mobile communication systems, offering types and examples of multiband antennas used in mobile communication. He also investigates wideband antennas for mobile communication applications.

In Chapter 9, Aquino-Santos, Guerrero-Ibáñez and Edwards-Block analyse two emerging wireless technologies that can be employed in RVC communication, IEEE 802.1p and IEEE 802.16-2004, in an urban scenario. The work compares simulations of these two prominent technologies for roadside communication. Main findings show that IEEE 802.16-2004 technology outperforms IEEE 802.11p technology in terms of throughput, packet loss and end-to-end delay.

Minhas, Mahmood and Malik (Chapter 10) discuss the role of ad hoc networks in mobile telecommunications. Present trends in mobility are presented with respect to technological, application and social aspects, which inherently influence telecommunication network design, and future trends in technology and its usage are addressed.

Gotsis and Sahalos, in Chapter 11, introduce 'DoA-based Switching (DoAS)' and its incorporation into a 3G mobile communications framework, e.g. the Universal Mobile Telecommunications System (UMTS). They evaluate and compare DoAS to 'Typical Switching (TS)', which refers to the typical operation of a switched-beam system (SBS), which is determined by the highest uplink Signal to Interference Noise Ratio (SINR) or by the highest mean received power.

Chapter 12 introduces a discussion by Lo and Guan whose objective is to leverage and combine the benefits of Coordinated MultiPoint transmission and reception (CoMP) and relaying in order to improve the performance of cell-edge users, which is severely degraded by inter-cell interference. The authors suggest that the joint relaying and CoMP technique can yield performance gains beyond what can be achieved by using either of the techniques alone.

In Chapter 13, Tostes, Duarte-Figueiredo and Zárate highlight that Call Admission Control (CAC) is one of the most important systems for a cellular network. The authors describe conventional and unconventional CACs. The chapter presents both types and concludes that it is not possible to say which is the best: it depends on the user and operator demands.

He, Du and Hu (Chapter 14) propose water quality monitoring as an essential tool and foundation in water resource management and water pollution supervision and control by government administrations. The chapter focuses on water quality sensors based on ion-specific electrodes, water quality dynamic monitoring systems based on web-server-embedded technology and their application to rivers and freshwater detection, aquaculture production and hydroponic plant production.

Dr. Juan P. Maícas
Associate Professor of Strategy,
Universidad de Zaragoza,
Spain

Part 1

Economic Oriented

1

Privatization, Reforms and Firm's Performance in Mobile Telecommunication Industry

Chiraz Karamti[1,2] and Aida Kammoun[1]
[1]*Higher Institute of Business Administration (ISAAS)*
[2]*In Telecom ParisTech*
Tunisia

1. Introduction

Since the introduction of mobile telephony in the early fifties in Europe, US and Japan, the demand for this service exploded. Actually there are countries that have a market penetration of more than 100 per cent. This dramatic growth in the mobile telecommunications industry can be, at least partly, attributed to the growing trend toward privatization (defined as the sale of total or partial previously state-owned enterprises to private owners), market liberalization and deregulation.

In Europe, while most national markets were monopolies in the late 1980s, by today, most of them have three or more competing mobile networks. The telecommunications reforms reflect changes in technology that might mitigate the reliance on government interventions and affect our understanding of the effects of these interventions. Many several studies have shown the impacts of telecommunications reforms on the market outcome and firms' performance. On the one hand, privatization is said to increase the incumbent's operational efficiency by reducing political control; on the other hand, it may very well deter market competition since the incumbent is able to engage in anti-competitive behaviors. The public incumbent, instead, due to political oversight may suffer from inefficient operation in competing with the rivals. Many studies point out that the privatization will produce the greatest efficiency gains where competition replaces monopoly. When both private and public firms are exposed to the same competitive pressures and market signals, they are expected to yield similar performance in terms of allocative efficiency, regardless of their ownership structure.

The introduction of competition, breaking up or unbundling monopolies, and the privatization of state-owned telecommunications operators have become the main themes of telecommunications sector reform in developing and developed countries. These reforms might result in a falling of telecommunication prices, a significant expansion of telecommunications networks and a substantial improvement in productivity. This study attempts to uncover evidence on these effects. Several recent econometric studies have examined the effect of telecommunications reform on sector performance, especially for European countries. The majority of these studies consider that competition on its own, and complementarities between competition and privatization, are positively correlated with telecommunications industry performance.

Although there has been much empirical research on the effects of privatization, competition and regulation on the telecommunications sector, very little empirical work was interested in studying these effects on mobile telecommunication sector. This chapter studies the effects of telecommunication reforms (privatization, competition and regulation) on mobile operator's performance in the OCDE area.

2. The dilemma of privatization

Many economists, policy makers and corporate managers have long believed that private firms are more efficient than public ones. Privatization, defined as the sale of (total or partial) previously state-owned enterprises to private owners, is, so, assumed to increase the firm's efficiency and profitability because, on the one hand, the change in ownership structure shifts the privatized firm's objectives and the managers' incentives away from those imposed by politicians. The managers are then subordinate to them on monitoring and discipline of profit oriented investors. On the other hand, privatization may very well deter market competition since the incumbent is able to engage in anti-competitive behaviors. The public incumbent, instead, due to political oversight may suffer from inefficient operation in competing with the rivals.

As a result, since the late of 1980s, several countries have undergone partial or full privatization of their utility sectors, especially telecommunications. In fact, until recently, in most countries, telecommunications service providers were state owned, state operated, and often monopolistic. The telecommunications sector was viewed as the quintessential public utility. Economies of scale, combined with political sensitivity, created large entry barriers and externalities. Since the 1980s, policy makers gradually began to recognize that telecommunications systems are an essential infrastructure for economic development. As the economy broadens and becomes critically dependent on vastly expanded flows of information, telecommunications acquires strategic importance for economic growth and development. Besides, rapid technological innovations in the past three decades have significantly reduced economies of scale and scope in this sector, attenuating the economic rationale for a state-owned natural monopoly in the Telecommunications sector. The solution was privatization which aims to break the monopoly and improve the efficiency and performance of the telecommunication industry.

Theoretically, privatization affects the firm's performance through multiple channels. It might cause firms to operate more productively because managers are subjected to the pressures of the financial markets and to the monitoring and discipline of profit-oriented investors. In addition, the change in ownership structure of privatized firms shifts the firm's objectives and managers' incentives away from those that are imposed on them by politicians, toward those that aim to maximize efficiency, profitability, and shareholders' wealth. By going public, firms would have many entrepreneurial opportunities because they would not be subject to government control (D'Souza, Megginson, & Nash, 2007).

Furthermore, Hartley and Parker (1991) developed a conceptual framework based on property rights and public choice approaches, in order to show that privatized firms are more efficient than SOEs because profit motivation is absent for public firms. This is why many authors found that privatization leads to significant improvements in the availability and quality of telecommunications services. In fact, privatization leads to network expansion and modernization of Telecommunications services.

In contrast to the aforementioned literature, which concludes that ownership does matter under competitive environments, other researchers pay more attention to the role of competition rather than ownership per se. The reduction in government ownership is not, in fact, the only factor that improves the performance of privatized firms. The competitive environment and capital-market discipline also increase the efficiency of these firms (Castro & Uhlenbruck, 1997). In this context, policy makers suggest that competition can greatly improve monitoring possibilities and hence increase incentives for production efficiency. Thus, it follows that private firms are more efficient than SOEs in competitive environments. However, in noncompetitive industries or in industries with natural monopoly elements, the performance of privatized firms is ambiguous, and results from empirical studies are inconclusive (Boubakri & Cosset, 1998). Vining and Boardman (1992) argue that at low levels of competition, the differences between public and private ownership would be insignificant, as both types of firms would adopt similar rent seeking behavior. When competition increases, however, private ownership offers incentives and motivation for managers to proactively adopt profit-maximizing behavior. In addition, D'Souza and Megginson (1999) indicate that privatized firms that work in competitive industries are likely to yield solid and rapid economic benefits as long as there are no economy wide distortions that hinder competition. Parker and Hartley (1991) point out privatization will produce the greatest efficiency gains where competition replaces monopoly.

When both private and public firms are exposed to the same competitive pressures and market signals, they are expected to yield similar performance in terms of allocative efficiency, regardless of their ownership structure (Fare, Grosskopf, & Logan, 1985). In the same vein, Forsyth (1984, p. 61) states, "Selling a government firm makes no difference to the competitive environment in which it operates; ownership and competitive structure are separate issues." Newbery (1999) proposed that the emphasis should be placed on breaking up monopolies before privatization. Omran (2004) further indicates that, due to spillover and learning effects, the performance of state owned enterprises does not depart significantly from that of their privatized counterparts once they anticipate later privatization and competition in the sector.

It appears that the importance of establishing an institutional framework, i.e., regulation and competition, before privatizing firms has been emphasized. So, the sequence of the telecommunication reform might affect the outcome of market competition, that is, the time and extent to which the incumbent monopoly is shattered. When privatization comes before competition, a monopoly can attract foreign investment more easily, leading to successful privatization, because the returns from investment are guaranteed by the100% market share. In this sense, the state as an owner is tempted to delay competition in exchange for the higher capitalization value of the firm during privatization (Bauer, 2003, p.12). Even if competition is allowed in a later period of time, the firm is still able to consolidate its market share since it possesses the network effects inherent in its large net work and is more likely to engage in anti-competitive behaviors in this asymmetric market (Rey&Tirole, 2007). On the other hand, the cost for the competitive rivals to challenge the established incumbent, such as interconnection charges and negotiation costs, is so formidable that they have difficulty becoming significant market players.

Seen otherwise, many papers[1] suggest that privatization without a simultaneous introduction of competition will simply create private monopolies. Most economists therefore argue that privatization works better when there is competition that limits the

[1] See for example George Yarrow, Privatization in Theory and Practice, Econ. Policy 324 (1986); J. A. Kay & D. J. Thompson, Privatization: A Policy in Search of a Rationale, 96 Econ. J. 18 (1986); Vickers & Yarrow, Privatization: An Economic Analysis (1995); World Bank, Bureaucrats in Business (1995)

market power of the incumbent(s). The paper of Chorng-Jian Liu and al. (2009) does highlight a dilemma in telecommunication reforms, in that a not-yet-privatized incumbent under market competition can no longer dominate the market but is turned into an inefficient operation. Authors call for a rethinking of telecommunications development theory that overlooks the importance of the sequencing between privatization and liberalization. Indeed, the timing of privatization affects the speed and the degree to which a monopolistic market is transformed into a competitive one. Competition and privatization are thus seen as complementary. Besides, Product market competition is a potent force that improves performance in its own right. It tends to weed out inefficient firms, if they face hard budget constraints. The threat of bankruptcy may compel existing operators to be more efficient so as to minimize the probability of a corporate failure. Since state-owned firms rarely operate under hard budget constraints, the positive impact of market competition on performance is more likely to be present in privatized firms, further suggesting a complementarity between privatization and competition.

As a conclusion, the dilemma of privatization set a particular attention on how the degree of privatization and competition affects performance and how components of the policies interact with each other in shaping the reform outcomes. For instance, does full privatization improve the performance of a country's telecommunications sector more than partial privatization? Is privatization (or competition) alone sufficient in improving economic performance, or are privatization and competition complementary policies? And finally, how do privatization and competition affect performance measures?

Policy makers suggest that there is a strong presumption that privatization and competition in the telecommunications sector improve economic performance. Whether this presumption is true remains largely an empirical question.

3. Corporate performance: Theory and evidence

Privatization, seen as an important economic phenomenon, has attracted much attention from academic researchers and policy analysts. Studies document generally that moving companies from state to private ownership improves firm performance. However, majority of these studies show that privatization works better when it is accompanied by other major institutional and legal reforms. These two lessons represent important contributions to economic thought and help to support the emerging consensus among policy-makers about how best to use privatization as a tool for promoting economic development.

Begin first with a definition of the performance concept. Based on research long-rooted in the management discipline, performance can be defined as the accomplishment of a given task measured against preset standards of accuracy, completeness, cost, and speed. Firm performance is measured against standard or prescribed indicators of efficiency, effectiveness, and environmental responsibility (Duty or obligation to satisfactorily perform or complete a task) such as, cycle time, productivity, regulatory compliance… etc. **Efficiency** means the comparison of what is actually produced or performed with what can be achieved with the same consumption of resources (money, time, labor, etc.). It is an important factor in determination of productivity. **Effectiveness** is relative to the degree to which objectives are achieved and the extent to which targeted problems are resolved. In contrast to efficiency, effectiveness is determined without reference to costs and, whereas efficiency means "doing the thing right," effectiveness means "doing the right thing."

During the last two decades important structural policies have taken place worldwide in the telecommunication sector. A significant number of studies attempt to assess the consequences of the aforementioned changes. Two types of analysis are usually encountered in the relevant literature: empirical econometric analyses and descriptive analyses. These research papers mainly examine the consequences of the telecommunication market reform and the corporate restructuring of traditional telecommunication organizations, which were fully or partly privatized through public offer or through direct sale to one or more investors. Studies on the privatization and performance of telecommunication industry started in the early 1980s. Many papers investigated the effects of privatization and competition on the expansion and performance of telecommunication network.

Results from the study of Wallsten (2002) reveal the correlation between privatization, competition, regulation, and performances of telecommunication industry in 30 Latin American and African countries. Fink et al. (2002) examined the effects of national policy reform in the telecommunication sectors of 86 countries and found that both privatization and competition can lead to significant improvement in telecommunication performance.

Few studies analyze the impact of public enterprise reform on profitability, productivity, exports, budgetary impacts, crowding out of the private sector, etc. Moreover, many of the studies also suffer from basic methodological deficiencies. For example, using cross-sectional data, Foreman-Peck and Manning (1988) conducted total factor productivity analyses to compare the performance of British Telecom (BT), which was privatized in 1984, with the performance of five telecom firms in Europe. They concluded that British Telecom is apparently less efficient than the companies in Norway and Denmark, but more efficient than those in Spain and Italy. Their finding is inconclusive, however, since ownership is by state in Norway, but mixed in Denmark, Spain and Italy. This methodology is incapable of linking variations in performance with the change in the company's ownership.

Several sector specific studies have also been conducted on the outcome of reforming telecommunications services, albeit in developed economies (Takano, 1992; Oniki et al., 1992; Imai, 1994; Foreman-Peck, 1991). The study of Foreman-Peck (1991) examined whether the transformation in the telecommunications sector altered or improved performance over that of the previous state regime. Results suggest a substantial improvement in the productivity performance of the telecommunications industry after privatization. Takano (1992) examined the process, as well as benefits and losses stemming from the partial privatization of Nippon Telegraph and Telephone Corporation (NTT), a government monopoly producer of domestic telecommunications services in Japan. The study evaluated the benefits to four important actors: NTT proper, stockholders, users and government. Oniki et al. (1992) assessed the impact of deregulation on NTT through improved management and operations by estimating a translog variable cost function for 1983- 1989 fiscal years. According to the study, deregulation resulted in a cost reduction of 1.31 or 2.29%, depending on the specification of the cost function adopted. In the same vein, Imai (1994) estimated the cost reduction associated with the 1985 deregulation of international telephone services in Japan. The study estimated that NTT's unit cost fell by a wide margin after deregulation (54.5%).

Many studies in the telecommunications sector seek to explore the regulatory institutions of different countries using the new institutional economics. Levy and Spiller (1996) conducted a comparative analysis of the impact of core political and social institutions on regulatory structures and performance in the telecommunications industry in Jamaica, the United

Kingdom, Chile, Argentina and the Philippines. The study examines the relationship between regulatory outcomes and performance, and how each country resolved its regulatory problems.

Galal and Nauriyal (1995) explored the relationship among the outcomes of regulatory reforms, regulatory incentives and government commitment on the basis of the recent regulatory experience of seven developing countries: Argentina, Chile, Jamaica, Malaysia, Mexico, the Philippines and Venezuela. They attempt to link the performance of the telecom sector with the extent to which these countries successfully resolved the information asymmetry, pricing and contracting problems. Results show that the sector continues to suffer from under-investment and low productivity. Other countries had mixed results.

The majority of these studies were interested on telecommunication firm performance without differentiating between mobile and fixed telephone activity. However, the reform of the sector has caused the rapid increase in the proportion of penetration of novel telecommunication services, most notably mobile telephony and the Internet (Clarke and Gebreab et al. 2003, Ypsilantis 2002, Xavier and Ypsilantis 2001). It is important to note that the proliferation of main telephone lines, which appears more intense during the initial years of the reform, is reduced after the full liberalisation of the market due to the intense American Economic Review 91, 320–334.

Doove S., Gabbitas O., Nguyen-Hong D. and J. Owen, 2001. Price Effects of Regulation that develops in the mobile telephony market impacts positively on the levels of productivity (Fink and Mattoo et al 2003). Similarly, the increase in production, mostly expressed in terms of phone call flows, increases the productivity index. Moreover, the reduction in the number of employees in traditional telecommunications organizations promptly increases work productivity (Dia and N' Guessan et al 2002). Ypsilantis and Min (2001) as well as Sacripanti (1999) observe a greater reduction of prices in mobile telephony services due to the more intense competition in these markets. Ypsilantis and Min (2000) examine the percentage of successful calls and the proportion of access in order to examine the quality of mobile telephony services, and conclude that the reform is positively associated with the quality of services in mobile telephony. Nevertheless, in some cases, the quality of services on offer showed no indications of improvement, despite the reforms of the sector, thus remaining at the approximate level before the reform. Bernardo Bortolotti and al. (2002) use the number of licensed operators in the mobile (analogue and digital) telephony market as a proxy for product market competition in 25 national markets involved. They were interested in measuring the competitive pressure faced by the privatized companies, so they refer only to operators not owned by the incumbents.

The paper of Chorng-Jian Liu and al. (2009) explores the factors that hamstrung Chunghwa Telecom in competition against its rival entrants. The econometric analysis substantiates the fact that handset subsidies are the most effective instrument for mobile firms to gain market share. Chunghwa Telecom, due to its public ownership status, was nevertheless prohibited at first from adopting such a marketing strategy. The empirical results pinpoint the importance of the sequencing of reforms in telecommunications: a prolonged privatization could help to promote competition in the industry. Public ownership makes Chunghwa Telecom vulnerable to political intervention and operational inefficiency, which is a barricade to performance and competitiveness for the not-yet-privatized company in a liberalized market. Taiwan's case paves a shortcut to successful implementation of telecommunication reform in a timely fashion.

The paper of Zheng, S., & Ward, M.R., (2011) studies the effects of competition and privatization on Chinese telecommunications performance, using panel data. First, mobile service has become the dominant platform for service. Over the sample, mobile calling volume went from less than half to almost three times that of fixed service. Second, growing income levels contributed to this shift. Higher income is estimated to be associated with increased demand for mobile service and decreased demand for fixed service. Third, a significant portion of the mobile price reductions are due to greater within mobile platform competition. Fourth, there is some evidence that the movement toward private versus state ownership also contributed to this transition. Privatization is associated with lower mobile usage prices and higher usage levels. However, it is associated with higher fixed prices and reduced fixed demand.

4. Reforms and special issues on the mobile communications sector

4.1 Telecommunications policy reform

Three dimensions of public policy reforms are relevant and have been applied in developing and developed countries: a change of ownership, an introduction of competition, and a strengthened regulation.

1. The first telecommunication reform strategy implemented by states in renovating the sector is often to privatize the national telecommunications provider. By selling off a controlling interest in the national telephone company, political leaders hope to expose the organization to market pressures for efficiency and profit. However, privatization without a simultaneous introduction of competition will simply create private monopolies. Most economists therefore argue that privatization works best when there is competition that limits the market power of the incumbent(s). Competition is thus seen as a complement to privatization (Xu and Li, 2002).
2. Hence, the second strategy is to break the provider's domestic monopoly over the consumer services market. The objective of liberalization is to induce competition in prices, creating incentives to lower production cost and increasing product innovation (Nicoletti G. and Scarpetta S., 2003). Since the beginning of 1998 a number of European Union member countries opened their mobile telecommunication markets to full infrastructure and service competition by allowing competition for public voice infrastructures and services[2]. Despite this market openness, many governments maintain the two roles as industry regulator and players by holding shares or directly competing on the mobile market. This may, on the one hand, very well deter market competition since the incumbent is able to engage in anti-competitive behaviors. On the other hand, because of privatization, governments can no longer overtly affect company decisions. But they often appeal to the public interest in order to stay politically engaged via weak regulatory structures. As a result, regulation is considered as a form of state involvement (Latzer et al, 2006).
3. Consequently, the third common reform is to insure the regulatory independence. This occurs when the regulatory body is separate from and not accountable to, any supplier of basic telecommunications services. Most OECD countries defined the "independence" of the regulator as a separation from day-to-day political interference,

[2] In addition to this opening of national telecommunication markets, was the agreement to liberalize international trade in basic telecommunications.

and independence of decision making based on powers vested in the regulatory body. However, for many years in many countries, the regulatory body is kept attached to the Ministry and the regulator is under the direct supervision of the executive of government. Hence, both scholars and international institutions advocate for the establishment of independent regulators, which means professionalizing the staff making decisions about telecommunications policy and appointing technocrats instead of political leaders to senior positions (Howard, Mazaheri, 2008). Some countries devote large resources to establish independent regulatory agencies, in compliance with the directives of the World Bank, the OCDE and the European Union.

Since there is evidence that policy reforms may have an impact on operators' performance, answering the questions above must involve assessing the effects of each of these three policy reforms on performance indicators.

4.2 Special issues on the mobile communications sector

Apart from the distribution of licenses[3] (for GSM and/or UMTS) and related questions on infrastructure sharing, several topics regarding mobile telecommunications were or still problematic for National Regulator Agencies (NRAs) and the European Union: roaming, Mobile Number Portability (MNP), Mobile Termination rates (MTR), universal services access requirements and mobile Virtual Network Operators.

4.2.1 International roaming charges

The first Regulation on international roaming services was published on 29 June 2007. The definitive text of Regulation (EC) No 544/2009 was published in the Official Journal of the European Union on 29 June 2009. Regulation on international roaming requires all operators in the EU to offer customers regulated voice and SMS retail roaming tariffs, which must comply with maximum price caps (known as the Eurotariff). The Regulation also provides that operators may offer alternative, i.e. unregulated, retail roaming tariffs alongside. Under the 2009 Regulation, the average wholesale roaming voice charge must be calculated on a per second basis, adjusted to take account of the possibility for the operator of the visited network to apply an initial minimum charging period not exceeding 30 seconds. This has led to a significantly lower surcharge in EU countries, from around 21% in Q2 2009 to around 6% in Q2 2010. Considering "Rest of World" retail voice roaming calls, typical prices are significantly greater than for calls wholly within EU/EEA. Overall, average Eurotariff retail voice roaming rates remained fairly near the regulated caps in many Member States. The Roaming Regulation does not seem to have had a significant impact on the pricing of other mobile services. Any waterbed effects would be expected to be small due to the fact that roaming revenue is a small part of overall mobile revenue (EU average of 4.2% in 2009).

---

[3] Government licensing policy in mobile telecommunications has various dimensions. First, the government needs to decide whether to set a single national (or international) standard, or whether to allow multiple technological systems to compete. Second, the government has to decide to how many firms will receive a license. This also involves an important decision with respect to the timing of first and additional licenses. Third, the government needs to decide how to grant licenses. In the early days of mobile telecommunications, licenses were often granted on a first-come-first-serve basis.

4.2.2 Mobile number portability

Mobile Number Portability (MNP) is a regulated facility which enables subscribers of mobile services to change their service provider whilst keeping their existing telephone number. Its purpose is to foster consumer choice and effective competition by enabling subscribers to switch between providers without the costs and inconvenience of changing telephone number. It is an important prerequisite for intensifying market competition[4] since it lowers switching costs, churn rates should be expected to increase[5]. The EU's Universal Service Directive requires member states to implement number portability for mobile services[6]. There are a number of countries where networks do not charge customers for porting numbers. For instance, in addition to Finland, MNP is typically free in the UK and in Ireland. In Belgium, only pre-paid subscribers pay for porting their mobile number. During the porting process, the ported number cannot handle incoming or outgoing calls.

The speed of porting is also heterogeneous across countries. While in some countries porting time is extremely short—porting takes only two and a half hours in the US—operators from other countries may need days, weeks or even months to port a number. More recently, Article 30(4) of the Citizens "Rights Directive" (2009) introduced a new requirement that consumers, *"having concluded an agreement"* shall have the number activated within one working day. The article also introduces a competence on Member States to impose sanctions on service providers, including a provision to compensate subscribers in case of delay in porting or abuse of porting by them or on their behalf.

4.2.3 Mobile termination charges

Call termination charges into mobile networks are currently one of the most crucial issues facing regulators in Europe. Call termination refers to the final completion of calls on a network, and in this case regards calls to mobile phones i.e. completion of calls in mobile networks which have originated in other fixed or mobile networks. A termination charge is a wholesale charge paid by the operator in whose network the call originates, to the operator of the network in which a call ends. The retail price paid by callers for a call from one network to a mobile network is broadly made up of two components: (a) the first operator's cost to originate and carry the call and (b) the termination charge paid by the first operator to the second terminating operator.

New regulatory measures, imposed on mobile termination markets by the Regulatory Framework of 2002, induced a global decrease of mobile termination rates. According to the European Regulators Group (ERG), the average decrease of Mobile Termination Rates (MTR) levels between 2004 and 2007 is about 26%, with important disparities between countries. Besides, the European Commission increasingly invites NRAs to make termination rates asymmetries disappear and to specify, meanwhile, the convergence conditions towards termination rates symmetry, with regard to both target level and time frame. The Commission considers that asymmetry, which refers to differences between MTRs of MNOs within the same member state, requires an adequate justification.

---

4 By today, almost in all European countries number portability is possible.

5 Unfortunately, there are no statistics on churn rates available for most European countries.

6 The United Kingdom and the Netherlands first implemented MNP in Europe in 1999. Countries such as pain (2000), Sweden and Denmark (all 2001), Belgium, Italy, Germany and Portugal (all 2002) followed suit. Most recently, Estonia implemented MNP due to regulatory intervention.

4.2.4 Universal services access requirements

The Universal Service Directive (USD, Article 3), requires to ensure that universal services are made available at the quality specified to all end-users in their territory, independently of geographical location, in the light of specific national conditions and at an affordable price. Among services included in the scope of the universal service, we can identify:

- Provision of access at a fixed location to the public telephone network;
- Special measures for disabled end-users to ensure access to and affordability of publicly available telephone services, including access to emergency services.

In addition to the services listed above, Member States may take the following measures:

- Specific measures to ensure that disabled end-users can also take advantage of the choice of undertakings and service providers available to the majority of end-users;
- provision of tariff options or packages to consumers which depart from those provided under normal commercial conditions (Article 9(2) USD)
- Provision of specific facilities and services allowing subscribers to monitor and control expenditure and avoid unwarranted disconnection of service
- Measures to cover different parts of the national territory.

Looking at the European market, one can conclude that, in general, it is not yet possible to provide the universal service at any location. Only in the densely populated countries with a high coverage level of mobile network such a possibility exists.

4.2.5 Mobile virtual network operators

The Regulatory Framework enables the operations of virtual operators and creates them a business opportunity. If the incumbent operators are not willing to open their networks voluntarily, the regulations help the NRAs to enforce the network access with reasonable terms. These terms have to be equal to the vertically integrated service operators of MNOs. In accordance with Finnish Communications Market Act. Section 23, MNOs with SMP can be imposed obligations regarding access to the MNOs network when necessary. These obligations are to give service providers the right to access the MNOs network.
In summary, the regulatory situation concerning different types of virtual operators is not yet harmonized between the EU countries.

5. Evidence from European countries

5.1 Data and variables analysis

Our empirical work relies mainly on several primary sources. Firstly, the industry data comes from the *ITU World Telecommunication Indicators* (2010) dataset. Secondly, the policy indicators were gathered from several web sites such as *OECD regulatory* database, *ITU World Telecommunications Regulatory* database, *Privatization Barometer* database and *POLCOIII* dataset. Note that, all mobile data used concern GSM mobile business.
We consider the mobile telecommunications markets in 31 European countries over the period 1993 to 2008, using network deployment (main mobile lines per 100 inhabitants), prices, output and quality as the dependant variables. We consider three main aspects of mobile telecommunications reform-privatization, competition and regulatory development- as explanatory variables. The set of explanatory variables, other than measures of the regulatory reforms, also includes control variables such as technological progress, demographic, political and macroeconomic indicators.

5.1.1 Mobile communications performance indicators and measurement issues

Traditionally, corporate performance at Mobile Network Operators (**MNOs**) tends to be evaluated by several Performance Measurement Indicators (**PMIs**) such as gross revenues, number of subscribers, ARPU, churn[7], quality of services, profits, as well as market share. However, in this study, the precise definition of the performance measures was dictated by the availability of data and also due to well known measurement issues. This is discussed below in analyzing the quantification of each performance indicator chosen in this study for the mobile communications industry.

5.1.1.1 Productivity

Productivity of service industries and especially mobile communications is hard to define. Indeed, unlike manufactured goods, services are characterized by a greater degree of heterogeneity, which makes aggregation difficult. As mentioned below, mobile telecommunications output may include the number of users serviced, the number of minutes of communication supplied, the range and the quality of services provided as well as the (generally immeasurable) network externalities. In analyzing performance, many studies on telecommunications consider both labor productivity (LP) measured as revenues per employee per year, and total factor productivity[8] (TFP) in order to assess productivity changes (Armando Calabrese and al., 2002). Some research has found that privatization leads to lower prices through the expansion of the network or improved labor productivity (Ros 1999, Li and Xu 2002, Fink et al. 2003). However the effects of privatization and competition were complementary. Pagoulatos and Zahariadis (2011) found that labor productivity is negatively affected by state ownership. Indeed, the aim of regulation in telecommunications is to meet social goals, avoid potential abuses due to predatory behavior, and stimulate competitive pressures to enhance consumer welfare. Hence, labor productivity is expected to increase as companies become more efficient in their quest for higher profits under external regulatory constraints. Note that any increase in size of employment will have a negative effect on productivity growth. The decrease in the number of employees, reducing the denominator, enhances the labor productivity indicators.

5.1.1.2 Employment

Studies show that state-owned companies tend to over-staff workers, pay high wages, and provide generous benefits. Therefore, it is argued that the effects of reforms, namely privatization, in the telecommunication sector on employment are likely to be negative since privatization reduces overstaffing; (Li and Xu 2002, Ypsilantis and Min 2001, Xavier and Ypsilantis 2001). However, cases of little or insignificant employment reductions (or even employment increase) exist. The main explanation is that generally, overstaffing usually

---

[7] The wireless industry is at an inflection point; marked by saturation, competition, stagnant revenue growth and increasing customer care and subscriber acquisition costs. As such, the ability to retain an existing customer has become critical to recapturing some of the revenue and margin sacrificed by customer acquisition programs and price promotions. With such figures, churn data, alongside subscriber acquisition costs, has become a key measure used by industry analysts to determine mobile operator performance.

[8] Total factor productivity picks up productivity gains that cannot be attributed to increases in the productivity of labor or capital usage alone. This residual productivity is attributed to the combined effects rather than to other factors.

occurs in clerical and administrative positions and not in the more technically skilled jobs such as in the telecommunications sector[9]. Indeed, in countries that carried out labor reforms early in the process, there was a minimal effect on employment post-privatization.
It is more likely that the studies reflect real differences in post-privatization employment changes between countries. However, the safest conclusion we can assert is that privatization does not automatically induce employment reductions in divested firms.

5.1.1.3 Output

One of the major obstacles in telecommunications comparisons is the measurement of output. Some studies measure output only in physical terms (for example, in the number of calls and access lines). Other studies weight physical output in terms of relative prices (for example, revenue or value of output per subscriber). Ariff, (2009), Nicoletti, (2000) and Pavlos C. Symeou, (2004) have defined the telecommunications output as the number of mobile subscribers, total revenue and outgoing telecom minutes. Heshmati and El-Rhinaoui (2009) uses as output the mobile traffic (minutes of use of all subscribers). Due to data constraints, a relatively narrow definition of output was adopted: the total revenue divided by the cost of a 3-minute local call on a mobile. In general, there is broad agreement that output increases more after reforms for telecom sector. However, results from different studies suggest that the effects of privatization are either complemented or overwhelmed by the effects of competition and/or regulation. Separating the effects is difficult, but the evidence suggests that privatization without strong regulatory support is less effective.

5.1.1.4 Prices

The weakest link in the data chain is on prices as these are notoriously difficult to know and to compare across economies. The MNOs offer increasingly complex and diversified products at lower and lower prices. Besides, changes in quality are reflected by improvements in physical and non-physical characteristics which induce several dimensions in the measurement issue (Karamti and Grzybowski, 2010). Furthermore, even though price differentials may reflect quality differences, most of the studies assume that the mobile service is provided at same quality by each mobile operator in the sample which makes the simple prices comparison irrelevant.
The tariff baskets are commonly approved as the most appropriate method for price comparisons among countries. Cross-country differences in observed prices may also reflect differences in price regulation. To account for some of these problems, OECD tariff baskets were supplemented with a measure of "average prices" in the mobile services: mobile revenues per subscriber (ARPU). However, because limitations in using price data for mobile services identified by previous studies (i.e. Magnien, 2002; Banerjee and Ros, 2004a) we were constrained, as many other studies, to proxy the price with the average revenue per minute (Sung, 2007).

5.1.1.5 Mobile density

The teledensity refers to the number of mobile phone lines per 100 inhabitants in a country (Ariff, 2009; Xu and Li, 2004). Despite the fact that the mobile telephony development is usually measured by the number of cellular subscribers in a country (Boon Lee and William

[9] Note also that requirements of a digital system would spell technological unemployment for older workers while younger computer-trained staffs are being recruited (Ure, 2003).

Shepherd (2000)), Hamilton, 2003; Gutierrez and Berg, 2000), we prefer taking density (subscribers per 100 inhabitants) since it is considered a better indicator of the development of the traffic of mobile network.

5.1.1.6 Quality

Quality is a multi-faceted concept which includes relatively objective features such as variety, reliability and serviceability as well as more subjective factors such as user satisfaction. The quality of service in mobile communication is defined here (the objective aspect) by a number of key indicators. They include technical faults, network availability, call set up access rate and call drop rate. Some papers conclude that the reform is positively associated with the quality of services in mobile telephony. However, very few quality indicators are available on a cross-country basis for the mobile communication services.
In our analysis and since such detailed information are very scarce, another feature of quality is considered, the "coverage". Many observers assume that with the liberalization of the sector which results in an increase of the number of mobile operators, the percentage of the population covered will also grow, since competition will induce lower prices and more affordable service. However, some examples demonstrate that the reverse can be true[10]. One possible reason for this counter-intuitive finding is that many mobile operators focus on the more profitable urban areas and lack the resources and/or interest to roll service out to rural areas, where the majority of people reside and where the social benefits of mobile connectivity are higher than in well-served urban areas. Besides, as more operators enter the market and competition intensifies, the utilization levels and profitability of many carriers drop, hindering their ability to invest in the network to expand further.

5.1.2 Telecommunications policy reform variables

- *Regulation Variables:* We used three dimensions of the regulation framework. First, we construct a dummy variable 'NRA' that denotes the establishment of an independent National Regulatory Authority in the sector. The variable NRA takes a value of 1 only if the authority is characterized autonomous and 0 otherwise. In order to take into account the dynamic effect of the regulatory framework, we include a count measure of the number of years since the establishment of an independent regulatory body, '*NRA_YEARS*'. Moreover, Mobile Number Portability (MNP) is a regulatory facility which is likely to affect retail prices, termination charges, price elasticity's, market shares, as well as entry and investment decisions. It is fair to say that most analyses on MNP have supported the notion that, on the whole, MNP intensifies competition in mobile telecommunications. The effect of MNP on the mobile sector performance has never been tested before. Thus, we include *'MNP'* a dummy variable which takes 1 when number portability in mobile networks is established in a country and 0 otherwise.
- *Privatization Variables:* The effect of change in ownership of the incumbent provider on performance is captured by a dummy variable *'PRIV'* which equals 1 when the firm has allowed for the first time, private participation in its operations and 0 otherwise. A variable **'State Ownership'** measures the percent of shares owned by the state. Besides, using a single point in time can only provide a limited impact of telecommunication reforms. Thus, besides a binary measure of whether or not a country has privatized the

[10] Despite the large number of operators (eight) in India, for example, population coverage lags significantly behind Jordan, with four operators; China, three operators; and the Philippines, three.

Variable	Definition
Performance variables	
Density	Mobile telephone density, measured by the number of telephone subscribers per 100 inhabitants.
Mobile staff	Reported number of employees.
Labor Productivity	The labor productivity, which is measured by real output per employee.
Output	The total revenue divided by the cost of a 3-minute local call on a mobile.
ARPU	Average revenue per user equals to the total revenue is divided by all subscribers.
Coverage	The percentage of the population covered by mobile telecommunication networks.
Reform Variables	
• **Regulation variables**	
NRA	The establishment of an independent regulatory body.
NRA_YEARS	Measure of the number of years since the establishment of an independent NRA.
MNP	Mobile Number Portability, a dummy for the introduction of number portability in mobile networks in each country;
POLCON III	Measures the quality of an economy's political system and rages between 0-100.
• **Privatization variables**	
State Ownership	Percent of shares owned by the state.
PRIV	A dummy equals 1 if the incumbent is privatized.
PRIV_YEARS	A count measure of the number of years since privatization.
• **Competition variables**	
COMP	A dummy equals 0 if the mobile telecommunications sector is served by a national monopoly operator, a value of 1 if the sector has two operators and a value of 2 if the sector has more than two operators.
New_entrants	Market share of the new entrants. The variable is the ratio of the number of mobile lines in operation owned by the new operator to the number of total mobile lines in the market.
MVNO	Mobile virtual network operator, a dummy for the launch of the first MVNO in each country;
MVNO_YEARS	A count measure of the number of years since the launch of the first MVNO in the country.
Control Variables	
Size	Takes the values of the size index.
Small	**A dummy takes 1 if economy is small based on the size index's median**

*) The data is available only until 2007.

Table 1. Variables Definitions

mobile sector, we use a count measure of the number of years since privatization, *'PRIV_YEARS'*.

- *Competition Variables:* To measure the degree of competition, we used a variables *COMP*, assigned a value of zero if the mobile telecommunications sector is served by a national monopoly operator, a value of one if the sector has two operators and a value of two if the sector has more than two operators in mobile market segment[11] (Wei Li, 2004). Besides, the market share of the new entrants captures competitive pressure in the mobile telephony market. The *'New_entrants'* variable, which concern only mobile network operators, is the ratio of the number of mobile lines in operation owned by the new operator to the number of total mobile lines in the market. However, the mode of competition has changed on the European mobile markets during the last ten years. Emergence of virtual operators, together with new content providers, has brought a large number of new players to the market. The traditional market structure of incumbent operators and their vertically integrated partners was so fragmented. Furthermore, the amount of competitors has increased and new kinds of competitors emerged. Indeed, the fast developing regulatory framework in European countries force MNOs to accept virtual operators, however, not all countries have changed their regulation to promote competition. Thus, the introduction of mobile virtual operators could be an interesting variable never taken before. Thus variable *'MVNO'* is a dummy variable which takes a value of one when the first virtual operator is launched in a country. We also added the variable *'MVNO_YEARS'* which is a count variable representing the number of years since the launch of the first MNVO.

5.1.3 Political and institutional variables

The regulation quality is very important in assessing company performance (Henisz, 2002). Indeed, the narrower the regulatory regime, the greater the political involvement is likely to be in company management. More political involvement translates into a blurring of market based performance since politicians seek to satisfy national and special interest needs that go beyond the company's "welfare." Political interference is in this case is more costly to the privatized company. The effect of the magnitude of institutional endowments on firms' performance is gauged by the variable *'POLCON III'* developed by Henisz (2002). This variable ranges between 0-100. Smaller values illustrate an economy with lower economic freedom, narrower institutional endowments, and higher political risks.

5.1.4 Country economic indicators

Firms in small economies have traditionally been assumed to encounter substantial difficulties improving their performance. They are characterized inter alia, by limited capacity which prevents them to exploit high economies of scale. This is particularly true for firms in sectors with high fixed and sunk cost such as telecommunications. Symeou (2009) analysis of the liberalization of small and large European countries finds that competition as an end in itself is less relevant to the success of liberalization in small economies. This can be explained, on one level, by the fact that market dynamics in small economies limit the prospects for efficient entry. On other level, because the number of operators required

[11] S. Zheng & M.R. Ward (2010) used the Herfindahl-Hirschman Index (HHI) as measure of competition intensity. Wallsten (2001) used the number of mobile operators. Instead, Barros and Seabra (1999) employed a dummy variable which drew a distinction between monopoly and non-monopoly markets.

generating the expected outcomes of liberalization efficiently is much smaller than in large economies. We assume then, that there exist a relationship between economy size and firm performance. Thus, following Symeou (2004, 2009), an index for smallness suggested by Jalan (1982) is adopted which combines population, income and geographical measures. These indicators (GPD, population, arable area) are used to measure the country's economy '*Size*' variable which is an index constructed as follow:

$$Size\ Index = \frac{100}{3}\left(\frac{P_i}{P_{max}} + \frac{A_i}{A_{max}} + \frac{Y_i}{Y_{max}}\right)$$

P_i, A_i and Y_i are population, arable area and GDP of each country respectively;
Finally, "**Small**", is a dummy variable which takes 1 if economy is small based on the size index's median.

5.2 Econometric analysis

5.2.1 Panel model

The approach taken in this study is to examine how indices of market efficiency such as network deployment, work productivity, prices and quality of services offered and employment have been effected by the introduction of competition, by the privatization of the traditional telecommunications organizations and by the establishment of independent regulatory authorities over both time and countries.

To do so, our empirical analysis consists in the specification and estimation of equations for prices, subscriptions, output, quality, employment and labor productivity. In each of these equations we consider three main aspects of telecommunication reform-privatization, competition and regulatory development as explanatory variables.

The current study employs the approach used by earlier studies. Each equation was estimated using two regression models: a random effects specification and a fixed effects specification. The fixed effect specification assumes that county-specific effects are fixed parameters to be estimated, whereas the random effect model assumes that countries constitute a random sample.

The general model we refer to can be written as follow:

$$Y_{it} = \alpha_i + \beta R_{it} + \delta X_{it} + \mu t + \varepsilon_{it} \qquad i = 1, \dots, N \quad t = 1, \dots, T$$

where Y is the set of variables used to proxy performance, *i* is an individual country, *t* is a period of time (1 year), α_i the country fixed effect that controls for country specific propensity to reform and other country specific unobserved factors. Explanatory variables include a set of reform variables (R_{it}) and set of control variables (X_{it}). Time (t) is a time trend[12] and is used to catch the temporal effect and reflects technological change and ε_{it} is the error term.

In order to account for dynamics in our data, we make use of the Differenced Generalized Method of Moments (DIF-GMM) developed by Arellano and Bover (1995) for analyzing panel data. However, fixed and random models systematically outperform these dynamic regressions[13].

---

[12] We have also experimented with including time dummies instead of the time trend, and the results are very similar. For ease in checking the tendency of the time trend and reporting the results, we therefore use the time trend specification.

[13] These results, not reported in this study, are available from the authors.

Generally, studies of the impact of reforms on enterprise performance encounter difficult issues namely the endogeneity bias. Indeed, the biggest potential problem is that competition, privatization and regulation may be endogenous to reforms. That is, reforms affect telecom performance, but performance may also affect reforms. A possible source of endogeneity is that unobservable factors affecting reform may also affect performance, e.g., managerial quality, that are correlated with both the dependent variable and with the included explanatory variables. The analysis deals with this issue by including country fixed effects. This permits control for a country-specific propensity to reform. The reform dummy variables, too, help control for a pro-pensity to reform, which could be correlated with performance changes.

5.2.2 Alternative specifications

For each outcome variable, we estimate the baseline equation under two different specifications and report the results in tables 2-4.

The two specifications tests for the whole performance measures are:

1. *Interaction between reform variables:* Theory suggests that simply privatizing a monopoly may not generate telecom improvements. Careful regulation is required to encourage a monopoly to improve its performance. To explore further the effects of regulation, we interact the regulation dummy with the privatization dummy. Then, following Wei Li and Xu (2004), we added an interaction variable between privatization and competition in order to estimate complementarities that may exist between the two reforms and estimate model 2 (M2). M2 allows us to explore separately the effects of competition, privatization, regulation and how they interact.
2. *Small size versus Big economies:* In order to analyze whether the effects of the reforms change with the economy' size across countries, we rerun the baseline regressions allowing the reform effects to differ between big and small economies (those with Size index lower than the median value). We do so by including interaction terms between the small size dummy variable with the reform variables and estimate model 3 (M3).

5.3 Regression results

In this section, we address the existence of relationships between the reform variables- privatization, completion and regulation- of the mobile communications services outcomes by running a set of regressions. Table A2 in the appendix show the fixed-effect and random-effect estimation results on which we build our testing procedure asking whether the reform variables, have a significant impact on the variables of mobile communications outcomes, namely, ARPU, labor productivity, employment, coverage, output and mobile density.

In addition to showing the estimated values of the parameters associated with the explanatory variables listed at the left, Tables 1-3 include three additional items. Firstly, we provide an F-statistic (F) for fixed-effects or Wald statistic (*Wald*) for random-effects for testing the joint significance of the explanatory variables. Secondly, we rely on the Hausman test while opting for random effects or fixed effects. Thirdly, we include the number of observations included in each regression (Obs.).

The results in table 2, Wu-Hausman specification test to discriminate between fixed and random effect models show that in most cases a fixed effects model is the appropriate model specification. Recall that fixed-effect models allow controlling for fixed unobserved heterogeneity and are therefore preferred to random models when estimating the relationship between privatization and telecommunications outcomes. Four of the six equations seem to present fixed effects. Only in ARPU and labor productivity regressions random effects seems more suitable.

Tables 2-4 present the results of estimating the baseline equations besides estimates under the two alternative specifications explained above for each of the six dependent variables.

1. *Output and pricing:* Did privatized companies restrict output and raise prices? (Li and Xu, 2004). Overall, the degree of market competition (proxied by the share of new entrants) and years after the establishment of MVNOs considered as new competitors (could also be interpreted as the effect of prospective competition according to Nicolletti, 2001) emerged as the main explanations for the cross-country and time variability in output. Estimates in table 2 show that privatization have a limited, negative and statistically insignificant, effect on output. It's true that call volumes often rise with network penetration; however the number of players on the market also arises, shirking the output of each operator. This result is confirmed first, by the negative and significant impact of number portability (MNP) on output. Unfortunately, we do not have information about churn rates; however, MNP can be taken as proxy since customers who want to switch to another operator generally prefer to keep their primer phone number. Second, the negative and highly significant coefficient on the interaction variable Priv×Comp confirms once again the negative impact of both privatization and competition on output. This finding is totally different of those found in previous studies. Li and Xu (2004) find that full privatization has a positive impact on real output and no evidence of complimentarily between privatization and competition on output expansion. However, the authors consider their results as puzzling and explain them by the fact that their data are not adjusted for changes in service quality. Estimates in table 3 show that overall, Privatization and the regulation indicators performed quite well, significantly improving the fit of the regressions. The estimates broadly suggest that countries having stronger actual competition and more regulated market tend to have lower prices. These findings are quite similar to those of Nicoletti (2000). Interestingly, political constrains seems to have important, positive and statistically significant impact on firms' ARPU. Finally, privatization seems to have larger and significant effects on firms' ARPU in small size economies contrary to Competition.
2. *Employment:* Table 3 presents estimates with employment in logarithm as dependant variable under the two specifications discussed above. Inspection of the results reveals that, consistent with the hypothesis, employment is reduced due to the privatization of traditional telecommunications organizations. The estimated effects of number of years since privatization are small but highly significant. Besides, employment decreased slightly but non-significantly with both actual and prospective competition (N_MVNOs). The greatest effect on employment stems from the new enterprises in the market where market structure is the main explanatory variable with small but positive and highly significant effect on employment. In column 2, estimates of the coefficient on the interaction variable between privatization and competition is negative, contrarily to regulation, the two coefficients are however small and statistically insignificant. Finally, competition seems to have larger effects on employment in small size economies contrary to privatization.

	LN(ARPU)				LN(OUTPUT)			
	(M1)		(M2)	(M3)	(M1)		(M2)	(M3)
	FE	RE			FE	RE		
PRIV	-.027	.119	.049	.090	-.002	.222	.223	-.036
PRIV_YEARS	-.009	.029*	.031*	.027*	-.053	.0044	-.044	-.058
State ownership	.005**	.002	.003*	.002	.005*	.0021	.005*	.005*
COMP	-.100	-.174**	-.078	-.179**	-.097	-.141	.174	-.102
MVNO	.152	.158	.144	.174	.090	.0400	.089	.105
MVNO_YEARS	.064**	.015	.028	.019	-.063*	-.071*	-.066*	-.058*
New_entrants	-.003	.000	-.000	.001	.018***	.026***	.016***	.018***
NRA	-.225*	-.277**	-.382**	-.268**	.296*	.440***	.206	.319*
NRA_YEARS	.012	.008	.008	.009	-.141**	-.078*	-.139**	-.133**
MNP	.289**	.363***	.347***	.377***	-.321*	-.318*	-.260	-.316*
POLCON III	-.558	-.176	-.284	-.080	2.364***	2.346***	2.345***	2.419***
SIZE	.019	-.001	-.001	-.001	-.024	.040***	-.036	-.020
YEAR	-.089**	-.122***	-.122***	-.128***	.435***	.261***	.442***	.424***
PRIV×COMP			-.122			15.862***	-.402**	
PRIV×REG			.246				.153	
PRIV_SMALL				.057				.079
COMP_SMALL				.030				.048
_CONS	6.276***	7.017***	7.06***	6.923***	18.162***		18.525***	17.993***
Observations	297	297	297	297	317	317	317	317
R-squared	0.425	0.112	0.094	0.112	0.834	0.4288	0.840	0.835
F	34.45***				109.32***		97.79***	94.44***
Wald		284.28***	325.04***	285.14***		1302.98***		
Hausman		9.96				45.49***		

* Significant at the 10 percent level.
** Significant at the 5 percent level.
*** Significant at the 1 percent level.

Table 2. The Impact Of Privatization, Competition And Regulation On Arpu And Output Using Fixed Effects And Random Effects

3. *Labor productivity*: Using labor productivity measured by real output per employee as the dependant variable, we estimate equation 1 under the alternative specifications and report the results in table 3. There we find that privatization increases labor productivity yet both statistically insignificant except under the second specification. This finding is ambiguous and difficult to explain since most of the previous studies confirm a positive and significant link between privatization and labor productivity. Only Nicoletti (2000) found a negative impact and explained that his result could depend on the limited concept adopted for privatization, which was defined as any initial sale of PTO shares, not necessarily implying loss of control by the state. However, these explanations could at best account for the lack of significance of this variable, certainly not a negative impact. We can adopt such explanation since our results suffer only from lack of significance. As found before for ARPU, privatization seems however to have large and significant effect in smaller economies contrarily to competition.
4. *Mobile density*: Table 4 reports the estimates of equation 1 under the two other specifications with Mobile Density as dependant variable. Focus first on the impact of privatization on network expansion. Estimates on columns 1-3 show that moving to private ownership is positively associated with the expansion of mobile network. However, unlike most of the studies under review (Li and Xu, 2004), the estimates here are small and statistically insignificant except in column 2. In our opinion, this can be explained by the fact that, in the studies mentioned above; authors consider growth in the mobile density as an expansion of the service coverage. We do not agree with such definition, since coverage is a geographic indicator of the mobile network expansion while density is simply the number of subscribers per 100 inhabitants. This indicator does not reveal the contrasting tele-densities in saturated urban areas and rural areas and could be then inflated by the number of subscribers in large metropolitan areas. Indeed, generally, statistics do not take into account the inadequacies in mobile phone coverage in the more rural and remote zones. Moreover, even though competition has prompted the mobile companies to improve coverage by adding new base stations, they generally avoid investing in difficult and less populated areas because of low revenue users in these zones. Most high end users are in urban areas only. This lead to an improvement in mobile density but more in urban zones and less in white areas where coverage is essentially non-existent. For all these reasons we consider the effect of privatization on mobile density could be limited. Regarding competition, as mentioned earlier, the market structure has a small but positive and statistically significant effect on mobile density. More interestingly, the increasing number of mobile virtual operators has a negative and statistically significant effect on mobile density. This confirms our former explanation. Even though MVNOs may have little or no network infrastructure of their own, they can focus on low penetrated rural markets which are outside the focus of that MNO, but unfortunately, this is not the case yet. In column 2, estimates reveal that the joint effects of privatization and competition are small and statistically insignificant. Surprisingly, the joint effects between privatization and regulation have a negative effect on density. Finally, competition seems to have larger effects on mobile density in larger size economies however the estimate is statistically insignificant. Finally, a sound institutional endowment consisting of a strong telecom regulatory body and a stable political system increase the level of main lines per 100 inhabitants.

	LN(EMPLOYMENT)				LN(LABOR PRODUCTIVITY)			
	(M1)		(M2)	(M3)	(M1)		(M2)	(M3)
	FE	RE			FE	RE		
PRIV	-.1222	-.0248	-.0450	.0156	.2055	.2323*	.2223	-.1169
PRIV_YEARS	-.0864**	-.0607*	-.0833**	-.0841**	.0574	.0366	.0388	.0289
State ownership	-.0005	-.0022	-.0004	-.0003	.0027	.0028	.0026	.0016
COMP	-.0008	-.0178	.1188	-.0903	-.1189	-.1092	-.2198*	.0606
MVNO	.0180	-.0044	.0166	.0163	-.1451	-.1461	-.1421	-.1178
MVNO_YEARS	-.0134	-.0178	-.0143	-.0089	-.0952***	-.0841***	-.0864***	-.0851***
New_entrants	.0177***	.0219***	.0165***	.0209***	-.0032	-.0041	-.0025	-.0087*
NRA	.0210	.0465	-.0378	.0684	.0275	.2165	.3012*	.1445
NRA_YEARS	-.0132	-.0023	-.0152	-.0055	-.1631***	-.0738**	-.0769**	-.0818**
MNP	-.453***	-.433***	-.4259***	-.4619***	.3094**	.2908*	.2686*	.3283**
POLCON III	.5977	.5540	.5839	.7166	1.346**	1.580**	1.5747**	1.4566**
SIZE	.0020	.0377***	-.0028	.0094	.0108	.0102	.0100	.0130*
YEAR	.2291***	.1619***	.2355***	.2062***	.1986***	.1271***	.1263***	.1449***
PRIV×COMP			-.17480				.1532	
PRIV×REG			.0995				-.1940	
PRIV_SMALL				-.3136				.8343***
COMP_SMALL				.2373**				-.3798***
_CONS	5.999***	4.701***	6.158***	5.5717***	11.385***	11.260***	11.268***	11.421***
Observations	317	317	317	317	317	317	317	317
R-squared	0.7914	0.4407	0.7935	0.7974	0.4257	0.0834	0.0802	0.0844
F	82.00***		71.59***	73.19***	16.02***			
Wald		102.31***				187.79***	194.09***	224.93***
Hausman		27.45**				6.72		

* Significant at the 10 percent level.
** Significant at the 5 percent level.
*** Significant at the 1 percent level.

Table 3. The Impact Of Privatization, Competition And Regulation On Employment And Labor Productivity Using Fixed Effects And Random Effects

	Ln(1+Density)				Ln(Coverage)			
	(M1)		(M2)	(M3)	(M1)		(M2)	(M3)
	FE	RE			FE	RE		
Priv	.068	.100	.254*	-.022	-.038*	.1134***	.0813**	-.0391*
Priv_Years	.019	.009	.026	.014	-.109***	.0001	-.1044***	-.1111***
State ownership	-.002	-.001	-.002*	-.002	-.0001	.000	.0001	-.0000
COMP	.057	.118**	-.055	.095	.012	.022*	.0781***	.0216*
MVNO	-.094	-.077	-.080	-.082	.001	.002	.0017	-.0013
MVNO_Years	-.138***	-.116***	-.142***	-.136***	.001	.0005	-.0002	-.0001
New_entrants	.015***	.013***	.018***	.014***	.001	.0002	.0005	.0000
NRA	.351***	.432***	.629***	.347***	.021	.047*	.0882***	.0159
NRA_Years	-.038	-.006	-.002	-.035	-.015	.000	-.0117	-.0136
MNP	-.302***	-.291***	-.327***	-.294***	-.004	-.005	-.0032	-.0044
Polcon III	1.182***	1.492***	1.261***	1.171***	.038	-.065	.0511	.0408
Size	-.005	-.003	-.005	-.006	-.0001	-.001	-.0013	-.0013
Year	.281***	.247***	.238***	.284***	.1247***	-.001	.1180***	.1269***
Priv×Comp			.141				-.0726***	
Priv×Reg			-.464***				-.0817**	
Priv_Small				.208				.0105
Comp_Small				-.072				-.0261*
_cons	.596	.295	.462	.663	4.251***	4.436***	4.192***	4.304***
Observations	317	317	317	317	188	188	188	188
R-squared	0.946	0.419	0.948	0.947	0.779	0.839	0..803	.776
F	383.59***		345.74***	333.34***	37.82***		41.04***	35.05***
Wald		426.69***				102.46***		
Hausman		171.43***				339.18***		

* Significant at the 10 percent level.
** Significant at the 5 percent level.
*** Significant at the 1 percent level.

Table 4. The Impact Of Privatization, Competition And Regulation On Teledensity And Coverage Using Fixed Effects And Random Effects

5. *Coverage*: To begin with, privatization has a negative and highly significant effect on coverage. This result seems surprising. However, as explained before, MNOs avoid investing in difficult and less populated areas because of low revenue users in these zones. This lead to an improvement in mobile coverage and density but more in urban zones and less in white areas where coverage is essentially non-existent. Additionally, the combined effects of a new regulatory environment and of privatization of the national telecommunications organization result in a more robust decline of the coverage. It's important to note that, these findings can't be compared with findings in previous studies on mobile telecommunications, since most of them do not consider this performance indicator. Only very few studies examine the correlation between quality and reform process on the mobile sector and characterize quality as the technical performance of the mobile networks. Other studies find that cellular coverage was largely developed in the decade since the privatization of mobile telecommunication services. Further, they conclude that formulation of a regulatory policy and the establishment of an independent regulatory body impact positively on the quality (Fink and Mattoo et al 2001, Galal and Nauriyal 1995).

6. Conclusion

This research studied three chief aspects of the process of telecommunications market reform. We analyzed the effects of privatization, competition and regulation on a comprehensive set of performances in the mobile telecommunications sector in 31 European countries during the period from 1993 to 2008. Using an econometric model, the results are quite mitigated since some of them are different of those from prior research.
To preview, the results show that competition is associated with increased penetration, and lower prices while privatization by itself is associated with few benefits. Privatization combined with an independent regulator, have most a negative impact on performance indicators. Moreover, smaller size economies appear to experience similar reform impact as larger economies on output and coverage. However, the impact of competition on mobile phone density and employment seems to be higher in small economies and lower for labor productivity contrarily to the privatization effect. Furthermore, in small economies, MNOs, appear to achieve higher ARPU more if the sector is privatized but competition bring it down. Moreover, in contrast to competitive pressure, privatization reduces employment. The opposite effects that privatization and competition on employment is confirmed by the negative joint effect in the estimates. Further, privatization have no identifiable impact on output, this surprising result may be explained by the fact that most of the previous studies attribute robust growth in output to total factor productivity not considered in this analysis. More important, competition is found to raise output contrarily to labor productivity. Consistent with this result, competition and privatization exhibit strong opposite effects on output, and complementarily, yet insignificant, on labor productivity. Similarly, we find that large portion of mobile network expansion can be attributed to both privatization and completion.
We used historical average revenue per user (ARPU) as an indicator of user willingness to pay and as proxy to price. As expected, even though the number of subscribers exploded in the last decade in the European mobile telecommunications networks, the ARPU fell under both competition and regulation pressures.

Regarding quality, surprisingly, privatization seems to decrease network coverage. One possible reason for this counter-intuitive finding is that when competition intensifies, more operators enter the market, the utilization levels and profitability of many carriers drop, hindering their ability to invest in the network to expand further. Overall, even though new technological developments and the digitization of the technological infrastructure in particular, exert a very substantial effect on coverage, in some cases, the quality of services, despite the reforms of the sector, remains at the approximate level before the reform. Finally, a common finding is that higher quality of institutional endowments has globally a positive impact on firm efficiency which is confirmed by our results and concerns both small and large economies. Besides enhancing firm efficiency, higher quality of institutional endowments and lower political risk in the economy may also reduce the apparent riskiness of the economy in the global market. Higher quality institutional endowments may even overcome the regulation authority role. This reinforces the growing impression in the literature on telecommunications policy that generic competition law might be sufficient to maintain a healthy competitive environment and makes industry-specific regulation unnecessary, as long as strong institutional foundations and low political risk are in place (Symeou, 2004).

7. References

Ariff, M., Cabanda E. and M. Sathye, 2009. Privatization and performance: evidence from telecommunications sector. Journal of the Operational Research Society", 60(10), 1315-1321. doi:10.1057/jors.2008.103

Armstrong M, Network Interconnection in Telecommunications. Economic Journal, 1998, vol.108, no. 448, sid. 545–64. Jean-Jacques Laffont, Patrick Rey and Jean Tirole: Network Competition: Overview and Nondiscriminatory Pricing. RAND Journal of Economics, 1998, vol. 29, no. 1, pp. 1–37. Jean-Jacques Laffont, Patrick Rey and Jean Tirole: Network Competition: Price Discrimination. RAND Journal of Economics, 1998, vol. 29, no. 1, pp. 38–56.

Banerjee, A. and A. J. Ros, 2004. Patterns in Global Fixed and Mobile Telecommunications Development: A Cluster Analysis. Telecommunication Policy 28: 107-132.

Bauer, J. 2003. The coexistence of regulation, state ownership, and competition in infrastructure industries. Quello Center Working Paper 03-2003.

Ben Naceur, S., Ghazouani, S., and M. Omran, 2007. The performance of newly privatized firms in selected MENA countries: The role of ownership structure, governance and liberalization policies. International Review of Financial Analysis, 16, p.332-353.Amess and Roberts, (2007).

Bishop, M. and J. Kay, 1988. Does Privatization Works?. London: London Business School.

Bishop, M. and J. Kay, 1989. Privatization in the United Kingdom: Lessons and experience. World Development 17, 643-657.

Boardman, Anthony E and Vining, R, Aidan, 1989. Ownership and Performance in Competitive Environments: a Comparison of the Performance of Private, Mixed, and State-Owned Enterprises. Journal of Law & Economics, University of Chicago Press, vol. 32(1), p. 1-33.

Bortolotti B., D'Souza J., Fantini M. and W. L. Megginson, 2002. Privatization and the sources of performance improvement in the global telecommunications industry. Telecommunications Policy 26 (2002), p. 243–268.

Boubakri, N. and J.C. Cosset, 1998. The financial and operating performance of newly privatized firms in developing countries. Journal of Finance, 58, p 1081-1110

Boubakri, N. and J.C. Cosset, 2002. Does privatization deliver? Evidence from Africa. Journal of African Economics, vol 11.

Boubakri, N., J.C. Cosset and O. Guedhami, 2004. The performance of newly privatized firms: Evidence from Asia. Pacific-Basin Finance Journal, january, p 65-90

Boubakri, N., J.C. Cosset, O. Guedhami and K. Fisher, 2005. Ownership Structure and the performance of privatized banks. Journal of Banking and Finance 29, p 2015- 2041

Boubakri, N., J.C. Cosset, O. Guedhami, 2005. Investor protection, corporate governance and privatization. Journal of Financial Economics 76,p 369-300

Boubakri, N., J.C. Cosset, O. Guedhami, 2005. Liberalization, Corporate Governance and the performance of newly privatized firms. Journal of Corporate Finance 11, p 747-946

Cabeza Garcia, Laura & Gomez Anson, Silvia, 2007. The Spanish privatisation process: Implications on the performance of divested firms. International Review of Financial Analysis, Elsevier, vol. 16(4), pages 390-409.

Calabrese A., Campisi D., and P. Mancuso. Productivity Change in the Telecommunications Industries of 13 OECD Countries, International Journal of Business and Economics, Vol. 1, No. 3 (2002) pp. 209-223.

Castro, J. and K. Uhlenbruck, 1997. Characteristics of privatization: Evidence from developed, less developed, and former Communist Countries. Journal of International Business Studies, 28, 123–143.

Chakraborty, C., 2003. Privatization, Telecommunications and Economic Growth in Selected Asian Countries: An Econometric Analysis. Communications and Strategy, 4th Quarter(52),31-48.

Cho Shin, Choi Byung-il and Choi Seon–Kyou, 1996. Restructuring the Korean telecommunications market – Evolution and challenges ahead», Telecommunications Policy, Vol. 20 – Issue 5 pp 357 – 373 (Elsevier Science)

Chorng-Jian Liu, YuntsaiChou, Shyang-HuaWu, Yi-ShinShih. The public incumbent's defeat in mobile competition: Implications for the sequencing of telecommunications reform, Telecommunications Policy 33 (2009) 272–284

Conway, P. and G. Nicoletti, 2006. Product Market Regulation in Non-Manufacturing Sectors of OECD Countries : Measurement and Highlights. Documents de travail du Département des affaires économiques de l'OCDE, n° 530.

D'souza, J. and W.L. Megginson, 1999. The financial and operating performance of Privatized Firms during the 1990s. The Journal of Finance, vol. (4).

Dewenter, K. and P.H. Malatesta, 2001. State-owned and privately-owned firms: an empirical analysis of profitability, leverage, and labour intensity. American Economic Review 91, 320–334.

Doove S., Gabbitas O., Nguyen-Hong D. and J. Owen, 2001. Price Effects of Regulation: International Air Passenger Transport, Telecommunications and Electricity Supply' Staff Research Paper, Productivity Commission, Commonwealth of Australia.

D'Souza, J. and W. Megginson, 1998. The Financial and Operating Performance of Privatized Firms During the 1990's. Mimeo, Department of Finance, Terry College of Business, The University of Georgis, Athens, GA.

D'Souza, J., Megginson, W., and Nash, R. C., 2005. Effect of Institutional and Firm-Specific Characteristics on Post-Privatization Performance: Evidence from Developed Countries. The Journal of Corporate Finance, , 747-766.

D'Souza, J., Megginson, W., and Nash, R. C., 2007. The Effects of Changes in Corporate Governance and Restructurings on Operating Performance: Evidence from Privatizations. Global Finance Journal, 157-184.

Dunnewijk T. and S. Hultén, 2006. A Brief History of Mobile Telecommunication in Europe. Working paper, United Nations University.

EC ,2006, European Commission 2006, Annex to the Communication from the Commission to the Council, the European Parliament, The European Economic an Social Committee and the Committee of the Regions, European Electronic Communications Regulation and Markets (11th Report) COM(2006)68 final

EC, 1997, European Commission, 1997. Directive 97/33/EC of the European Parliament and of the Council of 30 June 1997 on Interconnection in Telecommunications with regard to ensuring universal service and interoperability through application of the principles of Open Network Provision. (ONP) (OJ L199/32, 26.07.97).

EC, 2002, European Commission, 2002. Eighth Report on the Implementation of the Telecommunications Regulatory Package. COM(2002) 695

ERG (2005) Report on Transparency of Retail Prices (with implementation of Number Portability. Available at:
http://erg.eu.int/doc/publications/retail_prices/erg_05_52_transp_retail_prices_report.pdf.

ERG (2009) Report on Transparency of Tariff Information.

Fare, R., S. Grosskopf and J. Logan, 1985. The relative performance of publicly owned and privately owned electric utilities. Journal of Public Economics 26, no. 1, Feb., 89-106.

Farinós J.E., Viñas, C., Garcia J. and A.M. Ibáñez, 2007. Operating and Stock Market Performance of State-Owned Enterprise Privatizations: The Spanish Experience. International Review of Financial Analysis, Vol. 16, No. 4, p. 367–389.

Fink, Carsten; Mattoo, Aaditya; and Rathindran, Randeep. An Assessment of Telecommunications Reform in Developing Countries. Working paper. Washington, D.C.: World Bank, 2002.

Foreman-Peck J. and D. Manning, 1988. How well is BT performing? An international comparison of telecommunications total factor productivity. Fiscal Studies, Institute for Fiscal Studies, vol. 9(3), pages 54-67, August.

Foreman-Peck, J., 1991. The Efficiency Effects of Privatization and Liberalization: The Telecommunications Industry Under State and Private Ownership. Economics Series Working Papers 99124, University of Oxford, Department of Economics.

Forsyth P., 1984. Airlines and Airports: Privatization, Competition and Regulation. Fiscal Studies, volume 5, Issue 1, p. 61-75.

Galal Ahmed and Nauriyal Bharat (1995), «Regulating telecommunications in Developing Countries - Outcomes, Incentives and Commitment», World Bank

Galal, A., L. Jones, P. Tandon and I. Volgelsang, 1994. Welfare Consequences of Selling Public Enterprises. Published for the World Bank, Oxford University Press, NewYork.

Gruber, H., Verboven, F., 2001a. The diffusion of mobile telecommunications services in the European Union. European Economic Review 45, 577–588.

Gruber, H., Verboven, F., 2001b. The evolution of markets under entry and standards regulation – the case of global mobile telecommunications. International Journal of Industrial Organization 19, 1189–1212.

Grzybowski, Lucasz, 2005. Regulation of Mobile Telephony across the European Union: An empirical analysis. Journal of Regulatory Economics, vol. 28, nº 1, July 2005.

Gutierrez L. and Sanford B., 2000. Telecommunications liberalization and regulatory governance: lessons from Latin America. Telecommunications Policy, Vol. 24 - Issue 10-11, pp 865 - 884 (Elsevier Science)

Henisz, W. and B. A. Zelner, 2001. The Institutional Environment for Telecommunications Investment," Journal of Economics and Management Strategy 10(1): 123-47.

Heshmati A. and El-Rhinaoui R., 2009. Effects of Ownership and Market Share on Performance of Mobile Operators in MENA Region. TEMEP Discussion Papers 200921, Seoul National University; Technology Management, Economics, and Policy Program (TEMEP), revised Nov 2009.

Hiroyuki I., 1994. Assessing the Gains from Deregulation in Japan's International Telecommunications Industry. Journal of Asian Economics, Vol. 5, No. 3, 1994, pp. 381-398.

Howard P. N. and N. Mazaheri, 2008. Telecommunications Reform, Internet Use and Mobile Phone Adoption in the Developing World. In proc. World Developement Vol.37 No 7 pp 1159–1169

Ioannis N. Kessides, 2004. Reforming infrastructure: privatization, regulation, and competition. World Bank policy research report, 306 p.

Jerome, A. 1996. Privatization in Nigeria: Expectations, illusion and reality. In A. Ariyo, ed., Economic Reform and Macroeconomic Management in Nigeria. Ibadan: Ibadan University Press.

Karamti C. and L. Grzybowski, 2010. Hedonic study on mobile telephony market in France: price-quality strategies. Netnomics, 2010, 11(3), 255-289.

Kole, Stacey R. and J. Harold Mulherin, 1997. The Government as a Shareholder: A Case from the United States. J. L. Econ. 40, pp. 1-22.

La Porta R., López-de-Silanes F., Shleifer A., and R. Vishny, 1999. The Quality of Government. J. L. Econ. Organ., 15, pp. 222-279.

Latzera M., Justb N., Saurweinc F., and P. Slominski, 2006. Institutional variety in communications regulation: Classification scheme and empirical evidence from Austria. Telecommunications Policy 30 (2006) 152–170

Lee B., and Shepherd W., 2000. Output and Productivity Comparisons of the Transport and Communication Sectors of South Korea and Australia, 1990 to 1998. School of Economics and Finance Discussion Papers and Working Papers Series 081, School of Economics and Finance, Queensland University of Technology.

Levy, B and P. Spiller, 1996. A Framework for Resolving the Regulatory Problem. In Levy, B & Spiller, P (eds.) Regulations, Institutions and Commitment: Comparative Studies in Regulation, Cambridge, Cambridge University Press.

Littlechild S., 2005. Mobile Termination Charges: Calling Party Pays versus Receiving Party Pays" mimeo, Judge Institute of Management Studies, University of Cambridge, May 26.

Magnien. F., 2003. Mesurer l'évolution des prix des services de téléphonie mobile : une entreprise difficile. Économie et Statistiques n°362, Juin 2003.

Maicas, J.P. et al, 2009. Reducing the level of switching costs in mobile communications: The case of Mobile Number Portability. In Telecommunications Policy, vol. 33, n° 9, October 2009.

Majumdar, Sumit K. 1996. Assessing Comparative Efficiency of the State-Owned, Mixed, and Private Sectors in Indian Industry. Pub, Choice, 96, pp. 1-24.

Megginson W. L., Nash R. C. and M. Van Randenborgh, 1994. The financial and operating performance of newly privatized firms, an international empirical analysis. The Journal of Finance Vol. 49, No. 2, P. 403-452.

Megginson, W., Nash, R. C. , Netter J., and A. Poulsen, 2004. The Choice of Private Versus Publc Capital Markets: Evidence from Privatizations. The Journal of Finance, p. 2835-2870.

Nash, R. C., Megginson, W., and M. Van Randenborgh, 1996. Does Privatization Work? Journal of Applied Corporate Finance, 23-34.

Nash, R. C., Megginson, W., and M. Van Randenborgh, 2001. The Financial and Operating Performance of Newly Privatized Firms: An International Empirical Analysis. Privatisation and Corporate Performance, The Journal of Finance, 397-446.

Newbery David M., XXXX. Privatization, Restructuring, and Regulation of Network Utilities. Review of Industrial Organization, Volume 19, Number 4, p. 497-500.

Nicoletti G. and Scarpetta S., 2003. Regulation, productivity and growth: OECD evidence. Economic Policy, CEPR, CES, MSH, vol. 18(36), pages 9-72, 04.

Omran, M., 2004. The performance of state-owned enterprises and newly privatized firms: Does privatization really matter? World Development, 32(6), 1019–1041.

Oniki, H., Oum, T., Stevenson, R. and Zhang, Y., 1994. The productivity effects of the liberalization of Japanese telecommunications policy. The Journal of Productivity Analysis 5, 63-79.

Öniş, Z., 1991. The Evolution of Privatization in Turkey: The Institutional Context of Public Enterprise Reform. International Journal of Middle East Studies, 23:163-176.

Pagoulatos, G. and N. Zahariadis, 2011. Politics, labor, regulation, and performance: lessons from the privatization of OTE. Hellenic Observatory papers on Greece and Southeast Europe, GreeSE paper no. 46. The Hellenic Observatory, London School of Economics and Political Science, London, UK.

Pan A. Yotopoulos, The (rip) tide of privatization: Lessons from Chile, World Development, Volume 17, Issue 5, May 1989, P. 683-702.

Parker D., and K., Hartley, 1991. Do changes in organization status affect financial performance? Strategic Management Journal, 12(8), 631–641.

Pentzaropoulos G.C, and D.I Giokas, 2002. Comparing the operational efficiency of the main European telecommunications organizations: A quantitative analysis. Telecommunication Policy, 26(11), p. 595-606.

Pinto B., Belka M., and S. Krajewski, 1993. Transforming State Enterprises in Poland: Evidence on Adjustment by Manufacturing Firms. Brookings Papers Econ. Act., pp. 213-261.

Ramandaham, V., 1989. Privatization in Developing Countries. New York: Routledge.

Rey P., and J. Tirole, 2007.A prime ronfore closure. In Mark Armstrong & Rob Porter (Eds.), Handbook of industrial organization, Vol.III(pp.2145–2220).

Ros A., 1999. Does Ownership or Competition Matter? The Effects of Telecommunications Reform on Network Expansion and Efficiency», Journal of Regulatory Economics Vol. 15 Issue 1 pp 65-92, January, (Kluwer Academic Publishers)

Singh J.P., 2000. "The Institutional Environment and Effects of Telecommunication Privatization and Market Liberalization in Asia." Telecommunications Policy, 24, 885-906.Kessides (2004)

Stennek J. and T.P. Tangerås, 2007. Better regulation of mobile telecommunications. European Policy Analysis, February, Issue 1-2007.

Symeou P.C., 2011. Economy size and performance: An efficiency analysis in the telecommunications sector. Telecommunications Policy Volume 35 Issue 5, June, 2011

Takano Y., 1992. Nippon Telegraph and Telephone Privatization Study. World Bank Discussion Paper, 1818 H. Street, NW, Washington, DC 20433.

Takashi Y. and D. Ypsilantis, 2001. Regulatory Reform in the Telecommunications Industry in Ireland. OECD

Thatcher M., 2001. The Commission and national governments as partners: EC regulatory expansion in telecommunications 1979-2000, Journal of European Public Policy, 8:4 August, pp. 558-584.

Tian, G. Lihui, 2000. State Shareholding and Corporate Performance: A Study of a Unique Chinese Data Set," working paper, London Business School, London.

Ure J., 2003. Telecommunications Privatisation: Evidence and Some Lessons. Working Paper, Telecommunications Research Project, University of Hong Kong.

Vining, Aidan R and Boardman, Anthony E, 1992. Ownership versus Competition: Efficiency in Public Enterprise. Public Choice, Springer, vol. 73(2), pages 205-39, March.

Wallsten S., 2000a. An econometric analysis of telecommunications competition, privatization, and regulation in Afric. Working paper, Palo Alto: Stanford University.

Wallsten S., 2000b. Telecommunications privatization in developing countries. The real effects of exclusivity periods. Working paper, Palo Alto: Stanford University.

Wallsten, S., (2002), "Does sequencing matter? Regulation & privatization in telecommunication reform", World Bank Working Paper No. 2817.

Xavier P. and D. Ypsilantis, 2000. Regulatory Reform in the Telecommunications Industry in Spain. OECD

Xavier P. and D. Ypsilantis, 2001. Regulatory Reform in the Telecommunications Industry in Czech Republic. OECD

Xu, Lixin Colin, 2002. The Impact of Privatization and Competition in the Telecommunications Sector around the World. World Bank, Washington, D.C.

Ypsilantis D., 2002. Regulatory Reform in the Telecommunications Industry-Canada. OECD

Ypsilantis D., 2002. Regulatory Reform in the Telecommunications Industry - United Kingdom. OECD

Ypsilantis D., and M. Wonki, 2000. Regulatory Reform in the Telecommunications Industry in Korea. OECD

Ypsilantis D., and M. Wonki, 2001. Regulatory Reform in the Telecommunications Industry in Italy, OECD

Zheng, S., and M.R., Ward, 2011. The effects of market liberalization and privatization on Chinese telecommunications. China Economic Review, Volume 22, Issue 2, June 2011, Pages 210-220.

2

A Consumer Perspective on Mobile Market Evolution

Laura Castaldi[1], Felice Addeo[2],
M. Rita Massaro[1] and Clelia Mazzoni[1]
[1]*Second University of Naples*
[2]*University of Salerno*
Italy

1. Introduction

In 2006 we performed a wide research on consumer behaviour in the Italian mobile communication market (Mazzoni et al, 2007). Using a multidimensional segmentation framework (LAM model), we identified three demand clusters according to consumers' *lifestyles, use motivations* and product *attributes*. One of the main findings was that two clusters out of three were characterized by a propensity to an integrated and service-oriented use of mobile communication. In other words, some consumers conceived mobile phones not only as simple communication devices, but also like technologically advanced multipurpose tools.

Nowadays these results seem to foresee the increasing importance that services are assuming in consumers' preferences. Indeed, since 2006, many changes have occurred in mobile market, among them the *servitization* phenomenon, leading handsets manufacturers towards an extension of their value chain on service delivering. Mixing both good and service components in their offerings, they are integrating phone devices with numerous software applications.

While previous studies focus on the supply side of servitization, its implications on consumers is less investigated.

In this chapter we therefore aim at analyzing the impact of servitization on mobile phones demand and on the LAM model's three dimensions.

The chapter is therefore organized as follows. Firstly, we present the LAM multidimensional segmentation model and its theoretical rationale, contextualizing it within the current state of literature on the subject of market segmentation with particular attention to the system of used variables (section 2). Results of 2006 research are described in section 3. Industry evolution since that time and consequent changes in the mobile market value are pointed out in section 4. Basing on a literature review, we then focus on servitization, underlining its market and consumer implications (section 5). Finally, the effects of servitization on the LAM model's three dimensions are briefly considered (section 6). Conclusion brings some considerations about further research (section 7).

2. The LAM model

Market segmentation is an activity of demand analysis leading to the identification of different groups of consumers (*segments*) as much as possible homogenous internally and heterogeneous with each other with respect to some relevant variables/characteristics.
Since Smith's 1956 seminal article, it has become the subject of attention on the part of academia and firms, with a wide variety of topics being investigated, ranging from its conceptual basis to the methodology to be adopted (Dickson & Ginter, 1987; Fabris, 1972; Frank, Massy, & Wind, 1972; Green, 1977; Haley, 1968, 1971, 1984; Saporta, 1976; Wedel & Kamakura, 2003; Wind, 1978; Yankelovich, 1964).
The stream of research on market segmentation has its theoretical foundations in the reflections upon imperfect competition theorized in the studies of Edward Chamberlin (1933) regarding monopolistic markets and of Joan Robinson (1933) on imperfect competition. In economic theory the moment of discontinuity as compared to the general equilibrium of perfect competition is marked by the configuration of systems in which firms stimulate the tastes of their potential customers through a differentiated supply, supposing that purchasers exhibit heterogeneous demand functions in relation to some significant variables (price sensitivity, interest in specific product attributes, brand notoriety, individual lifestyle and so on)[1].
Nowadays segmentation is well known for its different, and in some ways contradictory, approaches: on the one hand, the segment has become increasingly micronized and marketing increasingly personalised based on a fragmented consumer base (Collesei, 2000, 59); on the other, economies of scale are sought after on global markets, configuring transversal demand segments (Mazzoni, 1994) independent of the geographic location of consumers that belong to them[2].
In literature, since the works of Frank, Massy and Wind (1972) and Saporta (1976), the multiplicity of variables usable for demand segmentation has lead to an array of taxonomies. A useful one is offered by Wedel and Kamakura (2003, pp. 7-16) who, basing on some previous classifications, distinguish *segmentation bases* (i.e., sets of variables used for market segmentation) according to their *generality/specificity* and *observability/non-observability* (Figure 1).
Segmentation bases are *general* or, on the contrary, *product-specific* when, respectively, independent or dependent on the characteristics of products/services and on consumption/purchase circumstances; moreover, they are *observable* if directly detectable or *unobservable* when they can only be deducted.
The adoption of different segmentation bases leads to the identification of diverse demand segments. According to Wedel and Kamakura (2003, pp. 4-5, 16), depending on consequent segment *identifiability* (i.e., ease of distinction among segments), *substantiality* (targeted segments must represent a large enough portion of the market), *accessibility* by means of promotional and/or distributional actions, *stability* in time, *homogeneous responsiveness* to marketing efforts and *actionability* (consistence with firm goals and core competencies), the

---

[1] From here the normative approach to segmentation, based on the premise that demand functions need to be interpreted and categorised (Dickson & Ginter, 1987) in the attempt to maximise the effectiveness of management intervention through companies' marketing mix politics, to which consumers react differently.

[2] On cross-national segmentation studies see, among others, Agarwal, 2003; Aurifeille et al., 2002; Bolton & Myers, 2003; Hassan et al., 2003; Steenkamp & Hofstede, 2002.

efficacy of segmentation bases varies from *very poor* to *very good*. In general, the most effective bases are product-specific unobservable, although the interpretative capacity of segmentation analyses is improved by the combination of a number of different segmentation bases (Wedel & Kamakura, 2003, pp. 16, 341-342).

	General	**Product-specific**
Observable	Cultural, geographic, demographic and socio-economic variables	User status, usage frequency, store loyalty and patronage, situations
Unobservable	Psychographics, values, personality and life-style	Psychographics, benefits, perceptions, elasticities, attributes, preferences, intention

Fig. 1. Classification of segmentation bases (Wedel & Kamakura, 2003)

In 2006, in order to explore the Italian demand for mobile phones, we adopted a three-dimensional segmentation model, simultaneously using different segmentation bases.

The choice of the industry was driven by the fact that in the dynamic mobile telecommunication market firm competition is based on changes spurred by innovation, leading to a broad and varied offering of products (cell phones, smart phones and nowadays tablets) and services (from traditional voice services and common alarm, schedule, calculator and videogames functions to data management utilities and innumerable and constantly growing applications). More importantly, mobile telecommunication is one of the most surprising mass consumption phenomena of the last decades, which transversal diffusion has embraced all social classes. This is especially true in Italy, with the highest mobile penetration in EU (154,47% in 2009) and mobile voice traffic above the fixed one (53,3% vs 47,7% in 2009) (European Commission Digital Agenda Scoreboard, 2011)[3].

The numerousness and heterogeneity of Italian cell phone users have therefore made mobile telecommunication an ideal market for the adoption of a multidimensional segmentation model based on the joint use of three dimensions: lifestyles, product/service attributes and use motivations (Mazzoni, 1995).

The *lifestyles* dimension gives general indications on values and psychological characteristics of individuals, besides providing socio-demographic indicators, spending behaviours and mass media exposure[4]. It aims at investigating the individual's reference universe, so as to be familiar with his/her social values and his/her actual behaviour as a consumer and as a user of communication means.

---

[3] Italy is a front-runner in mobile broadband penetration of dedicated data services, stood at 10,2% as of January 2011 (EU average at 7.2%) (European Commission Digital Agenda Scoreboard 2011).

[4] The use of this dimension is fairly well-established both in literature and practice. Literature on lifestyles for market segmentation considerably developed in the 1970s (Gunter & Furnham, 1992; Kamakura & Wedel, 1995; Michman, 1991; Plummer, 1974; Wells & Tigert, 1971; Wells, 1974, 1975; Ziff, 1971 and some more recent works: Gonzalez & Bello, 2002; Vyncke, 2002; Yang, 2004).

As to the other two model dimensions, unlike the traditional benefit segmentation (Haley, 1968, 1971, 1984) as well as some more recent works (Ratneshwar et al., 1997; Wu, 2001), we distinguish between *product/service attributes* preferred by consumers and individuals' *use motivations*. Indeed, the two dimensions give different information: *use motivations* represent needs that induce purchase while *attributes* are indicative of the characteristics of the product/service that influence consumers' choice among the various models and brands in the market.

As consumer segments are described through the contemporaneous use of the three dimensions, they will be graphically identified by a parallelepiped (Figure 2).

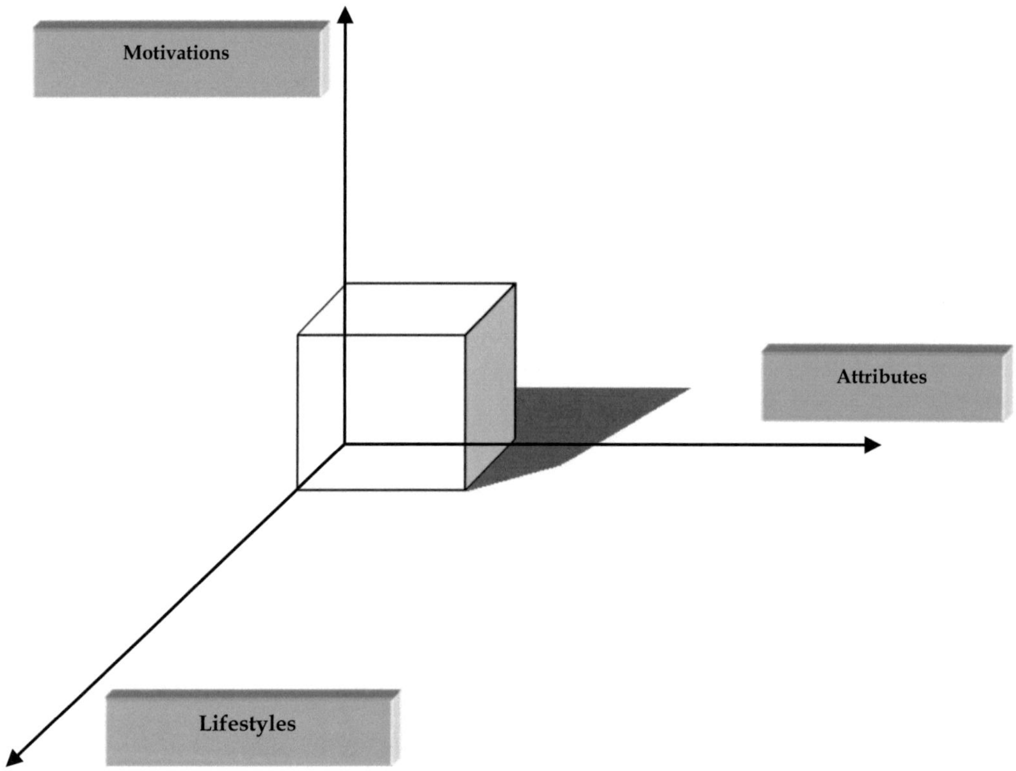

Fig. 2. Representation of a segment within the LAM model (Mazzoni, 1995)

The *LAM model* (lifestyles-attributes-motivations) aims to integrate analytical perspectives which are often used alternatively, one being based on subjective consumer characteristics and the other on the benefits/attributes sought in the product.

It uses three out of the four basic categories into which matrix in Figure 1 is subdivided: lifestyles are *general unobservable bases*, individual's socio-demographic data - falling, as stated above, within the definition of lifestyle that we adopt - are *general observable variables*, finally attributes and motivations dimensions are *product-specific unobservable bases*.

An evaluation of the LAM model based on the parameters proposed by Wedel and Kamakura, leads to the consideration that it should be an effective model for demand

segmentation. Its interpretative effectiveness is related not only to the use of unobservable product-specific segmentation bases, but also – as already said - to the joint adoption of different variable sets.
Table 1 shows the effectiveness of segmentation bases according to Wedel and Kamakura's six evaluation criteria, on a five-point scale ranging from very poor (--) to very good (++), thus highlighting the efficacy of the proposed model on all parameters (see boldface). Indeed, LAM model's general segmentation variables (lifestyles dimension) – in particular those observable (i.e., socio-demographic data) – are effective with respect to identifiability, substantiality, accessibility and stability parameters, while those product-specific and unobservable (i.e., attributes and use motivations, falling under the voice "benefits" in Table 1) lead to the identification of responsive and actionable consumer segments.

Variables/Criteria	Identifiability	Substantiality	Accessibility	Stability	Responsiveness	Actionability
1. General observable	++	++	++	++	-	-
2. Product-specific observable						
Purchase	+	++	-	+	-	+
Consumption	+	++	+	+	-	+
3. General unobservable						
Personality	±	-	±	±	-	-
Lifestyles	±	-	±	±	-	-
Psychografics		-	±	±	-	-
4. Product-specific unobservable						
Psychographics	±	+	-	-	++	±
Perceptions	+	+	-	-	+	
Benefits	±	+	-	+	++	++
Intentions	±	+	-	±	-	++

Table 1. Evaluation of segmentation bases (Wedel & Kamakura, 2003)

3. An application of LAM model on mobile communication market

The LAM model has been applied in several empirical studies covering different markets (Mazzoni, 1995; Mazzoni & Leone, 2001; Mazzoni, 2006; Mazzoni et al., 2007). Here we briefly synthesize the results of an application of the model to the analysis of the consumer behavior in the Italian mobile telecommunication market in 2006. This short summary is useful because the methodological path and the results of this application will be the starting point for the further research developments that will be discussed at the end of this chapter.
Multidimensional segmentation of mobile market was pursued using a post-hoc and descriptive segmentation method: *post-hoc* means that "the type and number of segments are determined on the basis of the results of data analyses" (Wedel & Kamakura, 2003, 17), while *descriptive* refers to the analysis of "the associations across a single set of segmentation bases, with no distinction between dependent or independent variables" (*ibidem*).

One of the most significant features of our research design was the choice of a mixed method approach (Cresswell, 2003; Tashakkori & Teddlie, 2003): this relatively new methodological perspective aims at widening and, possibly, enhancing the analysis of a phenomenon by integrating both quantitative and qualitative techniques to collect and analyze data (Bernard, 2000; Morse, 2003, 189-208).

The mixed method research design of our research could be described as a *model 1* design (Steckler et al. 1992) or a *sequential exploratory design* (Cresswell, 2003): in our research, qualitative data collection and analysis Precedes the quantitative ones, and the results of the former are used to develop measurement tools to be applied in the latter.

Put specifically, data collection was accurately performed through a sequence of: focus group, pre-test and CATI survey.

In the first research stage, two focus groups were organized, each with 10 participants selected according to their socio-graphic characteristics, their phone usage, and their propensity to use technological devices. The qualitative analysis of focus groups interactions allowed research group to gather important methodological information about the unit of analysis[5] and the variables to be inserted into the conceptual map of the research and therefore in the questionnaire. In particular, qualitative analysis confirmed some preliminary hypotheses of research group about the future development of Italian mobile market. In fact, in the first five years 2001-2005, Italian companies operating in mobile market pushed towards the integration of video and mobile communication, offering products (mobile phones) and services that should have fostered the diffusion of mobile TV and videophone calling. But they were pointing at the wrong target: Italian consumers did never fully appreciate video extension of mobile communication. From the analysis of focus groups emerged that customers considered videophone calling too invasive, while mobile TV was seen as not very useful, uncomfortable and too expensive. Besides, the integration of mobile communication with the Internet and online services were in the embryonic stages at that time, so even if research group recognized this integration as a possible development of mobile communication, it was decided not to analyze it in depth.

The use of mobile phone as sort of personal computer was took into account by researchers too: "using mobile as a palmtop", a name that sounds anachronistic now, was one of the items to be rated by respondents as a motivation to use mobile communication. However, even in this case, researchers decided to limit the items related to this potential palmtop use of mobile phone, considering the state of art of mobile technology and its market penetration at that time. Moreover, as pointed out by qualitative analysis of focus groups interactions, using mobile phone as a palmtop was perceived by consumers as limited to certain categories of people, i.e., manager or business men, a very different situation from the current state of play of mobile market, characterized by the increasing diffusion of smartphones and mobile devices (see section 4).

So researchers' considerations about the too early stage of the integration between mobile communications and other media were supported both by qualitative and quantitative data analysis. However, in light of the fast changes in mobile market during the last five years, the integration between mobile communications and the Web has to be considered one of the starting point to redefine the multidimensional segmentation model that will be discussed in section 7.

---

[5] At the beginning of the research process, research unit analysis was thought to be made up of people aged at least seventeen. Qualitative analysis showed that people begin to use mobile communication much earlier.

All the suggestions emerged from focus group analysis led to a preliminary version of the questionnaire that was administered as a pre-test on 100 people, randomly extracted from the research population. The size of the sample is unusually large for a pre-testing: according to literature a few cases could be sufficient (Sheatsley, 1983, p. 226). This large pre-test sample was chosen in order to have sufficient data to perform a reliable multivariate analysis and evaluate correctly the goodness of the techniques used in the segmentation procedure. Pre-test results gave other indications to refine the choice of variables.

The result of this mixed research process was an accurate variables' selection for each dimensions of the LAM model, product attribute, use motivation and lifestyle. These variables were then operationally defined into the items of the final version of the questionnaire, administered with the CATI (Computer Aided Telephonic Interview) technique.

The variables of attribute and motivations dimensions (Tables 2 and 3) were measured with a Cantril scale technique, while the multidimensional nature of lifestyles dimension required a more elaborate operational definition (Table 4).

variable	operational definition and description	measurement
	*interviewees were asked to specify the **importance** attached to each attribute*	
economic		
price	price of mobile phone	
promotion	promotional offers	
tariff	cost of calls and other services	
physical		
handiness	convenience for cell phone use	
battery life	duration of cell phone battery	
screen visibility	quality of cell phone display	
durability	solidity of cell phone over time	
signal reception	quality of signal reception	each variable is measured using a Cantril scale
aesthetical		
aesthetics	cell phone design	
personalization	possibility of personalising the mobile phone with covers, ring tones, etc.	
brand reputation	mobile phone brand reputation	
technological		
advanced services	availability of technologically advanced services (data management, Internet, etc.)	
accessories	quantity of available accessories (earphone, speakerphone kit, etc.)	
other functions	availability of other functions (photos, music, videophone, etc.)	

Table 2. Description of *attributes* variables (Mazzoni et al., 2007)

variable	operational definition and description	measurement
	*interviewees were asked to specify the **importance** attached to each motivation*	
relationships		
family	It allows me to communicate with my friends	
friends	It allows me to communicate with my family	
SMS/MMS	I use it to send SMS/MMS	
entertainment	I use it to entertain myself (music, games, photos, Internet, etc.)	
affiliation		
trendiness	I like to be trendy	
group	It makes me feel part of a group	each variable is measured using a Cantril scale
security		
work	It is necessary for my work	
privacy	I can safeguard my privacy	
expenses	I can control/contain my telephone expenses	
personal security	It makes me feel safe	
information and entertainment		
additional functions	I find the additional functions useful (calendar, calculator, alarm, etc.)	
advanced services	I'm interested in the advanced services (videophoning, Internet, etc.)	
photo camera	I can use it as a photo camera	
palmtop	I use it as a palmtop (agenda, data management, etc.)	

Table 3. Description of *motivation* variables (Mazzoni et al., 2007)

Questionnaire was administered with CATI technique to 1067 Italian citizens, aged between 14 and 65selected through a random sampling procedure[6].

According to the goals of LAM Model, data analysis consisted in a segmentation procedure based on the sequential application of two multivariate techniques, factor analysis and cluster analysis. The sequential combination of these two techniques is a common practice in methodological literature as it provides an extreme synthesis of data (Di Franco, 2001).

Factor analysis was performed to identify latent factors beneath each set of variables; cluster analysis allowed to group respondents according to those factors.

Factor analysis showed that a three-factor solution best represented each dimension (Table 5).

[6] Random sampling procedure was proportionally stratified: strata were built considering the distribution of Italian population by region, gender and age. Sample size was calculated with a 3% standard error.

variable	operational definition and description	measurement
	each variable required a different operational definition	
socio-graphic		
sex	Gender	dichotomy
age	Years	open ended questions
qualification	higher qualification obtained by interviewees	classification
marital status	current marital status of interviewees	
occupation	current occupation of interviewees	
residence	province in which interviewee lives	
values and interests	*interviewees had to specify the **importance** attached to each item*	each variable is measured using a Cantril scale
bodycare	importance attached to care of the body	
culture	importance attached to culture	
environment	importance attached to environmental respect	
family	importance attached to family	
friendship	importance attached to friendship	
love of country	importance attached to love of country	
personal success	importance attached to personal success	
politics	importance attached to politics	
religion	importance attached to religion	
social commitment	importance attached to social commitment/voluntary work	
solidarity	importance attached to social equality/ solidarity	
sport	importance attached to sport	
work	importance attached to work	
media usage	*interviewees were asked to specify **how often** they perform the following activities:*	each variable is measured using a 4 point Likert scale
cinema	going to the cinema	
theatre	going to the theatre	
radio	listening to the radio	
video games	playing with video games	
book	reading a book	
dailies	reading dailies	
magazines	reading magazines (weekly, monthly)	
sports dailies	reading sports dailies	
Internet	surfing the Internet	
mobile phone	using a mobile phone	
computer	using the computer	
television	watching television	

Table 4. Description of *lifestyles* variables (Mazzoni et al., 2007).

Motivation dimension is represented by the following three factors:
- *integrated use*: the choice of mobile devices is driven by the need for a tool allowing efficient communication and time management;
- *info-entertainment*: this factor highlighted the alternative use of mobile phone (this use was emergent at the time research was performed, and it is widespread nowadays) as a device for gaming, music and photo;
- *relationships*: it is the conventional and basic use of mobile phones as simple tools to communicate, in other words, mobile devices are seen as an extension of the traditional phone.

The three factors representing attributes dimension were:
- *practical aspects*: this factor regards those basic attributes, allowing an easy and efficient use of the mobile phone (ease of use, battery life, screen visibility, and durability);
- *state-of-the-art*: it underlines the preference attached to the attributes of technologically advanced mobile phones (availability of technologically advanced services, quantity of available accessories, availability of other functions);
- *service convenience and quality*: this is characterized by economic and quality attributes connected, on the one hand, with the pricing policy of handset manufacturers and network operators, and on the other hand, with the good functioning of the phone in its basic function, i.e., signal reception.

dimension	*factor's name*	*explanation*
motivation	integrated use	sophisticated needs for efficient communication and time management
	info-entertainment	need for amusement, information availability, and the desire to keep up with the times
	relationships	simple need for usual communication and for more traditional mobile services
attributes	practical aspects	basic attributes of mobile phones allowing simple and immediate use
	state-of-the-art	more advanced characteristics of mobile phones
	service convenience and quality	economic and quality attributes
lifestyles	Connected	multiple uses of modern media and a great interest in sport
	Committed	demanding cultural consumptions, principles of culture and social participation, and scant interest in the care of the body and sport
	Traditionalist	basic information consumption and traditional values

Table 5. Factor Analysis Results

Finally, lifestyles factor analysis gave these three factors:
- *connected*: characterized by a variety of cultural consumptions (Internet, video games, cinema, radio, etc.), this first factor is positively correlated with the sport value and negatively correlated with social commitment/voluntary work value;
- *committed*: positively correlated with cultural consumptions (such as reading books, going to the theatre, reading newspapers), and with social and cultural values; on the contrary, it is negatively correlated with care of the body and sport values;

traditionalist: this factor is positively correlated with traditional values, such as family, friendship and work, and negatively correlated with that of social commitment/voluntary work and social equality/solidarity. The nine factors above shown were subsequently used as criteria variables in a non-hierarchical cluster analysis, performed using K-means algorithms and Euclidean distance.

Cluster analysis results led to the identification of three clusters that could be considered as demand segments for mobile market.

According to the different combination of factors and to the socio-demographic information characterizing each clusters solutions, demand segments were named *techno-fun*, *value-driven* and *basic users*.

Figure 3 shows their position in multidimensional space.

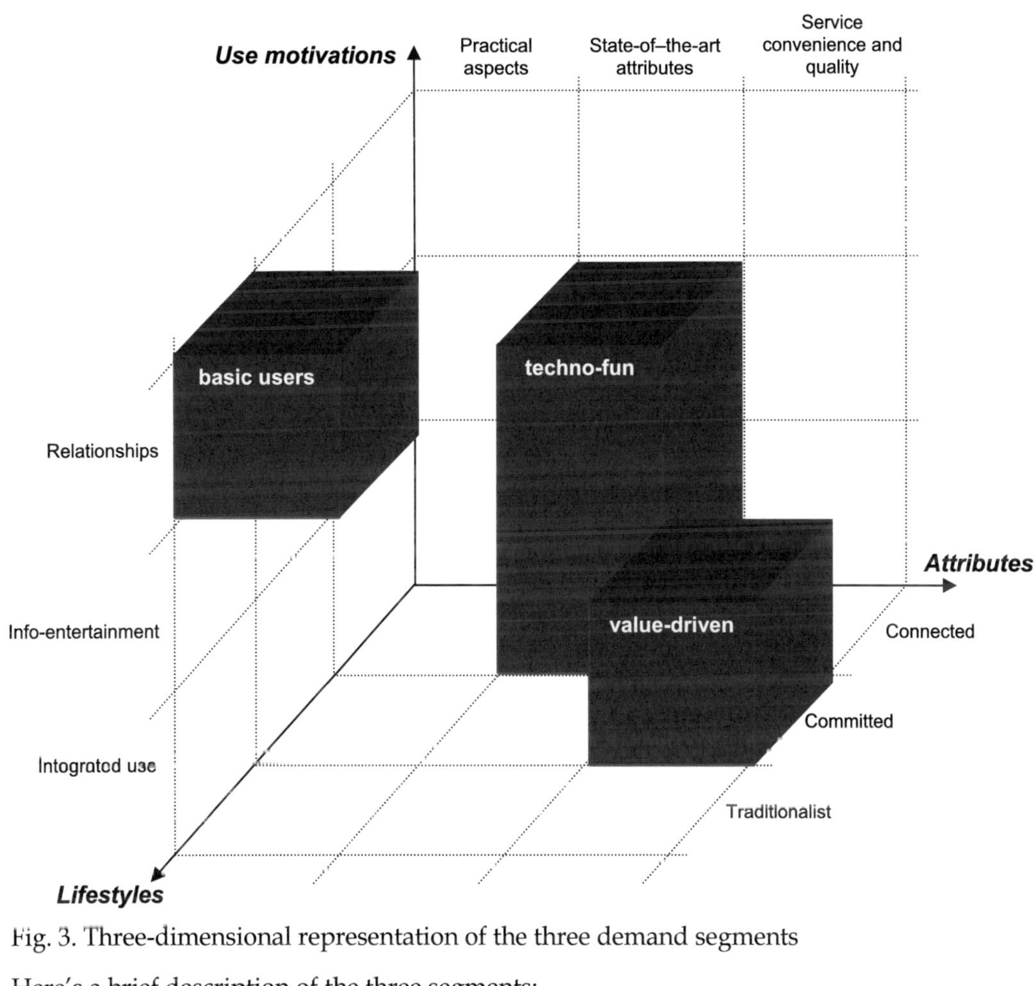

Fig. 3. Three-dimensional representation of the three demand segments

Here's a brief description of the three segments:

- the *techno-fun* segment: consumers belonging to this cluster (24.3% of respondents) are sensitive to media integration and technology, mainly with an entertaining attitude. Techno-fun people are mostly males (66.4 %), with an average age of 26 (in fact they are mainly students, unmarried and have completed secondary education). They have a connected lifestyle (multiple consumers of new media) and they buy a mobile phone (often more than one) motivated by integrated use (take advantage of additional functions and advanced services) and info-entertainment (self-entertainment and the desire to keep up with the times). Techno-fun consumers pay particular attention to the state-of-the–art attributes: presence of technologically advanced services, availability of other functions (photos, music, videophone) and accessories, while they show a limited interest in price, handiness and aesthetic characteristics. Considering the recent development of mobile market, this cluster is the most interesting because it embodies the direction of the change. These consumers, mostly young, conceived and used mobile phone as a multimedia devices and they were enthusiastically willing to explore all the possibilities offered by the integration between mobile communication and other media. People belonging to this cluster now have grown older and they have probably bought a smartphone or a tablet (or maybe both of them). In other words, they were a driving force that has led the market evolution to the current paradigm shift in mobile communication: the always connected style. Probably this segment has grown up to become the biggest one among mobile market consumers;

- the *value-driven* segment: it includes consumers (38.3% of respondents) driven by rationality and functionality criteria. In their daily life, so in purchasing and using mobile phone too, they try to maximize the value of their choices. The segment is composed mainly of married and single women (58.4 %), aged 35 (average) with a high educational level; they are mostly clerks and show principally a committed lifestyle. When purchasing a mobile phone, they are driven by an integrated use motivation and they search for service convenience and quality attributes (costs, promotional offers and signal reception), showing low interest in mobile phone design and aesthetics. Put specifically, consumers belonging to this cluster show a rational evaluation of mobile phone: it is conceived as an useful and technologically advanced tool, necessary not only to communicate, but also to manage time and organize daily life. So, the mobile phone purchasing choice is based on an accurate evaluation of cost/quality ratio. Value-driven segment plays a key role in our analysis: it is an "adult" cluster in the sense that these consumers have an aware and focused consumption of mobile phone. They are permeable to technological innovation, but they tend to subordinate it according to their personal needs. In a certain sense, a value-driven individual is what a techno-fun may become in the future;
- the *basic users* segment: consumers belonging to this segment (37.4% of respondents) share an essential use of mobile phone. They have a traditionalist lifestyle, based on primary values (family, friendship and work) and basic consumptions. They are women (59.8 %) with a low qualification and an average age higher than the other clusters (48 years), usually married. They are self-employed workers, retired people or housewives. Mobile phone for this consumers has to perform exclusively its primary function: communicate (motivations are connected to their social relationships). So they search for those attributes linked to practical aspects, that is, to all the features allowing a simple and efficient use of mobile phone: ease of use, battery life, screen visibility and durability. It is not unlikely to forecast that this segment is destined to shrink as the diffusion of integrate mobile communication is wide spreading among new generation of consumers.

Research findings proved significant differences among mobile phone consumers according to the three dimension of the LAM model proposed by Mazzoni. Each cluster had very

peculiar characteristics related to lifestyles, motivations and attributes. The three market segments got in touch with mobile devices in different times and ways, and therefore have a different approach to mobile communication.
Even if results seem to be good and reliable, epistemological cautions and methodological limitations should never be forgotten: segmentation is a representation of a reality, constructed under the conceptual and methodological choices of the research group. So, when interpreting cluster analysis results, one must be careful, avoiding to fall into the fallacy of reification, as Wedel and Kamakura clearly state: "in applying models to segmentation, one should recognize that every model is at best a workable approximation of reality. One cannot claim that segments really exist or that the distributional form of unobserved heterogeneity is known. Segmentation is a marketing concept that is used to approximate the condition of market heterogeneity by positing diverse homogeneous groups of customers. It has proven to be a very useful concept to managers, and we conjecture that it will continue to be so far some time" (2003, p. 329).
However, a broad representation of consumers could be very useful for those interested in the evolution of mobile market: academic scholars, companies and firms. Moreover, conceiving segmentation results (i.e., the consumers clusters) as something not fixed, but open to be discussed and eventually changed, made LAM multidimensional segmentation model a flexible analytic tool that could be refined according to the development of the market.

4. The evolution of mobile market towards a user-centred structure

When we conducted our research the mobile telecommunication market appeared rather different from today.
At that time, mobile telecommunication business was MNOs-centred. Indeed, also because of their direct and strong relationship with consumers, mobile network operators (MNOs) were leading firms, ruling, connecting and coordinating the other value-creating network operators in order to generate a high end value. Thus MNOs directly delineated the structure and managed the network, defining operators' connection modalities and communication codes. MNOs also stimulated innovation, determining its speed and direction. More than mere data carriers they were mobile media companies, exercising not only control over provided content and services but also influence on handset manufacturers as to obtain the diffusion of cell phones enabling the easy fruition of advanced, innovative services (Sorrentino, 2006, pp. 53-60).
In summary, as shown in Figure 4, MNOs were central in the mobile telecommunication value network (Kothandaraman & Wilson, 2001; Li & Whalley, 2002; Maitland et al., 2002; Stabell & Fjeldstad, 1998; Tilson & Lyytinen, 2006), managing the relationships with all network agents and conveying to end-users other operators' valuable products and services.
Moreover, at that time, mobile handset business was a stable oligopoly mainly dominated by operators that made successful entry decisions at the beginning of the mobile era: Nokia, Motorola, Samsung, Sony Ericsson and LG Electronics (ENTER & IDATE, 2007, p. 21; West & Mace, 2007, p. 2).
The year 2007 marked the beginning of a new era for mobile telephony, following the introduction of feature-rich and easy to use cell phones, first of all Apple's iPhone[7] (went on sale on June 29, 2007).

[7] Over 500,000 units were sold on the first weekend immediately after its launch (Laugesen & Yuan, 2010, p. 91).

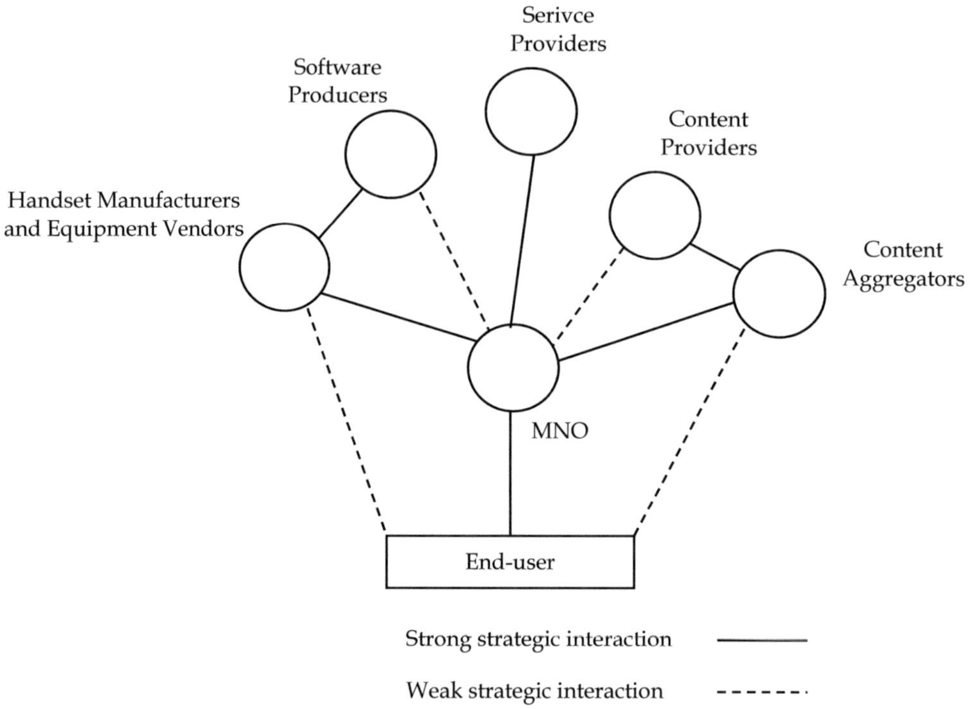

Fig. 4. Mobile telecommunication value network (Sorrentino, 2006)

Since that time many iPhones have been sold, making of Apple - a new comer - the fourth largest brand in the worldwide mobile phone market, with 16.9 million mobile devices sold to end-users in the first quarter of 2011 (Table 6).

	2007			1Q11	
Company	Units	Market Share (%)	Company	Units	Market Share (%)
Nokia	435,453.1	37.8	Nokia	107,556.1	25.1
Motorola	164,307.0	14.3	Samsung	68,782.0	16.1
Samsung	154,540.7	13.4	LG	23,997.2	5.6
Sony Ericsson	101,358.4	8.8	Apple	16,883.2	3.9
LG	78,576.3	6.8	RIM	13,004.0	3.0

Table 6. Top 5 worldwide mobile terminal sales to end-users 2007-1Q 2011 (thousands of units) (Gartner Press Release, 2008; 2011)

The technological innovation of Apple's iPhone has changed consumers' needs, has moved competition to a new level, has ruled out of market operators unable to adapt to change and has originated a *gateway to entry*[8], favouring the access of new firms often operating in other industries and accelerating the media convergence phenomenon.

[8] On *gateways to entry* see Yip, 1982.

Despite the lack of experience with mobile telecommunication, Apple could draw upon its core competencies (Prahalad & Hamel, 1990) in product design, innovation and marketing, personal computer hardware and software, online distribution systems and network management[9]. It could therefore deliver a converged handset able to provide traditional voice, data, entertainment and mobile Internet services (West & Mace, 2007, p. 2).

The iPhone success is strongly due to the high quality of browsing experience (Eaton, 2009; Laugesen & Yuan, 2010, p. 94).

While in the past it was largely believed that there was a need for a new version of the Internet in order to make it appropriate for use on cell phones (because of limited mobile data speed, small screens and no keyboards on handsets), Apple - sustained by infrastructural innovations such as the development of 3rd generation standards and the wireless Internet connection - created a mobile capable of delivering (in a user-friendly way) the wired Internet, thus leveraging its already-mature ecosystems. The iPhone improved mobile browsing experience having a large touchscreen, the Safari standard browser based on that developed for its personal computers (rather than a rewritten one), a graphical user interface with intuitive scrolling, panning and zooming designed specifically for touchscreens, and no physical keyboard (Eaton, 2009; Laugesen & Yuan, 2010, p. 94; West & Mace, 2010, pp. 275-276). This way it proved that the killer application for the mobile Internet was the same as for the wired Internet, i.e., a web browser: as the browsing experience became similar to that on PCs, mobile Internet usage increased dramatically (AdMob, 2010; West & Mace, 2010, p. 279).

Moreover, basing on its iTunes competencies, Apple developed the App Store, an online marketplace to deliver its own as well as third-party applications. It was launched in July 2008 and in the first month users downloaded more than 60 million apps (Wingfield, 2008) to arrive to 10 billion downloads out of the 350,000 apps available as of January 2011 (Apple Press Release, 2011).

As it controls many of the assets related to the value proposition and has a direct and strong relationship with customers, the business model adopted by Apple through the iPhone represents what has been named a *system integrator platform* (Gonçalves et al., 2010).

As illustrated in Figure 5, while customers pay MNOs for network access, they buy handsets and mobile content and services through the system integrator platform (i.e., Apple). Indeed the latter produces and delivers third-party products/services[10].

Several competitors – already operating in the mobile market as well as new comers (e.g., Google) – have followed (imitated and emulated) Apple's strategy. Indeed mobile phone manufacturers have acknowledged the rising importance of content and services for end-users (Cusumano, 2010, p. 22) and therefore recognized the value of delivering devices enabling the fruition of personalized services[11].

---

[9] Software developers were crucial for Macintosh adoption and diffusion as well as content providers and third-party add-on suppliers for iPod success.

[10] System integrator platforms' success rests upon the ability to simultaneously attract application developers and end-users. Therefore they allow and encourage third-party developers to use their platform in order to increase its value as well as end-user service offering (Gonçalves et al., 2010, pp. 67, 69-70).

[11] This is moving competition from physical attributes (e.g., handiness, battery life, signal reception, etc.) to soft components of handsets, such as operating systems, graphical user interfaces, online marketplaces.

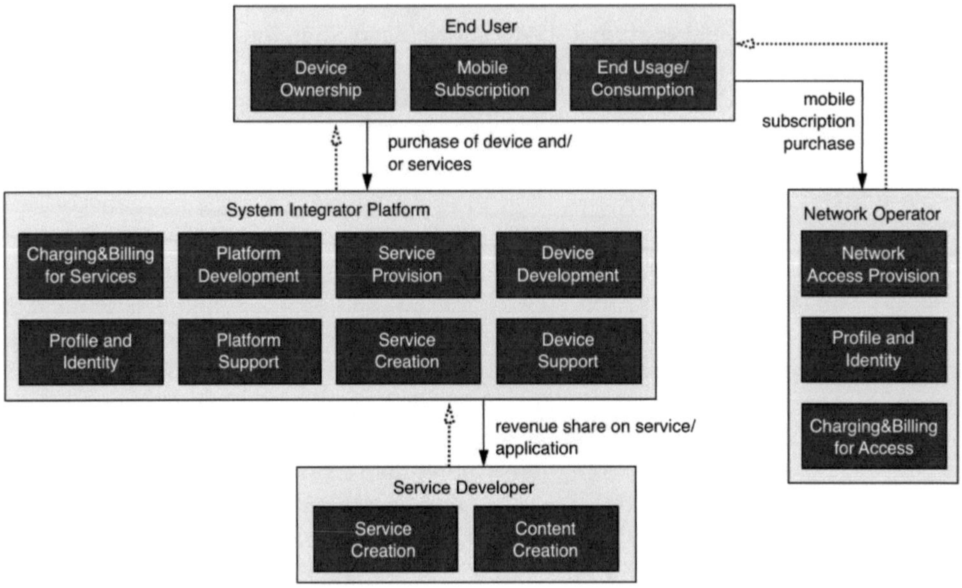

Fig. 5. System integrator platform (Ballon, 2009, as cited in Gonçalves et al., 2010)

As a consequence, today's handsets are not simply phones in their traditional definition but better mobile computing platforms for voice communication and content and services fruition, i.e., what has been named a *mobile ecosystem* (Mitchener, 2009).

The huge amount of available software applications and accessories (often third-party and subject to approval procedures) allows vast personalization of handsets through hardware and multimedia content (news, games, music, utilities, VoIP software, social networks, etc.), determining devices' functional capabilities.

The emergence of Apple in the mobile telecommunication market has therefore changed the original characteristics of the mobile phone business, broken down its traditional boundaries and moved power from MNOs to handset manufactures, hardware and software developers and consumers, thus leading to the reconfiguration of the value network (Laugesen & Yuan, 2010; Vogelstein, 2008). In particular, end-users have assumed a central role as they can directly define, thus personalize, the characteristics of their devices. This is also due to the fact that, because of the ease of access to the wired Internet through new mobile phones, all of its existing applications, content and services are immediately available for consumers, through a Wi-Fi connection or MNO service (where MNO operates as mere bit-pipe)[12].

This is making obsolete many of the services previously offered to consumers by MNOs and paradoxically is likely to set back these operators' role to that prevailing of voice and data carriers, complemented by the most recent of bit-pipe for the Internet connection.

Today's mobile telecommunication business thus appears Internet-based and user-centred, as consumers can easily surf the Web and create their own personalized mobile telecommunication product/service

[12] "Now, in the pursuit of an Apple-like contract, every manufacturer is racing to create a phone that consumers will love, instead of one that the carriers approve of" (Vogelstein, 2008).

In the new value network, as shown in Figure 6, handset manufactures are directly connected with end-users, to whom they convey content and services. Therefore, more than being influenced by MNOs, nowadays handset manufacturers exercise an influence on them as they represent the gateway to new consumers[13].

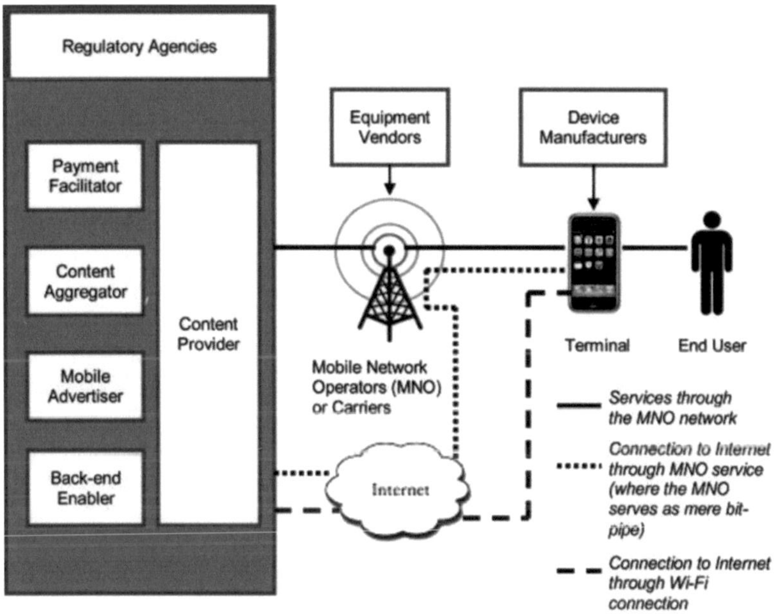

Fig. 6. Mobile value network (Seal, 2010)

Mobile telecommunication business therefore is user-centred nowadays, as consumers can directly shape the characteristics of the product/service on their exact needs.
As a consequence, content, services and innovative processes are not MNOs-directed anymore but better a direct consequence of end-users' characteristics and needs.

5. The advent of service economy

As already said in section 4, during the years and particularly since 2007 mobile market has undergone evolutions. Among the several changes that have occurred, in what follow we focus our attention on *servitization*. Through a literature review, the attempt is to define the servitization phenomenon (section 5.1) and describe its implication on company business and activities (section 5.2). Finally (section 5.3), how servitization takes shape in mobile market is described.

5.1 Servitization

The expression *service economy* refers to the increasing economical importance that service sector - compared to other sectors as agriculture and industry - has been taking in the industrial economy. Strict meaning, service economy (Fuchs, 1968; Gustafsson & Johnson,

[13] For the period July 2008-June 2009, 40% of newly activated iPhones brought AT&T new subscribers, accounting for 48% of the operator's new subscriptions during the same year (West & Mace, 2010, p. 279).

2003) is also used as synonymous of servitization (Baines et al., 2007). The servitization is the evolution of the offering from a material product to one which is inseparable from services. Since products and services become one offering, manufacturers shift and extend their business from goods selling to services delivering.

Boundaries among what was conceived as the material part (product) and the intangible part (service) of supply are vanishing; indeed, as a product offering is enriching with services, inverting this perspective, *productization* is leading the inclusion of material component in service offering. Thus many authors argue that there is no more reason to distinguish among tangible and intangible components, as offerings in service economy are composed by a variable mix of products and services often named *product-service-system (PSS)* (Manzini et al., 2001; Baines et al., 2007; Mont, 2002).

A PSS offering consists of three parts (Goedkoop et al., 1999):

1. *product*: a tangible commodity manufactured to be sold;
2. *service*: an activity (work) done for others with an economic value;
3. *system*: a set of elements including their relations.

As the mix and the combination of product and services and their relations can vary, there are different types of identified PSS. In a study on the state of art of PSS research (Baines et al., 2007) emerges that literature converges on the existence of three PSSs types:

1. product-oriented offering: selling the product to customers while adding after sales service (i.e., assistance, maintenance and repair);
2. use-oriented offering: selling the availability of a product without giving to customer its ownership (i.e., leasing a product without selling it);
3. result oriented offering: it is sold the product result instead of a product (i.e., selling photocopies instead of photocopier, or washed laundry instead of a washing machine).

Tukker (2004) reports eight types of PSSs representing subcategories clustered by their economical and sustainability potential and characteristics (see Figure 7).

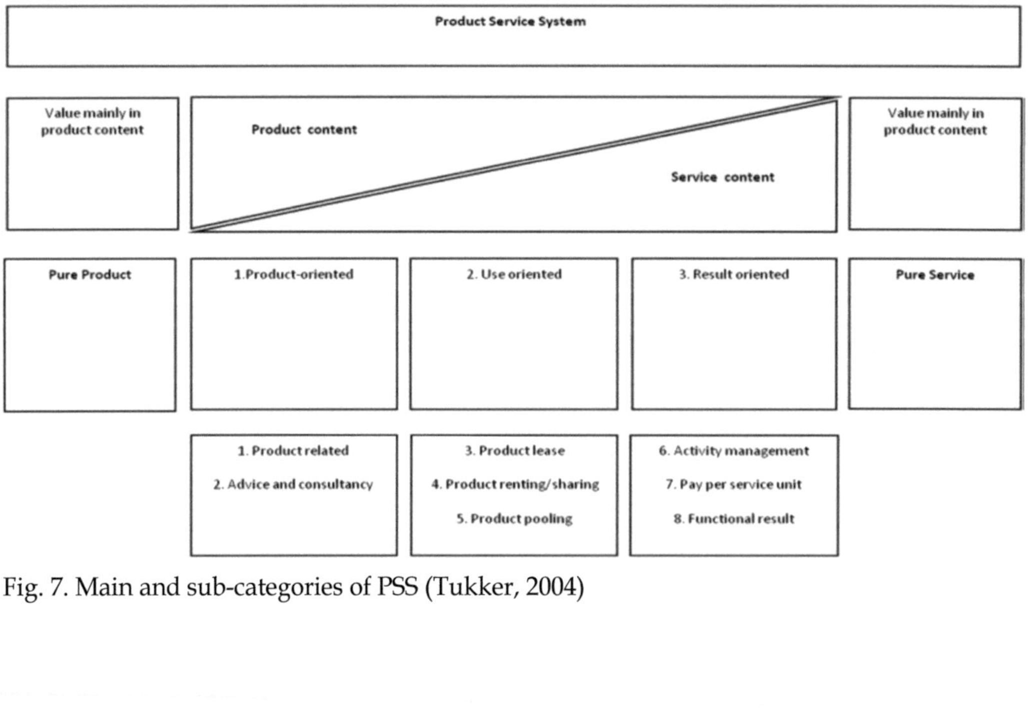

Fig. 7. Main and sub-categories of PSS (Tukker, 2004)

Beyond many ways to cluster, a product service system can be put on a continuum where on one side of the spectrum there is the traditional manufacturer who adds service to its product and, on the opposite side of the continuum, there is a product that is just a vehicle for services and value delivery (Wong, 2004, as cited by Baines et al., 2007).

5.2 Moving towards services: business implications

Servitization may affect several aspects of firm's business model: operations, organizational structure, costs and investments structure, marketing management and on a broader scale, the overall strategy and its focus.

From an operational point of view, companies need to move up the value chain in order to couple manufacturing activities with service providing. This shift implies a reorganization of firm's structure and activities, while efforts in the acquisition of new managerial and technical skills are necessary.

Business servitization also compels companies to make changes in the behavioural process and in the organizational culture; indeed, despite the high revenue expectation stemming from extending a business to services, firms have low propensity towards the servitization of their own business. As a previous study shows (Gebauer & Fleish, 2007), companies seeing service as 'evil', hinder supplementary services providing because their managers lack of motivations.

Enriching a product offering with services might also require to change the centre of the value proposition and, consequently, the business strategy. Indeed, through the adoption of the most radical servitization model towards the end of the spectrum, business focus will necessarily shift from the product to those services strictly linked with it. While in a product oriented approach manufacturers projects and sell just products adding few services, in a more servitized approach, the strategy is plotted out on many services for which product is a complementary good, while services become the basis on which competitive strategy is drawn.

Companies also gain competitiveness by costs reduction and energy saving; in fact, it is not by chance that in the literature, the ecological implications leaded by servitization of manufacturing are one of the most frequently investigated aspects (Baines et al., 2007; Gebauer & Fleish, 2007; Tukker, 2004). It is clear that while moving along the product-service continuum from the product-oriented type up to the result-oriented type, these positive effects become stronger and stronger. Surely, servitization is a growth strategy that offers opportunities to reach a competitive edge by increasing revenues.

Nevertheless, the major motivation pushing managers to servitize firms' business seems to be the willingness to match customers' needs (Baines et al., 2010). Servitizing a business adds value to product and supports firm competitiveness since it allows to deliver a high customized offering that is tailored on customer' needs and enables to create a strong consumer relationship (Tukker, 2003). A mass-market product like the mobile phone becomes extremely customizable by the complementary services that can be integrated into it: software updating allows customers to entail the mobile phone functionality on their unique needs. Therefore, it should be clear that extending business to service also involves changes in the marketing strategy that might switch from a transactional approach to a relational one, as services can be sold and delivered over a period of time (Baines & Lightfoot, 2011). Authors argue that servitization fosters the rise of a new *service dominant logic*, where marketing becomes a process of doing things in interaction with customer who is a co-creator and a co-producer of service (Vargo & Lush, 2004).

5.3 Combining service and product in mobile market industry: ecosystem

During 1990s, featured by a product stagnant demand in many manufacturing industries, moving downstream of value chain towards providing services was a winning strategy for many firms. Companies became aware that most of opportunities lay in services that have been promising high margins and sales despite few assets are required for their delivery (Wise & Baumgartner, 1999). Therefore, servitization is not a new phenomenon, but nowadays the integration of intangible components and service into manufacturing good is involving also mobile market.

Service and product in mobile market has been always strictly linked, since some services provided by the MNOs (e.g. Multimedia Message Service, Internet connection, etc.) are accessible only if mobile manufacturers enable these functionalities on their mobile device and vice versa. Even though this relation is also intuitive, empirical researches support these statements. For example, Ono and Tang's research (2010) shows that handset and network mobile service are chained up in their evolution since the diffusion of a network mobile service is wider if handset entailed on this service will be broadly spread in the market. Although convergence in mobile industry does not concern the integration of handset with network service, it cannot be denied that the offering in mobile market is also enriching of intangible components named contents, applications, software and generally speaking services. "The increased offering of fuller market packages or 'bundles' of customer focused combinations of goods, services and knowledge" (Vandermerwe & Rada, 1988, as cited by Baines and Lightfoot, 2011) is totally involving mobile industry as well. The cornerstone of this evolution towards servitization is often placed in June 2007 when the iPhone was launched on the market.

Apple's strength mainly lay in its ability to set the standard for mobile web surfing and change consumers' browsing experience. While other industry players as MNOs and content suppliers were trying to improve networks or to adapt web contents to the handset use, Apple just made surfing more similar to the wired Internet experience by introducing a wide touch screen, scrolling and zooming functionalities (West & Mace, 2010). With the enhanced opportunity to have an easy access to Internet, the "walled garden" (*ibidem*, p. 283) of contents managed and provided by service companies was already moving in manufacturer hands. Thus, when Apple made its entrance on the mobile market, this company was not just a device manufacturer, but also a service provider through the iPhone applications development. In June 2008, with the launch of App Store, although Apple gives to third parties the application development, the company still keep the control of services and contents by its portal (Laugesen & Yuan, 2010). Services and application enrich iPhone users' experience and add value to it, becoming source of its competitive edge. This innovation strategy has been so successful that has been emulating also by other competitors like, for example, Samsung with its web market.

At confirming that servitization of business is taking off in mobile industry as well, there is the more and more importance of the *ecosystem*. An ecosystem is composed of products and services complementing a device, this concept is rooted in the idea that the value generation is not solely in a product, but it is also generated by its surrounding environments. Mitchener (2009) reports three different ways to use this term. Firstly and broadly, an ecosystem is a set of environments with which a device interact (e.g., car's radio and speakers). Secondly, the ecosystem is made of software, content and services complementing the device (i.e., applications for iPhone and iPod). Finally, ecosystem is also referred to the accessories and gadgets built by third parties with schemes for approval and accreditation. Anyway, all these material and immaterial related offerings have the power to add value to the original manufacturer's device and to extend the user's experience. So more compulsory is becoming manufacturer's need to consider a product not as a standalone business but with all real and potential complementary innovations: the value and its several forms take shape not just in a product but in an broader ecosystem.

In a consumer perspective, applications and services add value to customers' handset and are becoming the main source of value creation. Thus also mobile manufacturer companies are changing their strategy focus: switching from the product, for which applications and services are designed, on applications which are becoming an increasing and valid source of new value creation. Mobile companies are also aware that value is moving downstream toward customers, thus in order to add further value, they try to place on the market as much as high-personalized offerings. They reach this goal engaging customers in developing their own applications and by delivering them development tool and allowing them to distribute their software through firm's web store.

Definitely in mobile industry, competition field seems to be moved from mobile handset to its service, both in suppliers and consumers' perspective.

6. The impact of servitization on market demand

Considering that the mobile telecommunication market and its value network are changing quickly under the spur of many technological innovations, new challenges or opportunities stem from the exploitation of innovations in mobile devices (section 4). The service economy, that implies the shift of manufacturers from good selling to service delivering, is one of them. Mainly since 2007, with the iPhone introduction, *servitization* has been an extending trend among mobile phone suppliers, as they try to mix in their offering both good and service components, integrating phone devices with increasing software and applications (section 5).

In a supplier perspective, this shift has an important impact on economical aspects, stemming from selling a variety of complementary services for products. Nevertheless, servitization also brings implications in the operation management and in the innovation strategy and compels providers to revise their business models (section 5.2).

But what is happening in the consumer perspective? A mass-market product like the mobile phone has become extremely customizable through the complementary services that can be integrated into it: software downloading allows customers to entail mobile phone functionality on their unique needs.

Servitization has meant a shift of market demand from the hardware component of cell phones to the possibility to buy devices as "terminals" enabling the fruition of personalized services.

It has therefore impacted on the LAM model's three dimensions (section 2), i.e., on the needs that induce purchase (use motivations), on the characteristics of the offering influencing consumers' choice among the various models and brands in the market (attributes), and on the characteristics of individuals (lifestyles).

As to the *attributes* dimension (Table. 2), servitization has mainly affected the following variables:

1. *handiness and screen visibility*, as the size of cell phones and the characteristics of their displays directly affect Web browsing experience and the ability to access content: as opposed to previous miniaturization, handsets are becoming increasingly bigger and with ample touchscreens enabling panning, zooming and scrolling;
2. *personalization*, as the possibility of personalising mobile phones is progressively moving from their accessories and tangible features (e.g., covers) to services and software applications;
3. *brand reputation*, as brand image nowadays depends on cell phones' operating system (OS) more than on the reputation of handset manufacturers since the former determines end-users' access to services and external ecosystems[14];
4. *advanced services*, as nowadays many more technologically advanced services are available to consumers (e.g., online marketplaces, Wi-Fi connection, m-banking, etc.).

Moving to the *use motivations* dimension, the main impact of servitization has regarded the *information and entertainment* variable (Table 3). Indeed nowadays - given the integration of mobile communication and the wired Web and the high and continuously increasing number of available services - this variable should be better explored to consider the increasing relevance assumed by content and software applications among purchasing motivations for mobile phones.

Finally, as to the *lifestyles* dimension (Table 4), given media convergence and the rising importance of services for end-users, it should be better explored the *media usage* variable to consider the relevance assumed by the Internet and social networks.

By the end, as servitization is leading to the increasing importance of offering's intangible component and therefore also to low storage and distribution costs, the LAM model could allow to point out micronized segments, thus representing the base for increasingly personalised marketing policies to reach niche markets.

We will consider and study in depth the above-mentioned impact of servitization on the model's dimensions for the design and implementation of further empirical research.

7. Conclusion

Empirical researches discussed in this chapter proved that LAM model could be fruitful applied in market segmentation; in particular LAM model seems to be very suitable in analyzing mobile market communication. However, as pointed out by many epistemologists (Popper, 1963; Kripke, 1972; Marradi, 1990), when analyzing a classification one must be careful and not fall under the essentialist fallacy: to think that classifications are almost immutable because they reach the essence of things. A segmentation technique, is essentially a form of classification, so its results could not be considered fixed (Wedel &Kamakura, 2003), especially when segmentation techniques are applied to a fast changing phenomenon like the mobile market communication one.

Therefore, we think that could be interesting to update the multidimensional segmentation model according to the latest evolution of mobile market, above all integrating servitization perspective into the main LAM conceptual dimensions. In other words, our research will try to conceptualize, and operationally define the progressive shift from the product (smarthphones) to the service (web applications). Moreover, attributes and motivations will

---

[14] This could explain why in the first quarter of 2011 Android (followed by Apple's iOS) dominates the smartphone operating system business (Gartner Press Release, 2011).

be significantly redefined to take into account the new tendencies of the consumption of mobile communication. For example, technological attributes will receive a higher consideration than in the previous research (Mazzoni et. al. 2007), while the integration between mobile communications and other media, above Web services, will be deeper analyzed in the motivation dimension.

Our research will maintain its mixed method approach, because, as showed in paragraph 3, the integration of different methodological perspectives enhance the quality of research design and findings. So, data collection will be performed integrating a qualitative technique (focus group) with a quantitative one (questionnaire). Probably, we will also use online market research technique for quantitative data collection, such as a web questionnaire, because they have many advantages: low costs, fast data collection, a potentially global reach, ease to compile (Murthy, 2008; Migliaccio et al., 2010).

Data analysis will be performed following two steps; firstly, we will adopt the same multidimensional segmentation procedure used in previous researches: factor analysis and cluster analysis applied in sequence. This will guarantee a certain degree of comparability between the results of the new research and those of the previous ones. Obviously, we will not compare the results from a substantial and theoretical point of view, as the LAM model will be composed by different variables, but we will evaluate the descriptive power of LAM model. In other words, we will test if LAM model is still able to perform a good synthesis of the actual state of play of mobile communication market.

Secondly, we will test different market segmentation procedures, giving a predictive orientation to LAM model. Put specifically, we will apply finite mixture regression models, because these statistical techniques seems to be very helpful in overcoming limitations and constraints of the usual descriptive market segmentation techniques (Wedel & Desarbo, 2002; Sarstedt, 2008).

8. References

AdMob (2010). *May 2010 Metrics Highlights*, 11.07.2011, Available from http://metrics.admob.com/wp content/uploads/2010/06/May-2010-AdMob-Mobile -Metrics-Highlights.pdf

Agarwal, K. M. (2003). Developing Global Segments and Forecasting Market Shares: A Simultaneous Approach Using Survey Data. *Journal of International Marketing*, Vol. 11, No.4, pp. 56-80

Aurifeille, J. M., Quester, P.G., Lockshin, L. & Spawton, T. (2002). Global vs. International Involvement-based Segmentation: A Cross-national Exploratory Study, In *International Marketing Review*, Vol. 19, No. 4, pp. 369-386

Anderson, J. & Narus, J. (1995). Capturing value of Supplementary Services. *Harvard Business Review*, Vol. 73, No. 1, pp. 133-141

Apple Press Release (January 22, 2011). *Apple's App Store Downloads Top 10 Billion*, 28.06.2011, Available from http://www.apple.com/pr/library/2011/01/22Apples-App-Store-Downloads-Top-10-Billion.html

Baines, T. & Lightfoot, H. (2011). Towards an Operations Strategy for the Infusion of Product Centric Services into Manufacturing, In: *Service System Implementation*, H. Demirkan, J. C. Spohrer, Krishna V. (Eds.), pp. 89-111, Springer, New York, NY, USA

Baines, T., Lightfoot, H., Benedettini, O., Whitney, D. & Kay, M. (2010). The adoption of servitization strategies by UK-based manufacturers. *Proceedings of the Institution of Mechanical Engineers, Part B: Journal of Engineering Manufacture*, 224, pp. 815-829

Baines, T., Lightfoot, H., Evans, S., Neely, A. & al. (2007). State-of-art in product-service system. *Proceedings of the Institution of Mechanical Engineers. Part B: Journal of Engineering Manufacture*, 221, pp. 1543-155

Bernard, H. R. (2000). *Social Research Methods: Qualitative and Quantitative Approaches*, Sage, Thousand Oaks, CA, USA

Bolton, R. N. & Myers, M. B. (2003). Price-based Global Market Segmentation for Services. *Journal of Marketing*, Vol. 68, No. 3, pp. 96-107

Chamberlin, E. H. (1933). *The Theory of Monopolistic Competition*, Harvard University Press, Cambridge, MA, USA

Collesei, U. (2000). *Marketing*, Cedam, Padova, Italy

Cresswell, J. W. (2003). *Research Design: Qualitative, Quantitative, and Mixed Methods Approaches* (2nd), Sage Publications. Thousand Oaks, CA, USA.

Cusumano, M. A. (2010), Platforms and Services: Understanding the Resurgence of Apple. *Communication of the ACM*, Vol. 53, No. 10, pp. 22-24

Di Franco, G. (2001). *EDS: esplorare, descrivere e sintetizzare i dati*, Franco Angeli, Milano,Italy

Dickson P. R. & Ginter J. L. (1987). Market Segmentation, Product Differentiation, and Marketing Strategy, In: *Journal of Marketing*, Vol. 61, No. 2, pp. 1-10

Eaton, K. (2009). iPhone King of Mobile Web with 50% Share, In: *FastCompany.com*, 28.06.2011, Available from http://www.fastcompany.com/blog/kit-eaton/technomix/iphone-success-shows-what-people-want-mobile-net

ENTER & IDATE (2007). *Mobile 2007*, ENTER, Madrid, Spain

European Commission (2011). Digital Agenda Scoreboard 2011, Available from http://ec.europa.eu/information_society/digital-agenda/scoreboard/docs/regulatory/it_reg_dev_2011.pdf.pdf

Fabris, G. (1972). *Il comportamento del consumatore*, Franco Angeli, Milano, Italy

Frank, R. E., Massy, W. F. & Wind, Y. (1972). *Market Segmentation*, Prentice Hall, Englewood Cliffs, NJ, USA

Fuchs, V. R. (1968). *The service economy*. Columbia University Press for National Bureau of Economics Research, New York, NY, USA

Gartner Press Release (February 27, 2008). *Gartner Says Worldwide Mobile Phone Sales Increased 16 Per Cent in 2007*, 11.07.2011, Available from http://www.gartner.com/it/page.jsp?id=612207

Gartner Press Release (May 19, 2011). *Gartner Says 428 Million Mobile Communication Devices Sold Worldwide in First Quarter 2011, a 19 Percent Increase Year-on-Year*, 11.07.2011, Available from http://www.gartner.com/it/page.jsp?id=1689814

Gebauer, H. & Fleisch, E. (2007). An investigation of the relationship between behavioural process, motivation, investments in the service business and service revenue. *Industrial Marketing Management* , Vol. 36, No.3, pp. 337-348

Goedkoop, M., van Halen, C., te Riele, H. & Rommens, P. (1999). *Product Service-Systems, Ecological and Economic Basics*. Report for Dutch Ministries of Environment and Economic Affairs, PRe Consultants, Amersfoort Netherlands

Gonçalves, V., Walravens, N. & Ballon, P. (2010). How About an App Store? Enablers and Constraints in Platform Strategies for Mobile Network Operators, *Proceedings of Ninth International Conference on Mobile Business/Ninth Global Mobility Roundtable*, pp. 91-99, Athens, Greece, June 13-15, 2010

Gonzalez, A. M. & Bello, L. (2002). The Construct "Lifestyle" in Market Segmentation: The Behavior of Tourist. *European Journal of Marketing*, Vol. 36, No. 1, pp. 51-85

Green, P.E. (1977). A New Approach to Market Segmentation. *Business Horizons*, Vol. 20, No. 1, pp. 61-73

Gunter, B. & Furnham, A. (1992). *Consumer Profiles. An Introduction to Psychographics*, Routledge, London-New York

Gustafsson, A. & Johnson, M. (2003). *How to create competitive advantage through service development and innovation*. Jossey -Bass, San Francisco, USA

Haley, R. I. (1968). Benefit Segmentation: A Decision-oriented Research Tool. *Journal of Marketing*, Vol. 32, No. 3, pp. 30-35

Haley, R. I. (1971). Beyond benefit Segmentation. *Journal of Advertising Research*, Vol. 11, No. 4, pp. 3-8

Haley, R. I. (1984). Benefit Segments: Backwards and Forwards. *Journal of Advertising Research*, Vol. 24, No. 1, pp. 19-25

Hassan, S. S., Craft, S. & Kortam, W. (2003). Understanding the New Bases for Global Segmentation. *Journal of Consumer Marketing*, Vol. 20, No.5, pp. 446-462

Kamakura, W. A. & Wedel, M. (1995). Life-style Segmentation with Tailored Interviewing. *Journal of Marketing Research*, Vol. 32, No. 3, pp. 308-321

Kothandaraman, P. & Wilson, D. T. (2001). The Future of Competition. Value-Creating Networks. *Industrial Marketing Management*, Vol. 30, No. 4, pp. 379–389

Laugesen, J. & Yuan, Y. (2010). What Factors Contributed to the Success of Apple's iPhone? *Proceedings of Ninth International Conference on Mobile Business/Ninth Global Mobility Roundtable*, pp. 91-99, Athens, Greece, June, 2010

Li, F. & Whalley, J. (2002). Deconstruction of the Telecommunications Industry: from Value Chains to Value Networks. *Telecommunications Policy*, Vol. 26, No. 9 10, pp. 451-472

Maitland, C. F., Bauer, J. M. & Westerveld, R. (2002). The European Market for Mobile Data: Evolving Value Chains and Industry Structures. *Telecommunications Policy*, Vol. 26, No 9, pp. 485-504

Marradi, A. (1990). Classification, Typology, Taxonomy. *Quality and Quantity*, Vol. 24, No. 2, pp. 129-157

Mazzoni, C. (1994). La segmentazione "trasversale" del mercato europeo, *Sinergie*, Vol. 33, gennaio-aprile

Mazzoni, C. (1995). *La segmentazione multidimensionale dei mercati*, Cedam, Padova, Italy.

Mazzoni, C., Castaldi L. & Addeo, F. (2007) Consumer behavior in the Italian mobile telecommunication market. *Telecommunication Policy*, Vol. 31, No.10-11, pp. 632-647

Mazzoni, C. & Leone S. (2001) La segmentazione dei lettori: approfondimenti sulla base di un modello Multidimensionale, In: *I sistemi locali nell'editoria giornalistica. Il caso della Campania*, C. Mazzoni & A. Rea (Eds.), ESI, Napoli, Italy

Michman, R. D. (1991). *Lifestyle Market Segmentation*, Praeger, New York.

Migliaccio M., Addeo F., & Rivetti F. (2010). Market Knowledge Exploration and Web 2.0: Initial Empirical Evidence on Hotel Chains, *Proceedings of IFKAD (International Forum on Knowledge Asset Dynamics)*, Matera, Italy, 24-26 June

Mitchener, J. (2009). Perfecting the ecosystem, In: *Engineering and Technology*, 9.May.2009-22.May.2009, Available from http://kn.theiet.org/magazine/ , 70-71

Mont, O. (2002). Claryfing the concept of product service system. *Journal of Cleaner Production*, Vol. 10, No. 3, pp. 237-245

Morse, J. M. (2003). Principles of Mixed Methods and Multimethod Research Design. In: *Handbook of Mixed Methods in Social and Behavioural Research*, A. Tashakkori & C. Teddlie (Eds.), Sage Publications, pp. 189-208,Thousand Oaks, CA, USA

Murthy, D. (2008). Digital Ethnography An Examination of the Use of New Technologies for Social Research. *Sociology*, vol. 42, pp. 837-855

Ono, S. & Tang, P. (2010). The role of mobile handsets in advanced network service evolution: evidence from Japan. *Telecommunication Policy*, Vol. 34, No. 8, pp. 444-460

Plummer, J. T. (1974). The Concepts and Application of Life Style Segmentation.*Journal of Marketing*, Vol. 28, No. 1, pp. 33-37

Popper, K. R. (1963). *Conjectures and Refutations: the Growth of Scientific Knowledge*. Routledge & Kegan, London, UK

Prahalad, C. K. & Hamel, G. (1990). The Core Competence of the Corporation. *Harvard Business Review*, Vol. 68, pp. 79-91

Ratneshwar, S., Warlop, L., Mick, D. G. & Seeger, G. (1997). Benefit Salience and Consumers' Selective Attention to Product Features. *International Journal of Research in Marketing*, Vol. 14, No. 3, pp. 245-259

Saporta, B. (1976). Les ambiguïtés du concept de segmentation, In: Revue Française du Marketing, No. 63, pp. 51-74

Sarstedt, M. (2008). Market segmentation with mixture regression models: Understanding measures that guide model selection. *Journal of Targeting Measurement and Analysis for Marketing*, Vol. 16, No. 3, pp. 228-246

Seal, K. C. (2010). A Framework for Understanding Mobile Value Offering through Multi-Country Studies, In: *Handbook of Research on Mobile Marketing Management*, K. Pousttchi & D. G. Wiedemann (Eds.), pp. 129-156, Business Science Reference, Hershey, PA, USA

Sheatsley, P. B. (1983). Questionnaire Construction and Item Writing, In: *Handbook of Survey Research*, P. H., Rossi, G. D. Wright & A. B., Anderson (Eds.), pp. 195-230, Academic Press, New York, NY, USA

Smith, W. (1956). Product Differentiation and Market Segmentation as Alternative Marketing Strategies, In: *Journal of Marketing*, vol. 21, no.1, pp. 3-8

Sorrentino, F. (2006). Innovazione e Creazione di Valore nel Settore della Comunicazione Mobile, In: *Scelte di Consumo e Reti del Valore nella Comunicazione Mobile*, C. Mazzoni (Ed.), pp. 37-63, Carocci, Roma, Italy

Stabell, C. B. & Fjeldstad, Ø. D. (1998). Configuring Value for Competitive Advantage: on Chains, Shops, and Networks. *Strategic Management Journal*, Vol. 19, No 5, pp. 413-437

Steckler, A., McLeroy, K. R., Goodman, R. M., Bird, S. T., McCormick, L. (1992). Toward Integrating Qualitative and Quantitative Methods: An Introduction. *Health Education Quarterly*, Vol. 19, No.1, pp. 1-8

Steenkamp J. E. M. & Hofstede, F. T. (2002). International Market Segmentation: Issues and Perspectives. *International Journal of Research in Marketing*, Vol. 19, No. 3, pp. 185-213

Tashakkori, A. & Teddlie, C. (2003). *Handbook of Mixed Methods in Social and Behavioural Research*, Sage Publications, Thousand Oaks, CA, USA

Tilson, D. & Lyytinen, K. (2006). The 3G transition; Changes in the US Wireless Industry. *Telecommunications Policy*, Vol. 30, pp. 569-586

Tukker, A. (2003). Eight Types of Product-Service Systems: Eight Ways to Sustainability? *Innovating for Sustainability International Conference of Greening of Industry Network*, San Francisco, October 2011

Tukker, A. (2004). Eight types of Product Service System: Eight Way to Sustainability? Experiences from Suspronet. Business Strategy and the Environment, Vol. 13, No. 4, pp. 246–260

Vargo, S. & Lush, R. (2004). Evolving to a New Dominant Logic for Marketing. *Journal of Marketing*, Vol. 68, No.1, pp. 1-17

Vyncke, P. (2002). Lifestyle Segmentation. *European Journal of Communication*, Vol. 17, No.4, pp.445-465

Volgelstein, F. (2008). The Untold Story: How the iPhone Blew Up the Wireless Industry, In: *Wired Magazine*, Vol. 16, No. 2, 28.06.2011, Available from http://wired.com/gadgets/wireless/magazine/16-02/ff_iphone

Wedel, M. & Desarbo, W. S. (2002). Market Segment Derivation and. Profiling via a Finite Mixture Model Framework. *Marketing Letters*, Vol. 13, No. 1, 17-25

Wedel, M. & Kamakura, W. A. (2003). *Market Segmentation. Conceptual and Methodological Foundations*, Kluwer Academic Publishers, Boston, MA, USA

Wells, W. D. (Ed.) (1974). *Life Style and Psychographics*, American Marketing Association, New York, NY, USA

Wells, W. D. (1975). Psychographics: A Critical Review. *Journal of Marketing Research*, Vol. 12, No.2, pp. 196-213

Wells, W. D. & Tigert, D. J. (1971). Activities, Interests and Opinions *Journal of Advertising Research*, Vol. XI, No. 4, pp. 27-35

West, J. & Mace, M. (2007). Entering a Mature Industry through Innovation: Appe's iPhone Strategy, *DRUID Summer Conference 2007*, 28.06.2011, Available from http://www2.druid.dk/conferences/viewpaper.php?id=1675&cf=9

West, J. & Mace, M. (2010). Browsing as kill app: explaining the rapid success of Apple's iPhone. *Telecommunications Policy*, Vol. 34, pp. 270-286

Wind, Y. (1978). Issues and advances in segmentation research. *Journal of Marketing Research*, Vol. 15, No. 3, pp.317-337

Wingfield, N. (August 11, 2008). iPhone Software Sales Take Off: Apple's Jobs. *The Wall Street Journal*, 28.06.2011, Available from http://msl1.mit.edu/furdlog/docs/2008-08-11_wsj_iphone_app_sales.pdf

Wise, R. & Baumgartner, P. (1999). Go Downstream: The New Profit Imperative in Manufacturing. *Harvard Business Review*, Vol. 77, No. 5, pp. 133-141

Wu, S. (2001). Benefit Segmentation: An Empirical Study for On-line Marketing. *Asia Pacific Journal of Marketing and Logistics*, Vol. 13, No.4, pp. 3-18

Yan X., (2004). 3G licensing in Hong Kong: The debate. *Telecommunications Policy*, Vol. 28, No.2, 213–226

Yankelovich, D. (1964). New criteria for market segmentation. *Harvard Business Review*, Vol. 4, No.2, pp. 83-88

Yip, G. S. (1982). Gateways to Entry. *Harvard Business Review*, Vol. 60, No. 5, pp. 85-92

Ziff, R. (1971). Psychographics for Market Segmentation. *Journal of Advertising Research*, Vol. 11, No.2, pp. 3-9

3

The Role of WAC in the Mobile Apps Ecosystem

Zeiss Joachim, Davies Marcin and Pospischil Günther
FTW. Telecommunications Research Center Vienna
Austria

1. Introduction

The Wholesale Applications Community (WAC) was founded to change the overall market for mobile applications. WAC intends to achieve this by introducing open standardized technologies based on W3C widgets and/or OneAPI definitions. In addition, WAC provides complimentary commercial models. This will allow developers to deploy an application across multiple devices and across multiple operators. Developers should not need to negotiate with each of them. WAC provides the commercial prerequisites.

WAC's objective is to commercialize products for its member companies. Open Web standards are utilized in support of this commercialization effort as long as such adoption does not impact the required time to market. The WAC widget specification is therefore based on W3C and OMTP standards to the greatest extent possible.

As already mentioned, WAC widgets utilize Web technologies. If a developer understands HTML and CSS to design a good looking website and has used JavaScript to create functionality and services for a user, he/she will also be able to write mobile apps in WAC. The widget packaging format is based on the W3C Widget Packaging specification and introduces some extensions to meet WAC requirements, such as specifications for billing. WAC widgets can optionally utilize a comprehensive handset API. A code signing security system ensures that widgets can only access APIs that are suitable to their level of trust.

This chapter gives an understanding of WAC and compares it to existing mobile app eco-systems and technologies. We investigate the details of WAC regarding its API definition, runtime implementation, technologies used, and SDK-based development. We answer the questions on what makes the difference about developing WAC apps and how developers and users can both benefit. Finally, recommendations are given to industry on how to use WAC and how to further develop WAC in upcoming releases of the standard.

2. Mobile app development overview

The current mobile platform space suffers a lot of fragmentation. Developing software for mobiles generally can happen in a native way (like Objective C for iOS) or by using a virtual platform that offers (theoretical) portability.

Today, the most relevant native development approaches are:

- iOS using Objective-C, Xcode and Interface Builder
- Native Qt libraries and SDK from Nokia: may survive only for embedded systems but not for app development, they are still too complicated although much better than the basic Symbian APIs

- Java based virtual machine programming in Android based on the Dalvic VM, Linux kernel and Eclipse Plugin SDK

NET or Silverlight type of development for Windows Phone 7 which didn't gain momentum yet

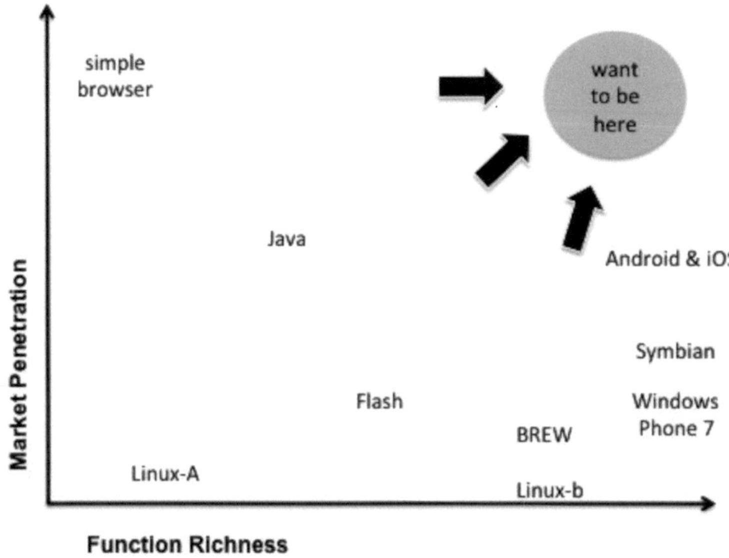

Fig. 1. Mobile Platform space (source: Fraunhofer Fokus)[1]

Virtual development approaches include:

- Web based development, WebOS: the scene and stage based paradigm in conjunction with a W3C DOM compliant model and Javascript development is promising and innovative. The Mojo framework provides access to device functionality, but native development is offered as well
- Java ME, has become a runtime for apps on middle to low-end mobile devices
- Flash: its future on mobile devices is unclear and technology might become obsolete in turn of HTML5 supporting browsers and web runtimes
- Desktop based web development (widgets): Yahoo! widgets on Windows (former Konfabulator runtime) and Dashboard on Mac OS X show the importance of including graphics to coding activities
- Other widget systems: Standardization as with W3C widgets, Opera Widgets, etc.

2.1 Frameworks

WAC needs to compete with the following frameworks for native or web-based mobile app design and implementation:

- COCOA and Quartz for iOS, or OpenGL ES
- Android Java APIs
- W3C widget based runtimes
- Generic Javasscript libraries: jQuery, ExtJs, jqTouch, SenchaTouch

[1] Linux-A and Linux-B denote different Linux Variants/Distributions

- HTML5 supporting runtimes
- WebGL supporting runtimes
- PhoneGap or Titanium apps (see Section 3)

For the iPhone, Apple offers a series of well known frameworks grown and proved in years of software design for Desktop computers and notebooks running the OS X operating system. Cocoa APIs for operating system task interaction and Quartz for GUI implementation have been adapted to the needs of mobile devices (iPhone and iPad), especially for multi touch events (a feature WAC does not support as of today). Although Objective-C as the used programming language is not so common, the frameworks are very well designed, easy to use, and the MVC (Model View Controller) software design pattern has been excellently implemented. The latter is one of the main reasons of the huge success of iPhone apps. Also the backwards compatibility to the C language and the support for Open GL allows rather easy porting of existing game software to the iOS platform.

Android decided for using Java as the main implementation language including most of the standard SDK APIs known for desktop computers. In addition, apps benefit from garbage collection and a virtual machine called Dalvik has been designed to fulfil the particular needs and to cope with performance restrictions of mobile devices. The use of Java attracts a huge designer community familiar with Java programming on the desktop and encourages the switch from designing Java programs on the PC to implementing apps for Android phones and tablets.

WAC is based on the W3C widget standard with all its good and bad implications. Being standardized, it can be easily implemented for all runtime programs and sandboxes already supporting W3C, e.g. the Opera browser or widget runtimes. However, there is more to be done – libraries like jqtouch or SenchaTouch need to be supported or integrated to allow for a state of the art user experience known from native apps. HTML5 is partially, WebGL not supported at all for WAC runtimes. This means that game development and sophisticated graphical user interfaces are hard or even impossible to implement with WAC.

2.2 App stores

Currently industry experiences a tremendous growth of the mobile application market. While mobile applications were distributed via individual channels in the past, developers and users now clearly turned to centralized catalogues of software offerings for their devices that are managed and provided by trusted entities.

The WAC store and distribution platform may compete with the following prominent stores:

- AppStore form Apple

The Apple AppStore is designed to meet the various application requirements of iPhone and iPod users. Being a storehouse with mobile applications that can be downloaded and installed, the AppStore also recommends related applications that can be rated by users. Apps are tested and certified by Apple before they are included in the AppStore catalogue. Except for developer testing, there is no (official) alternative way to install apps on the iPhone. Thereby Apple ensures a consistent user experience and avoids security issues. As of July 2011 there are 425,000 apps available with 15 billion downloads since the AppStore's inception. Due to its success, Apple recently started an Application store for desktop computers as well.

- Android market

The Google Android Market allows users to browse through and download third party applications for Android mobile phones. The store has grown from 2,300 applications in 2003 to 80,000 applications in 2010 with over 1 billion downloads in total. As there is no centralized testing of apps, Android systems currently suffer a little from poor quality apps and the fact that developers can make money mostly out of in-app advertisements but not with selling them in the store as such.

- OVI from Nokia

Although the destiny of Symbian remains unclear, the download rate has reached 5 million per day in July 2011. It seems to be successful for emerging markets like India, China or Turkey.

- Blackberry App World

Developed by Research In Motion (RIM), the Blackberry App world is all about BlackBerry and the various applications. It was previously known as the BlackBerry Application Storefront. Although RIM initially announced that the store would only be available in the United States, United Kingdom, and Canada, it is available across 70 countries today. The store has around 25,000 applications in its catalogue and over 2 million applications are downloaded daily. The store is offered in multiple languages (English, French, Italian, German, etc.) and allows the user to download free and paid applications. Unlike other application stores it offers latest news as well as a support system to take care of one's questions and concerns.

2.3 Software development kits

The following SDKs may be considered for Web-based app development:

- Eclipse plugins, e.g. for Android and JIL, WAC
- Online SDKs such as ARES for WebOS
- Standalone target specific SDKs (Xcode, Interface Builder, Visual Studio)
- Widget specific SDKs (template based with graphical tools) like Dashcode
- WebDesign SDKs such as DreamWeaver

These SDKs are considered in the following section regarding the differentiation of WAC from its competitors.

3. Differentiation of WAC

All technologies integrated in WAC have been used before by other initiatives, e.g. the usage of web design features by WebOS, offering widgets as apps by Opera, Apple or Yahoo. Application stores are already out there as well.

So what makes the difference?

Well, it's the combination of all of that, the device and network operator independence, the billing methods and in that sense the possibility to integrate network operator assets like network based GPS, call control, in-app billing or identity management. This is what the WAC consortium should focus on. Unfortunately, WAC 3.0 supporting OneAPI standards is not finished yet.

Properties of the WAC API and runtime are:

- Based on W3C widget standards
- Detailed certification based app functionality management (maybe a bit too complex / over engineered)

- Integration of remote APIs and Telco assets (based on GSMA OneAPI specifications)
- Operator independence
- Device/Manufacturer independence
- Entirely web based, i.e. apps run in a HTML rendering engine supporting CSS and Javascript, based on W3C Standardization; WAC specific JavaScript APIs are additionally available.
- Integration of Telco assets, remote APIs: e.g. for identity management

Two of the main competitors in web-based mobile cross-platform development are PhoneGap and Titanium mobile, which are discussed briefly in the following sections.

3.1 PhoneGap

PhoneGap (for details please see reference PhoneGap) is an open-source mobile framework that supports the following important platforms:

- iOS
- Android
- Blackberry
- Palm webOS
- Symbian WRT

As in WAC, developers write code in HTML/CSS/JS and deploy it to their target platform(s) (HTML5/CSS3 supported). However, no native code is produced, the final app is wrapped into a native "web view'"object (like in the desktop projects Fluid or Prism).

The currently supported API features of the PhoneGap platform can be found under reference (PhoneGap).

3.2 Appcelerator titanium mobile

Titanium mobile from Appcelerator is another open-source mobile framework for iOS/Android (Blackberry RIM to come). Developers use web technology (HTML5/CSS3 support) but Titanium's plug-in architecture also allows coding modules natively (Objective C or Java) to extend the app with native functionality. Titanium offers an integrated IDE and one of the biggest advantages is that native code is generated, so apps are faster and can make use of native UI controls. Another useful feature is that apps can also be ported to the desktop with the Titanium desktop edition.

3.3 Comparison with WAC

Table 1 compares PhoneGap and Titanium Mobile with WAC in terms of platform support, SDKs, UI controls, documentation & community, and whether a runtime is needed. It should be noted that native UI controls can be „emulated" in PhoneGap and WAC by using Javascript libraries such as SenchaTouch. Finally, the performance of all three is most likely best with Titanium (as native code is produced), however no hands-on test was performed.

Although WAC claims to be platform independent in the future, right now only Android based runtimes exist. Other platforms like PhoneGap or Titanium mobile are already there and well accepted by the web design community. Additionally, as WAC claims its own application store it is currently unclear if WAC apps could be offered on iOS devices, because of licensing issues. Developing under PhoneGap or Titanium mobile does not create this issue as these SDKs integrate into the platform specific stores.

	WAC	PhoneGap	Appcelerator Titanium
Platforms	n/a (depending on runtime support)	5 (6)	2 (3)
SDKs	Eclipse-based	Platform-specific	Integrated
Runtime needed	Yes	No	No
Native UI controls	No	No	Yes
Documentation & Community	Fair	Good	Good

Table 1. Comparison of WAC, PhoneGap and Titanium Mobile

4. Analysis of WAC APIs and SDKs

At the time of writing, there are three different SDKs for WAC widgets. One is provided directly by the WAC consortium, the two others are from Obigo (see reference Obigo) and Aplix (see reference Aplix). All three runtimes are similar tools that are based on the Eclipse platform and run on Android devices. In our opinion Eclipse may not be the best platform though. Being certainly a very powerful platform, it might be too complex for relatively simple widget projects and for standard web developers (which should also be attracted by the WAC initiative). A simpler and more tailored, graphical editor like for the ARES project (see reference ARES) might be better suited.

4.1 WAC standardization timeline

WAC 1.0 (December 2010)

- Based on subset of JIL 1.2.2
- Supports following W3C standards:
  - HTML 4.01, xHTML 1.1, CSS 2.1, SVG Tiny 1.2, Widget Packaging and Configuration, MediaQueries
  - Javascript
- Widget Security: based on W3C Widgets 1.0 (Digital Signatures)
- Handset APIs:
  - Accelerometer, Address Book, Application Launcher, Audio, Camera, Messaging (sending only) and Location

WAC 2.0 (aka "Waikiki", Jan 2011)

- Supports following W3C standards:
  - HTML 5 content parsing plus input element, canvas element, canvas 2D context, audio element, video element, contenteditable attribute
  - Javascript 1.5 (and JSON)

- CSS 2.1 and parts of CSS 3 (transforms, transitions, ...)
- DOM, XMLHttpRequest
- Support for popular libraries, e.g. jQuery
- Improved Security:
- Multi-level authentication/signatures: AppStores, Clients, Runtimes, Widgets...
- Policy-based access control
- Handset API enhancements:
- more methods per API, orientation, filesystem, calendar, tasks...

WAC 3.0 planned features (Sep 2011)

- Extended range of developer tools
- Billing support (+ in-widget billing)
- User identification
- Network APIs according to OneAPI specifications
- Advertising
- Feature phones

Technical details on WAC 3.0 are not available at the time of document writing (July 2011). Regarding user identification, Aepona presented a Message flow including prototype at the Mobile World Congress 2011 in Barcelona (based on OneAPI), but did not clearly identify the API, as the GSMA OneAPI suggests OAUTH.

Media transport or streaming is out of scope of WAC, SQLite is not supported and also ciphering is left to the developer to implement (not considering the possibility of communication with https, which is supported by the runtime). Cooperative multitasking is a must for WAC 2.0 runtimes but limited to the capabilities of the host system.

4.2 Creating a WAC widget

In the following we would like to go through the necessary steps to create a small widget that displays a camera preview window and offers taking pictures by pressing a button.

A simple WAC widget basically consists of three files, namely a HTML file that is the entry point of the widget and defines the layout, a Javascript file defining the logic, and, finally, a CSS file that specifies the visual properties of the widget.

We would like to start first with editing the HTML file (shown in Figure 2):

```
<html>

<head>
<meta http-equiv="Content-Type" content="text/html; charset=UTF-8">
<title>CameraApp</title>
<link href="CameraApp.css" rel="stylesheet" type="text/css"/>
<script type="text/javascript" src="CameraApp.js"></script>
</head>

<body onload="preview();" onUnload="destroy();">
    <object id="previewWindow" width="400" height="400"></object></td></tr></table>
    <div style="background-image:url(Default.png)"
            onclick="takePicture()" id="start" class="cameraButton"></div>
</body>

</html>
```

Fig. 2. HTML code for a WAC widget

The head of the file points to the CSS and Javascript files we will edit later. In the body we define the preview() method to be loaded upon runtime. We also define a preview windows and a camera button that launches a takePicture() method.
Figure 4 shows the Javascript code for the widget. The preview() method uses the WAC API to create a camera object. The window of that object is then assigned to the preview window defined in the HTML file before. The takePicture() method finally captures the image and stores it on the device. It makes use of a helper function generateFileName() that creates a unique filename based on the current DateTime. This was needed because the picture could not be stored if a file with the same name already exists.

```
#front {
    position: absolute;
    background-image: url("Default.png");
    top: 0px;
    left: 0px;
    width: 235px;
    height: 158px;
}

.cameraButton {
    position: relative;
    top: 300px;
    left: 5%;
    width: 235px;
    height: 158px;
}
```

Fig. 3. CSS code for a WAC widget

Finally, the CSS file (Figure 3) is relatively straightforward and just defines the styles for the button and the background. The SDK would now pack these three files together with content information in a zip file with filename extension wgt. The WAC app is finished and ready to be deployed (however, it should be noted, that a Publisher ID/certificate is needed for distribution in application stores).

4.3 Comparison with native App development frameworks

4.3.1 iOS

The primary programming language is Objective-C, the primary SDK is Xcode in combination with Interface builder. The SDK is free, but development is only possible on Mac Computers. iOS was designed to meet the needs of mobile environment, where users' needs are different than for a desktop system.
The iOS SDK contains the code, information, and tools needed to develop, test, run, debug, and tune applications for iOS. Xcode tools provide the basic editing, compilation, and debugging environment for code writing. Xcode is also the launching point for testing applications on an iOS device, and in iOS Simulator, a platform that mimics the basic iOS environment but runs on a local Macintosh computer.
In the meantime, iOS has become a mature mobile runtime environment for apps. This also applies to Xcode, which is very well tailored to the needs of developing function rich, highly appealing and simple to use mobile applications for the user. The possibilities of graphical user interface editing while separating the functional implementation in Objective-C gives some advantages compared to WAC widgets. This is because Xcode addresses the true

nature and power of Objective-C and COCOA UI libraries whereas the WAC SDKs try to mimic code writing development where they should address rather graphical oriented widget design.

```
var prevWindow;
var mCamera;

function preview() {
    prevWindow = document.getElementById("previewWindow");
    mCamera =  Widget.Multimedia.Camera;
    mCamera.setWindow(prevWindow);
}

function destroy() {
    mCamera.setWindow(null);
}

function generateFileName() {
    var currentTime = new Date();
    var day = currentTime.getDate();
    var month = currentTime.getMonth() + 1;
    var year = currentTime.getFullYear();
    var hours = currentTime.getHours();
    var minutes = currentTime.getMinutes();
    var seconds = currentTime.getSeconds();

    if (month < 10){
        month = "0" + month;
        }
    if (day < 10) {
        day = "0" + day;
        }
    if (minutes < 10) {
        minutes = "0" + minutes;
        }
    if (seconds < 10) {
        seconds = "0" + seconds;
        }
    fileName = year + "-" + month + "-" + day + " " + hours + "." + minutes + "." + seconds;
    return fileName;
}

function takePicture() {
    try{
    mCamera.onCameraCaptured= function(fullpath) {
    alert("Image path is:"+fullpath);
    };
    mCamera.captureImage("/sdcard/DCIM/Camera/" + generateFileName() + ".jpg",false);
    }
    catch(e)
    {
    alert("inside catch"+e.message);
    }
}
```

Fig. 4. Javascript code for a WAC widget

4.3.2 Android

The architecture of Android is based on the Linux kernel 2.6. It is responsible for memory management, process management and network communications. It also provides the hardware abstraction layer for the rest of the software and device drivers for the system.

Other important components are based on the architecture developed by Sun Microsystems (now Oracle), namely, the Java technology-based virtual machine Dalvik and its Android

Java class libraries. In order to program Android applications, the development system (m3-rc20a, published in November 2007) contains 1448 Java classes and 394 interfaces, of which 511 classes and 128 interfaces are Android-specific.
Applications for the Android platform are written exclusively in Java, taking advantage of the wide spread expertise on Java within the designer community. For speed-critical tasks Android apps may take advantage of many C or C++ written, native libraries under the hood. The catalogue includes codecs for media playback, a web browser based on WebKit, a database (SQLite) and an OpenGL based 3D graphics library.
In order to develop programs for Android, a recent Java SDK and also the Android SDK is required (e.g. Eclipse). First, the source code written in Java is translated with a normal Java compiler and then adapted by a cross-assembler for the Dalvik VM. For this reason, programs can in principle be created with any Java development environment.

4.3.3 WebOS

WebOS is a multitasking operating system for smart phones with a Linux-based kernel. Multiple applications may open and run simultaneously and can be browsed via a live preview, even videos may run in this preview mode. As iOS and Android, it is operated by finger gestures on a touch screen.
PIM data is not only stored on the device, but always synchronized with Internet services like, e.g. Gmail. Using a technology called Synergy, all information, e.g. from different calendar systems such as Exchange and Google Calendar, is summarized in a single application. Synergy links contacts, calendar events and e-mails from various sources (including the above-mentioned Exchange, GMail, Hotmail, Yahoo, Facebook).
With the HP app catalogue, the manufacturer offers (similar to iPhones and Android phones) an online service that allows to download and/or buy applications (Apps) for WebOS phones.
Palm HP offers several ways to develop applications for WebOS:

- SDK

The Mojo Application Framework SDK allows applications to be developed with HTML5, CSS and JavaScript. The SDK needs to be installed on a desktop computer.

- Ares

Ares is a new development environment for HP WebOS, which is now released in version 1.0. With Ares it is possible to develop applications directly in the browser. An installation of the SDK is no longer necessary. To test the applications in Ares, an emulator is integrated. Ares works with current Web browsers such as Firefox, Safari and Chrome.

- Plug-in Development Kit

Since March 2010 a "Plug-In Development Kit" offers the possibility of using C or C++ code in applications, which should ease porting of external applications from other platforms.
WebOs is a very modern and well thought through approach, which - in contrast to WAC - is taking full advantage of the introduced web technologies. It uses HTML5, CSS and Javascript to interact and synchronize with cloud and Web2.0 services whereas WAC simply offers HTML GUI rendering facilities, which are not even the strength of HTML. WAC should follow WebOS as it does with its Mojo application framework.

5. Recommendations for industry

The stakeholders to be kept in mind while offering application frameworks, SDKs, app stores and runtimes are mainly:

- Application developers
- Mobile device users
- Network operators

WAC has well considered the needs of network operators and partially the need of potential users, but so far it does not optimally attract and support app developers. As WAC is not primarily a platform for coding experts it should focus on the needs of web designers: a graphical toolset, a simple to use Javascript library for multi touch UIs, dpi resolution based GUIs, etc.

The registration process for becoming a WAC designer is currently rather cumbersome and introduces administrational hurdles not known for designers who signed up as an app developer at Apple or Google Android. With the App Market you simply sign up with your Google account and pay a minimal amount for the permission to submit apps. There is no need to fax or mail paper documents as required for WAC. Admittedly Apple or Google Android apps are limited with respect to using network APIs, hence some overhead on WAC-side for this additional functionality is understandable. Still the processes should be simplified.

In short, we summarize the key aspects that should be considered for the future WAC roadmap:

- Provide an SDK including a graphical toolset, like Dreamweaver or the like
- Integrate a framework á la SenchaTouch
- Offer an online SDK including development life-cycle support, like it is done for WebOS
- And in addition to the bullet above: Offer network functionality testing via network sandbox once WAC 3.0 is out
- Introduce user controlled device functionality access policies: In addition to certificate/configuration based functionality and privacy control interactive means of allowing or rejecting access to user data should be enabled
- Support for SQLite
- Support a dpi based solution for screen resolution scaling should be introduced. If this is not possible an automatic CSS adaptation by the device runtime should be applied. Currently it is not possible for WAC widgets to automatically adapt to different screen resolutions on different devices. In the worst case this means that for each resolution format a different widget needs to be written, which would be inacceptable for developers. How should this be represented in the WAC store?

And last but not least, in the times of cloud computing it should be considered to integrate the WAC APIs to any browser runtime so that the widgets (and their date) can be stored in the cloud and downloaded and executed on demand just as it is done for plain web sites or Web 2.0 services. This "running in the cloud" mechanism could be the main differentiator to prefer WAC widgets over native app development like its done on iOS, Android or Blackberry.

Finally, WAC runtimes need to be offered for all major platforms not only for Android. Simply because it is so easy to develop and deploy such software on Android phones, other platforms such as Blackberry and WebOS (and iOS?) should be supported as well.

The WAC activity is a promising start to enter the application market place. The operator and device agnostic approach together with the opportunity to offer Telco assets in applications might be very attractive to developers, users and network operators.
In our opinion the WAC activity in its current form (April 2011), however, will need some tuning to become successful. Currently it is not as innovative as its competitors (AppStore and Android Market) and does not fully leverage additional opportunities a Telco operator could give to users and developers.
The WAC 3.0 specification (due in September 2011), however, that supports network APIs for widget development may be an important differentiator from its competitors. Additional focus should be put on: A more tailored and user-friendly web-design type of SDK, a simpler sign up process for developers, more supported platforms or even the deployment of WAC APIs via mobile browsers and better support for state of the art user interface technologies.

6. Acknowledgment

The authors would like to thank the Telecommunications Research Center Vienna (FTW), partners of the APSINT project managed by FTW, and the COMET (Competence Centers for Excellent Technologies) programme of the Austrian Government for supporting the APSINT project.

7. References

APSINT http://www.ftw.at/research-innovation/projects/apsint
WAC-Dev http://www.wacapps.net/
WAC open collaboration http://www.wacapps.net/
WebOS http://developer.palm.com/
ARES https://ares.palm.com/Ares/about.html
PhoneGap http://www.phonegap.com
Titanium Mobile http://www.appcelerator.com/products/titanium-mobile-application-development/
Obigo http://www.obigo.com
Widget SDK fro Aplix http://widgetsdk.org/

4

Finding Services and Business Models for the Next-Generation Networks

Javier Martín López, Miguel Monforte Nicolás
and Carlos Merino Moreno
*Universidad Autónoma de Madrid / Almira Labs S.L.Madrid,
Spain*

1. Introduction

The growing demand in the use of mobile devices implies a huge investment by operators and infrastructure providers. However, these players have important questions when it comes to understanding how to finance these investments. ROI seems to be endangered by the new mobile ecosystem that has emerged during the last few years. Many factors threaten the potential of future income over the new infrastructures so there is huge need to find innovative ways of generating customer retention and traction, which in turn will lead to the generation of revenues that can build a profitable and healthy business around the next-generation networks.

This research should be carried out before any investment is planned to have a clear picture of the financial implications of the new deployments. There is a strong need to understand customers' new ways of technology consumption and plan how to provide this employing adequate services and business models.

As a result of the way that the mobile industry has been developed, investment in advanced networks must be acquired primarily through private companies, for example, the network operators. Other sectors, like health, civil engineering or even fixed telephony have historically enjoyed a stronger intervention from the public sector as governments take on the basic infrastructure for their countries population. However, this was not the case for the mobile telephone industry, which was regarded as a private initiative, in spite of the fact that there were many incumbent players at the beginning of the industries development. Even in those cases, the industry rapidly opened to competition and new entrants from the non-public sector came into this promising new industry.

Therefore, this established the basic condition for network evolution, all mobile industry investments must be aligned with clear business models to make them profitable. This has an immediate consequence; investments will not be forthcoming until there is a clear path to ROI. On the other hand, we are on the verge of a mobile infrastructure usage explosion. Customers demand more and more data services so that the old networks start to reach their limits or even collapse. Paraphrasing Shakespeare's Hamlet: 'To invest or not to invest? This is the question.' A natural response to this question should be a clear "YES" if we are to apply the industry standards from the 80's where unlimited booming was in place. But… things change.

In the present state-of-the-industry, the "who" that are making money out of the mobile networks has shifted from the operators exclusively to other companies like the device manufacturers and content providers who work outside the operators' networks, the main companies being Google and Apple. These companies are making huge profits by selling devices and services that rely on the operator's infrastructure with almost no financial paybacks. This means that the operator in is danger of becoming merely the transportation layer, instead of the value-added services provider, whose status in the industry is described as the "dumb pipes" (Wikipedia, 2011).

The network operators are at a crossroads. On the one hand, they clearly need to invest in the new networks to address the increase in customer needs, and to keep their position in the market. On the other hand, they are uncomfortable with the idea that those investments will be profited by other companies who will get the most out of the revenues generated without incurring any financial risks.

Some solutions are being drafted by the network operators, but they are receiving strong opposition from the rest of the players and even some of their customers.

These ideas revolve around making money directly from the infrastructure usage. One of them is the "tiered pricing" concept, already put in place by some US and European operators. The concept marks the end of the "eat-as-much-as-you-can" policy that has been in place for some years, and means that consumers will only be charged for the volume of data they really consume. (Telwares, 2011)

Other ideas which attempt to reverse the situation described above involve charging the content, software and devices companies for traffic crossing the network originated by them. For example, Telefónica announced their intention of charging Google for the traffic generated by consumers performing searches on their mobile devices. This move attempts to create a revenue sharing scheme between the infrastructure owners and content providers. (Boston, 2010)

All these initiatives are creating huge controversy as they are seen as an attack on the neutrality of the net, and on the freedom of the internet. In addition, companies like Google strongly oppose the idea of having to pay network operators for the volume of data requests sent to them by customers.

In a different approach to the problem, other industry players are trying to capitalise on a basic consensus in the industry. Most future network income will be derived from software. Device manufacturers like Nokia, Apple, Google and others have paved the way for a market of software and applications in the mobile market. The numbers themselves demonstrate the success of these initiatives, with almost 4250,000 applications available, and 15 billion downloads in four years of operation (Apple, 2011). New devices based on strong internet orientation and high usability have taken the market lead, shifting the industry from a hardware-based scheme to a software-based one. This success is based on both a new device concept oriented to the Internet more than to the classical telephony world, and a strong community of third-party developers who find easy ways of creating applications and selling them to the device users.

Operators have made a huge mistake in disregarding this market change. Their reaction has been slow and poor. They overlooked the device manufacturer's movements, thinking that they would never affect their position in the market until they realized the amount of money people were spending on applications in external App Stores. When they tried to react, they realised that they did not have the culture, development tools or teams needed to challenge the fast pace of new entrants coming from the pure software world. In addition, the financial crisis which started in 2009 stopped their R&D expenditure capabilities and VAS creation programs, resulting in an even lower innovation pace. Operators will need a whole new paradigm to recover the market initiative and to avoid the risk of just becoming a commodity player who offers infrastructures that will hardly be of any value in the long term. This chapter proposes a strategy that could help operators to regain the market lead.

However, the model of App Stores needs strong revision. Many questions have arisen about whether it really is a healthy market. After the initial hype has faded away, some doubts have been cast over the future of the App Store concept. The main concerns are:

1. Mobile device market fragmentation. There is a clear need to overcome the market fragmentation created by the existence of too many different mobile iOS and devices. (Rajapase, 2008))
2. The need for advanced, expensive handsets that is widening the digital divide between wealthy and non-wealthy people and countries.
3. The fact that services are created for the device, not for the people. The services should be driven towards customer needs more than just a display of technical skills.
4. Universal reach. Wireless technology has the potential to be the technology that helps to bridge the digital divide. Therefore, players should be moving in that direction instead of creating more complex systems every day. Technology should be simple and affordable to allow the inclusion into the digital society of those segments that are usually left aside: the disabled, the elderly, and those citizens from emerging countries. "No phone left behind" should be the driver of the industry.

The main subject of this chapter is to show how to find a healthy model around software for the telephony industry.

A model that combines the existing R&D needs, with a lower time-to-market of the products so that research investment is better justified. A model that shifts from the actual "Some develop services for some" towards "Millions develop services for everybody". A model that provides operators and third-parties a way of creating useful services that fit into the new networks, and which helps to monetise investments at the same time as helping to bridge the digital divide and incorporate everyone under the principle of "Universal Design " (UniversalDesign, 2011).

This model should take into account the different topics that compose the subject:

- Application stores and device stores, current status.
- Mobile Apps vs. Mobile Internet (services encapsulated in small code pieces vs. in-browser services replicating the PC/Internet experience)
- Sociology of the mobile user and aspirations.
- Limitations of the above models especially due to market and technologies fragmentation which bring a new digital divide.
- The need to search for technologies that bring the world of the mobile internet to the existing 5bn devices worldwide, and future growing numbers.
- Bridging the digital divide.
- Apps and services for everyone: kids, the elderly, the disabled and non technical people etc.
- The use of Cloud Computing in the mobile industry. The computing world is shifting towards a cloud-like schema, not centred in the PC, but in remote execution and storage environments. However, in the mobile world this strategy is aimed towards the devices and not a cloud-centred schema. How can this paradox be solved?

Finally, we will examine which of the new technologies mentioned in chapters of this book, and others appearing in the industry, have the potential to drive business and revenues.
As happened during the 3G industry-wide deployment, where video and video-calls were thought to be the new killer services which would drive mass-adoption of 3G, the next-generation also needs to find the technologies which will drive its adoption too. And to succeed where 3G failed.
Could plain high-rate data access do it, as many operators would like to think? Is there is a need to rely on new standards for services like RCS or WAC, improved voice technologies like VoLTE or HD-Voice, or again is it the time for video-based technologies?

2. The evolution of the telecommunication networks

Communication technologies have become the most ubiquitous service during the last decades. Among these technologies, mobile technology has become one of the fastest and broadest markets. Nowadays, billions of subscribers use a mobile device on a daily basis. Thus, the mobile telecommunications industry has nurtured an innovative market which has overcome the barriers of reaching every single person in the world.
Such barriers and limitations where deployed by the early lack of standards, which created boundaries within subscribers from different countries or even between different service providers in the same country. The first devices were based on early analogue systems, less flexible or able to be adapted to newer needs. One major drawback was their high cost, which moved service providers to bring competition and innovation into the industry. The tipping point, the digital era, allowed service providers to cope with their growing needs, allowing for better and diverse services. Nonetheless, digital communications brought major improvements into the transmission, switching and quality.

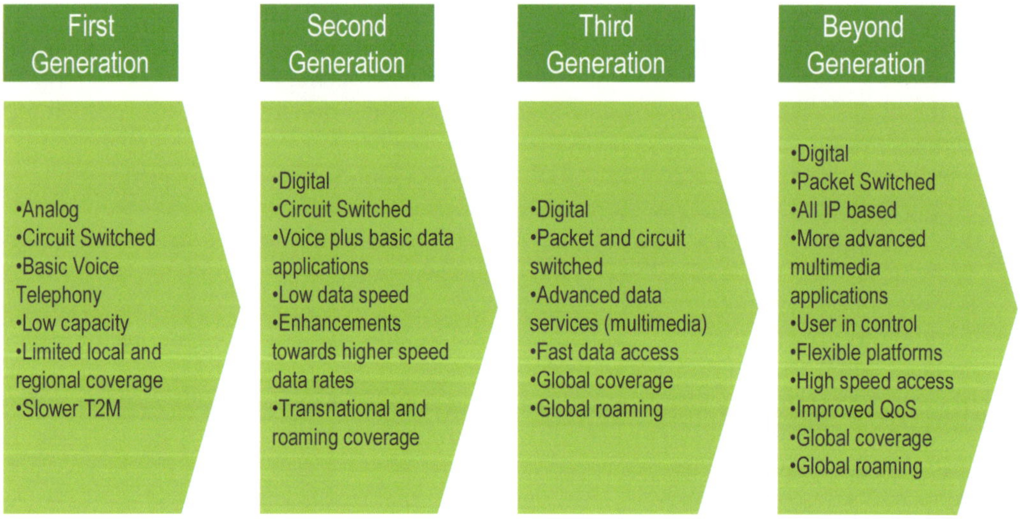

Fig. 1. The network evolution

At the beginning of the last decade, service providers were competing in newer markets as the technology evolved. The competitive landscape was changing year by year, forcing service providers to capture bigger stakes from the value chain. From pure voice communication, they moved into the data services. Based on the findings that SMS was a cash cow, service providers decided to expand their portfolio to think about communications in a holistic way. The fear of cannibalization among different services is not important, as it offers consumers comprehensive services, and the lifetime value of customers can be increased, CHURN can be reduced, and the overall value proposition of the operator increased tremendously.

With this idea in mind, service providers started working on the next generation network. One network that could cope with their service, and which offered a myriad of comprehensive services to increase the value of each subscriber.

2.1 The previous generations

During the 80s and 90s, communication service providers deployed networks based on analogue technologies. These technologies allowed for reliable and secure voice communication, bringing solid foundations for the next steps. Service providers were used to investing millions of dollars on equipment because they were making billions of dollars every quarter. The reason why the equipment was so expensive was that it was built by hardware engineers to allow millions of communications on a daily basis, relying on very expensive hardware equipment that was built specifically for this purpose. However, all that changed radically with the dot com crash, moving service providers to invest wisely in newer equipment.

Early network communications equipment and standards were driven by a small number of companies, such as Ericsson and Nokia. They invested millions of dollars on technology that could allow service providers to deploy nationwide voice networks. Strict requirements for building such a reliable communication network blocked newcomers who could bring the much needed innovation.

It wasn't until the introduction of i-mode by NTT DoCoMo (Fransman Martin, 2003) in Japan that mobile data services arrived on the international scene in earnest. The explosive growth of data services in Japan forced executives in carrier organisations to take the data services seriously—at a par with voice—and subsequently to make significant investments in the evolution of the data services market.

2.2 Deploying richer communication services

The introduction of i-mode by NTT DoCoMo (Fransman Martin 2003) in Japan was brought to the attention of the European services providers. Based on the explosive growth observed in Japan, mobile data services were an attractive market. Service providers' executives were forced to take the bid seriously and to invest in the evolution of their communications network. The deployment of mobile data services fostered consumer adoption, which created a self reinforcing effect because more connectivity led to more services and greater innovation.

This was the first attempt made by service providers to increase their value chain outside the basic voice services. During the last decade, service providers launched a myriad of services, which emphasised the fact that service providers were merely network service providers, and they needed to create an ecosystem of content and application partners.

2.3 The standardisation bodies

The telephone communication network had a key important requirement: breaking boundaries. Achieving such an objective was a matter of standardising every single protocol

used in the network. For that purpose, many governments and private parties created committees to provide a unique standard. Players such as the service providers, system integrators, manufacturers, content providers and IT providers joined such bodies to present their vision. The crucial factor then was the vision that each party brought to the body, influenced by factors such as region and place in the value chain, etc. (Grøtnes Endre, 2008)
Although necessary, it proved to be bureaucratic, leading to a slower time to market. The time that elapsed between each new release was too long, which proved fatal for bringing the right innovation. Bodies such as the ITU (International Telecommunication Union), the ETSI (European Telecommunication Standards Institute), the 3GPP (3rd Generation Partnership Program) covered standards ranging from radio transmission, network protocols and features. The key finding was that those bodies discussed not only technical matters, but a great deal of business strategy. Competitors where sitting at the same table discussing how to enable a new feature, which could potentially, increase the gap among them. Although some of the key features were hidden so that they could have a key advantage over their competitors.

2.4 The next generation networks

Earlier experiences showed service providers that it was urgent to come up with new ideas and to embrace innovation. The key point was to achieve faster time to market because the internet was seen as a threat to their business. Specifically, service providers were concerned about new ways of generating revenue streams which were different to those that they had. The deployment of broadband services proved to be the perfect stimulus for the mobile service providers. At one point, users could participate in chats, send electronic emails, and download videos or images.
The 3GPP standardisation body assumed the role of leading the way towards more open and simple communication protocols. Their vision was to embrace a packet network that could transport any content through end points. That required migrating from a circuit switched network, to a packet based network, whilst at the same time assuring the same quality that was delivered with the old network. The protocol used for building the network was the IP protocol (Ajit Jaokar & Chetan Sharma, 2010), used my billions of computers around the world. The IP protocol had the robustness required by service providers while providing more flexible schemas. The new core network was called IP Multimedia Subsystem (IMS) (Bertrand Gilles, 2007). The IMS was to be built around IP, which was developed by the Internet Engineering Task Force (IETF). The 3GPP requested IETF for their help with building the necessary protocols around the IMS idea. If the 3GPP required any ability, the IETF was responsible for delivering such protocol.
The primary goal of the architecture was to enable launching richer services that could potentially be used by any customer. The voicemail is a perfect example of such a service. It was very common to have a voicemail for the fixed phone and a voicemail for the mobile phone. Bringing convergence was easy, decoupling the access layer from the service layer, so that no matter how you accessed the voicemail, it worked for both scenarios. By separating the access layer, the transport layer, the control layer and the services layer, service providers could provide seamless access to their service despite which device the customer was using.
As a result, service providers could increase their service mix because they were not tied to any specific access layer. Therefore, you could watch a video from your mobile phone, chat from your fixed phone, etc.

2.5 The innovation trail

The service provider's strategy was to look towards other industries to learn from. At that time, the internet proved to be the best platform for launching new services. The internet was constantly evolving, bringing richer services to the end customers. This led to the conclusion that in order to bring innovation into the rigid mobile communications arena, a radical shift was needed. Building a worldwide communication network imposed certain conditions in terms of resilience and stability. Nonetheless, the internet at that time was a worldwide network; built from complex requirements that were able to cope with its demands. That reasoning allowed services providers to adopt the internet as their core protocol.

Adopting the IP family of protocols, service providers also brought the spark of innovation from the internet. Thanks to its architecture, the internet deploys its services at the endpoints, not at the core. The core is just responsible for the transport of data in the most efficient way. Therefore, small companies can innovate by creating compelling services at the endpoints. That schema was impossible with the core oriented architecture built by service providers.

The IMS is the perfect companion for service providers, as it can provide not only the quality needed, but also the innovation they are looking for. The trail lead by internet was used by the service providers as a guide for their networks.

2.6 The value of open systems

Early generations were built using proprietary hardware and software equipment, which blocked many companies from entering the market. With a few companies leading the innovation, service providers were totally dependent on their network providers. The shift made from circuit switched, towards packet switched networks enabled deploying standard equipment that any company could buy. However, there was still one barrier to overcome, and that was the proprietary software vendors.

The revolution came from the internet, as software was open, anyone could create applications over that software. The software required by service providers is now moving steady but slowly towards open systems. Consequently, innovation is open to any one with the desire to participate and bring in new ideas. The IT concept of SOA (Service Oriented Architecture), which is the based of the IMS network, is only possible if the protocol used to communicate is open.

2.7 Technologies and services built around NGN

The best way to show success is to lead by example. During the last years, many technologies have been created around the idea of IMS. Some of these technologies are based on the core, and others are based on the endpoints. This diversity enables service providers to come up with new and enticing services to offer their customers. The following are a few examples of very interesting technologies.

RCS: Rich communicate suite is an evolution of the typical basic services offered by service providers towards a complete and coherent service offering. Services such as instant messaging, presence and availability, video share, file share and others are defined at the handsets and the core network. The suite is built around many already standardised services and protocols, which allows faster definition and implementation. The success will come with the default inclusion of the client in every new handset sold.

LTE: Long term evolution is the next radio access for mobile devices. The UMTS (3G) provided speeds up to 2Mbps, then HSDPA provided up to 10 Mbps. LTE is a complete evolution in the way radio access works, providing speeds up to 300Mbps. LTE will enable mobile service providers to deliver richer multimedia services, with uncap video and audio.
HD voice: High definition voice provides double quality from its predecessor. During the last decades, the voice channel has not been updated to provide better definition. Therefore, the HD voice is not suited for the old analogue world, as it was designed for a narrow band voice. Once the IP shift has been deployed, HD voice will be a common service.
WAC: Wholesale Applications Community is a joint venture from many leading service providers to provide a unique set of APIs and storefronts to enable sales of handset agnostic applications. This initiative was started by JIL (Join Innovation Lab), which merged with WAC to provide a unique development platform. The idea behind it is to expose core network and device information to developers so that a developer ecosystem is created. That ecosystem will be equal to anyone deploying a WAC store.

3. The new R&D organization

Entrepreneurs have to face two different challenges when dealing with new technologies in new or existing markets. The first one is to become a renovator, which implies having a strong R&D to be able to create something new. The second is the business model. Most of the entrepreneurs do not have a specific profile, as they are not business development managers. Besides, business models hardly support any R&D effort. The need for immediate ROI wipes away the classical R&D models with long-term scope. Therefore, there is a need to find new ways of funding research activities.
This is the case of the technologies in the mobile industry which are opening the doors to new application scenarios that will revolutionize the way in which individuals have access to network capabilities anywhere, following the concept of "mobility". Now, the important and differentiating feature of this particular market is that delivery is not generated by the unilateral setting of a product or service derived from a company, but from the integration of interests from different industry players to establish "win-win" relationships where the management of reciprocity is the basis for viable solutions to end customers.
The group of agents that interact in this situation would be, in the first place, the network operators whose business is determined by the use or traffic that is generated on their network. This leads them to the need for multiple layers of value that are built to give meaning to their infrastructure. Therefore, one's business model must be very open to negotiation, even generating sponsorship of any R&D that may materialise in the context of the value proposition that reaches the end user, sometimes with their participation to create a system of "intelligent decisions". Later, we will detail the business model of these operators within the global system, related to the other agents that build that market reality.
Secondly, the technological start-up companies, who build the software that facilitates the further development of applications in the face of finding customers. This requires a bond of vertical integration forward and backward, i.e., approaches and liaison with operators and applications companies, they therefore, play the role of a "hinge" in the market. The business model will also be presented in a scheme shared with other agents.
And finally, content and applications companies generating opportunities and needs around a catalogue of applications that represent the "visible face" of the business described above, **and composing the "tip of the iceberg" (see Figure 2) where the accumulation of**

specific value proposition, with a layer of competitors nourished where creativity is crucial.

Fig. 2. The visible face

Furthermore, this scenario combines different types of organizations, i.e., large multinationals, large technology companies and the new technology-based companies phenomena arising from start-up or spin off, totally opposite in many aspects of their leadership and management, but similar in terms of the need to find common spaces of understanding that enable the development of value propositions in economic returns.
At the base are operators or network providers characterised by their large size, with an investment framework that supports high-volume. Because, there are not a large number of these operators worldwide they need geographical locations too large to meet the return of investment. The way to get the desired return is determined by the acquisition of shares in the business carried out by organisations that build technology and services based on the use of their network infrastructure.
Thus, we propose certain benefit rates that apply to the volume of usage of these networks, making it a partaker of the stakes associated with the success of all stakeholders seeking to develop businesses supported by such infrastructure.
Entering the business layer are agents who respond to the challenge of setting "base technologies". These are large organisations from several domains, usually general practitioners in implementing ICT in a multidisciplinary way with entrepreneurial projects, namely, new technology-based businesses, with value propositions in constant renewal, whose business model is eligible for funding the "interests" of the operators and the scheme of "fee" or revenue sharing as may be agreed with the companies that develop applications for the final customer in the market. **However, the importance of public support for this phenomenon and entrepreneurial risk, should be noted. The uncertainty of these**

companies' proposals, which need a longer process of maturation, means that operators are unable to take it "as it is".

Finally, the companies generating such applications are continuously trying to develop creative thinking based on both their talent in the use of several items: outside involvement, assuming the challenge of the launch, choosing the right powerful dynamic advocacy, communication, etc., and taking the risk of the entrepreneurial phenomenon. This is where the market cash is generated, the turnover that pulls the strings of agreements in the value chain, the basis for all the efforts that capitalise new planning.

These terms and conditions for viewing the "business" are the premise that should generate a structured approach to construct a formal system usually composed of elements that become a "business model". These may well relate to the reference named "business ontology" (Ostenwalder, 2004). However, this approach challenges some points of this document, namely:

a. Value proposition.

In this case the value proposition is tied around the needs that have been discovered in the context of mobility, by providing a variety of ubiquitous features that accompany the customer. Therefore, the needs of the professionals are increasing as they are linked to market needs, which is the reason why companies create compelling applications. However, one issue is the value proposition, and another is the supply, i.e., conditions that make it attractive and viable. It should support the whole chain of technology and infrastructure, making feasible some of the developments that the companies make to be in the innovation portfolio.

b. Elements that support the value proposition.

Among the elements that make the value proposition, we can distinguish key components, namely the constraints associated with the relational structure of the agents involved in the setting of the offer. Given the level of integration required, fundamental policies should be in place, forming cluster type development strategies and even setting up "joint ventures". These branches also require a management model for sharing information and knowledge to develop ways in which to coordinate the relational framework, establishing certain protocols and formal processes. Moreover, the relationship is considered a distinct expertise in which each party assumes a specific competency framework. There are three clear areas, on the one hand, infrastructure and investment, on the other, the technology base, and finally, the creativity around the applications.

c. Projection of the value proposition and stakeholders.

Looking into the market value, is important to note open and participatory dynamics that must be established within outsiders, trying to identify needs and specifically, segments where the feeling of belonging or clan, might allow the development of a branding and future development. To carry out this task all relevant information is channelled through different channels of communication that are inserted into the social networks, websites and networks and events, responding more to a relationship with information, and more people-centred experts.

d. Schemes of costs, revenues and results.

With all these variables, achieving a model that supports the revitalization of the value proposition is sustained in the cross-cutting activities that are determined by the cost structures that must be supported, namely, infrastructure and personnel etc… These activities are clearly linked to the flow of income, outside sales and within the "base technology companies". It is clear that this relationship of income and expenditure is associated with a control scheme that enables the progress of activities, the status of critical assets that sits on the competitiveness of the organization and its derivative in a control panel that can generate a scheme of "input-output" or "cause and effect" very enriching for directing and managing the task.

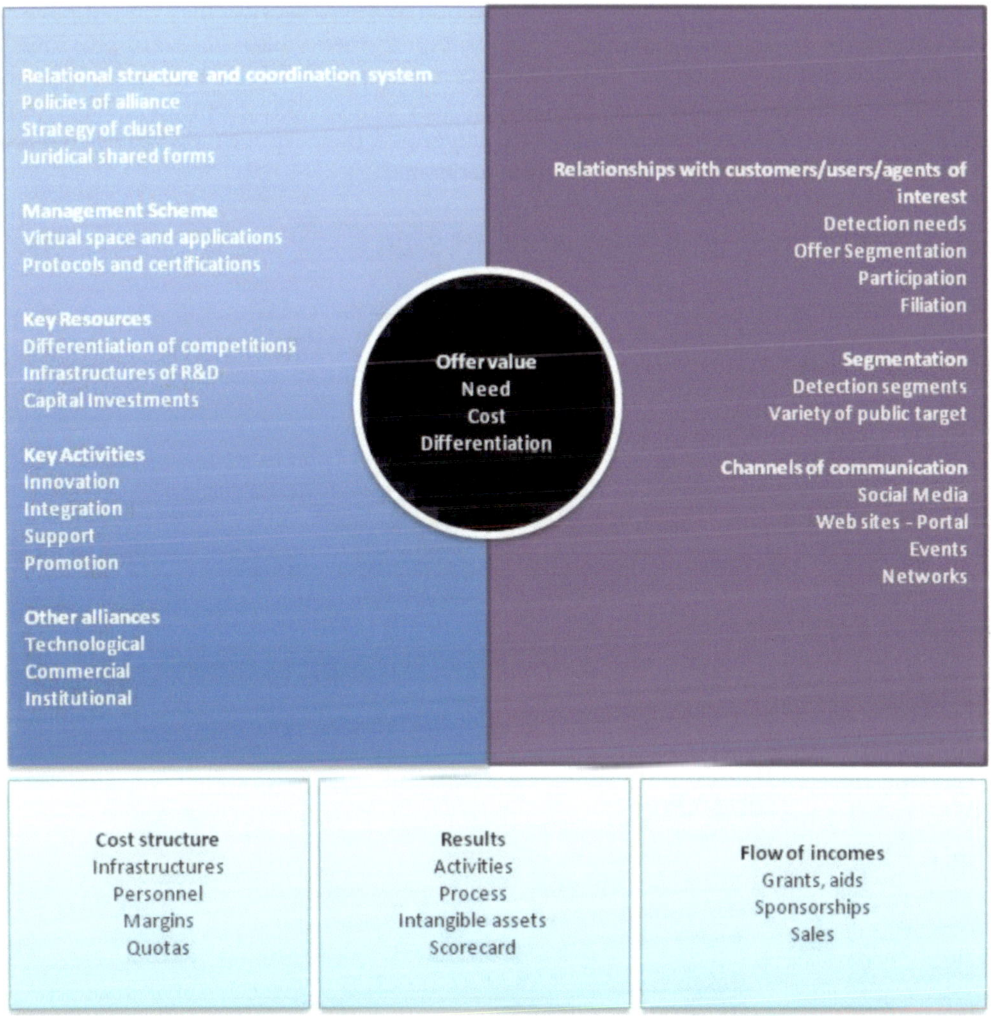

Fig. 3. Business Model Canvas

Therefore, the model provides a number of key considerations that must be taken into account when it comes to the strategic thinking that is required to organise the initial premise, and thereafter to realise the vision.

4. Finding the right services and business models

As described in the previous sections, the telecommunications industry needs to reinvent itself at this point in time to find new ways of maintaining its pre-eminent position in the market.

Now that we have reviewed the technologies and the current state of R&D, it is time to propose a new model, which in the authors' opinion, should be adopted and used by the industry in the years to come.

Application Stores have shown how the software market for wireless devices can be driven from outside the operators walls and gain a significant market share and the customers attention. Operators have shown little or no reaction to them, because they do not have the right tools to do so, and they have lost their position as the wireless software providers of preference for customers. Some of them have even abandoned the battle and assumed that they will be just data transportation media and make money from this traffic. Others have surrendered to the device manufacturers and integrated their Application Stores into their offers, and got a small portion of the revenues generated.

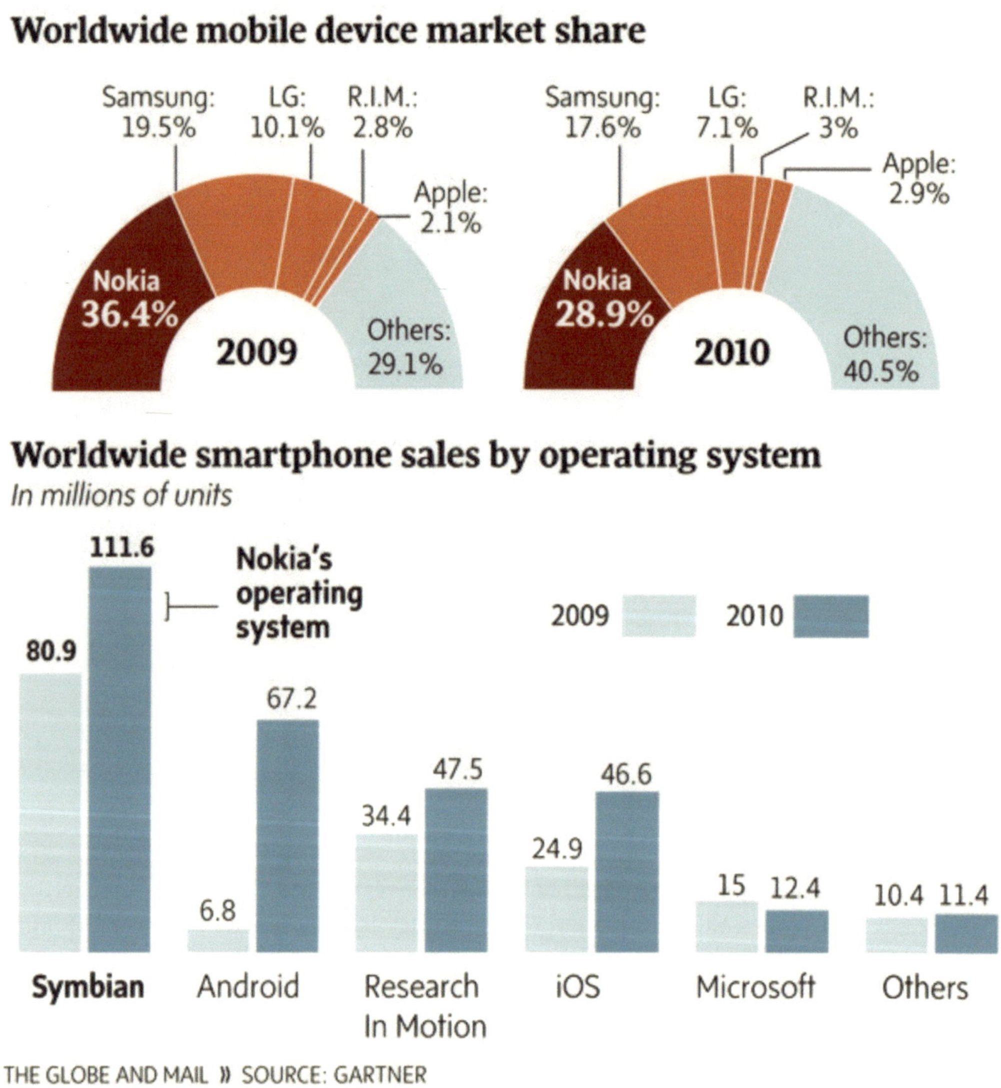

Fig. 4. Worldwide mobile device market share

Application Stores though, have a severe problem of fragmentation. As of March 2011 the new smartphones represented around 19% of the worldwide wireless devices market OnlineMarketing (2011). Growth is enormous and will continue in the coming years, but many factors point to the fact that smartphones will never take 100% the market. Sales figures point to around 200 million units being sold in 2011 (Celularis, 2011), (Moconews, 2011). But the wireless devices base is around the 5 billion mark. At this pace, it would take 25 years to achieve a full smartphone market. Therefore, and despite all the news coming from the industry, which talks about a smartphone world, the numbers show that this is not the case and there is still a huge market for non-smartphone devices that might shrink with time, but will never go away. Paraphrasing Geoffrey Moore in "Crossing the chasm" (Moore, 1991), the smartphone is a market of an Early Majority nowadays.

Additionally, many smartphone sales are being driven by huge subsidies from the operators, so that the numbers of smartphone sales and adoption are hugely distorted. We should not disregard the fact that although many customers do not really want to own a smartphone, they are not given an alternative. Therefore, what we are actually seeing is the birth of a new mobile consumer specie: the "dumb smartphone user". This kind of user owns a smartphone but hardly uses any of its advanced features. Besides the basic voice and texting features, they will possibly use e-mail and might access their social networks. But most of them do not seem to be willing to pay for the applications, being entangled with device downloads and technical tweaks.

One of the keys to the huge success of the Application Stores is the third-party developers' model. Developers have been given a clear model with development tools, certification process and access to an identified market. But again, some obstacles limit this model. First of all, competing against 350,000 applications is a severe problem. Reaching customers is not an obvious task, and a lot of money must be put in applications marketing and promotion, changing the initial low-cost model for a more costly one. Some applications have been a huge worldwide success, Angry Birds being the paradigm of this model as they have sold many millions of applications without any big promotional effort and their developers have become multi-millionaires. But the numbers show that this case is one in a million. The average application developers hardly make any money out of them. For a ten applications set, reports say that 1-2 are successful, 3-4 recover the investment or make some money, and 4-6 just do not make it.

As it can be seen from the figure below, developers wanting to create applications have an additional problem in the market fragmentation due to the presence of multiple operating systems.

Developers wanting to create a service have to choose to which operating system and therefore to which market share they are aiming. Developing multiple versions is an expensive exercise and raises the financial risk of the development. Multiple versions mean multiple developments to create and maintain, almost doubling development costs. Although some companies are creating development frameworks to develop applications that run on any operating system, experts in the field recognize the unfeasibility of the idea, both from a technical and a political point of view. There are some approaches, but the facts suggest that having a unique development that works in any device is just not possible, and will never be. Apple and Google, the main players in the domain will never get to an agreement of a common framework because their interests do not match. Apple especially believes in its own closed market and is not ready to communicate with others. But reaching broader markets is crucial to allow an easier monetisation of the applications.

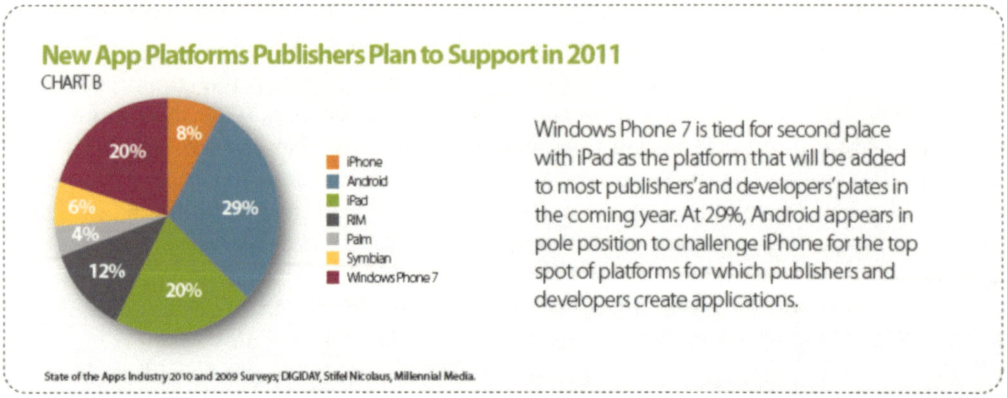

Fig. 5. App development platforms

We have reached the crucial point of the chapter, which has been the object of the R&D work developed by Almira Labs with the collaboration of the UAM over the last four years. It is: "How do I create a service that I develop once and reaches all or most of customers in the world? How do I overcome the fragmentation problem? Why do I have to choose a closed and limited market? How can I create applications that are useful for anybody regardless of their technical skill, budget, geography, cultural status or personal abilities? How do I overcome the R&D lower budgets in Operators? How do I develop rapidly low-cost services to match Internet industry pace?"

All these points have been addressed and the solutions found are proposed in the next section.

5. The case of Almira Labs

The Spanish regional government's current conviction about the benefits of supporting the creation and development of New Technology-Based Firms (NTBFs) is quite evident. An enterprising culture, training, funding, infrastructure and assessment services compose the entrepreneurship field and reveal a traditional strategic approach which has been followed for the last few years (see Figure 6), even more so when the consolidation undergone by science and technology parks is taken into account (MTI, 2007; European Commission, 2006; 2003; Rubiralta, 2003; Belso, 2004; GEM Project, 1999; Butchart, 1987).

This scheme shows a multidisciplinary reality which is shaped as inputs for the creation of a support plan for entrepreneurship and, more precisely, for what turns out to be the most attractive segment of companies for most regions - that is, NTBFs.

Thus, the reality of the NTBFs is among the strategic priorities of regional governments due to the impact on competitive updating around business networks and their potential for generating employment.

5.1 New technology-based firms

The conceptual framework and the characterisation of the collective of NTBFs have already been widely covered and studied (Bueno, 2003; Fariñas and López, 2006; Storey and Tether, 1998; Shearman and Burrell, 1988; Bollinger et al., 1983), in a dual perspective: on the one hand, the differentiation derived from the moment or stage in which the business project is, and on the other hand, the widespread consideration of the resources and capacities available, which involves the management of both the tangible and intangible, and the internal and external assets.

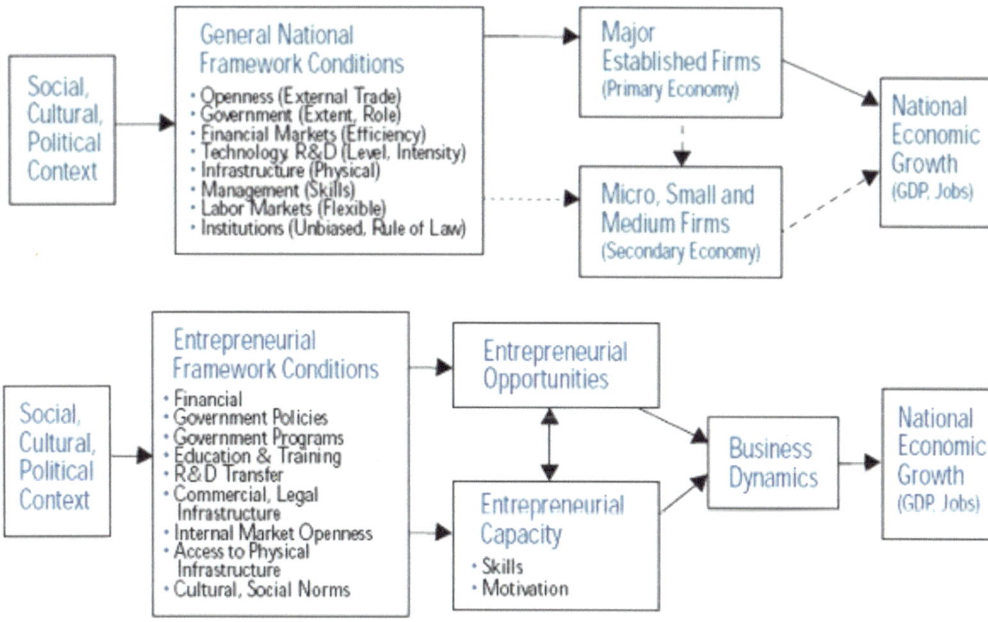

Fig. 6. Conceptual framework of the GEM Project for the analysis of the enterprising phenomenon *Source: GEM Project (1999)*

In turn, the temporal component is influenced not only by the traditional sequence idea-business plan-start-consolidation (Veciana, 2003), but also by the specific context derived from the sector to which the firm belongs.

In this case, the field of Information and Communication Technologies (ICTs) shows a clearly differentiated profile, especially regarding its maturation period and terms of supply. Undoubtedly, we face a reality of quick incubation that involves a rapid strategy creation with a clear orientation and the configuration of a dynamic organisational development (McQuaid, 2002).

Within the frame of ICTs, the life cycle of a new project shows a special, almost unique profile, especially for the following reasons:

- knowledge globalisation allows any firm or interest group with basic academic knowledge to have access to the latest novelties in applied engineering through the internet
- both barriers to information transmission and the difficulty of intellectual property registration increase investment risks
- the existing difficulty of establishing a project with wide objectives within a long-term period
- the low-cost competition from developing countries

All these factors force ICT business projects to gather the following basic features:

- a highly specialised niche with scarce competition
- an enterprising professional team highly specialised in the field and with previous experience

- the strategy of short-term projects/services and products, and R&D as a differentiating factor for medium- and long-term survival
- highly qualified personnel to assure productivity
- short time-to-market as a way to replace traditional methods of intellectual protection (entering the market first becomes a key factor).

In general, ICT projects appear as extensions of the entrepreneurs' work in previous companies, discarded working lines, etc., taking into account a small set of degree projects or PhDs studies.

The characteristics mentioned so far make the life cycles of projects extremely intense and concentrated within a short time period. Their average life oscillates between three and five years, and the most common objective is the sale of the product or service (and even the company) to a larger competitor, in order to apply the capital gain to a new business idea.

5.2 Almira labs as an NTBF in Madrid science park

Almira Labs is a company which was founded in 2006 with the clear vision of developing a technology allowing fast and low cost development of services for the telecommunication networks. After working in the industry for several years, the founders were convinced that developments for those systems should be made in a completely new way, bringing to the telecom space the practices of the software engineering world, with special attention to the SQA (Software Quality Assurance) principles, with technologies allowing the rapid development of services over open standards and programming languages that could be used from any device.

A new technology, Next-Generation Network (NGN) or IP Multimedia Subsystem (IMS), and a new programming standard called JAIN SLEE (JAINSLEE, 2009), based on Java, appeared to dramatically change the industry panorama. The services developed with this technology for the telephony network, work universally on the network nodes of any manufacturer, as long as they are compliant with the standards described above, either for mobile or fixed networks. This technology allowed the birth of a new concept: Fixed-Mobile Convergence or FMC (see Figure 7).

With this new paradigm, the number of possible service providers for operators increased, as it was no longer limited to the manufacturer of the network equipment acquired, but opened to service providers who were able to program with standard JAIN SLEE and had a good knowledge of the telephony network. Nevertheless, this situation still involved a certain degree of restriction for operators, since they were dependent on suppliers with certain specific knowledge of software development.

Almira Labs went a step beyond this. It positioned itself ahead of its competitors in this space, offering a graphical tool to create NGN services, reproducing the concept of a Service Creation Environment (SCE) that existed in the classical intelligent network for this new NGN paradigm. With this tool any non-technical member of a product engineering or marketing department can develop these kinds of services without programming, in a fast and efficient way and at a reduced cost.

To complete its offer to operators, Almira Labs has a catalogue of plug and play network services that were developed using its own technology and which can be deployed on any NGN node of any operator (see Figure 8).

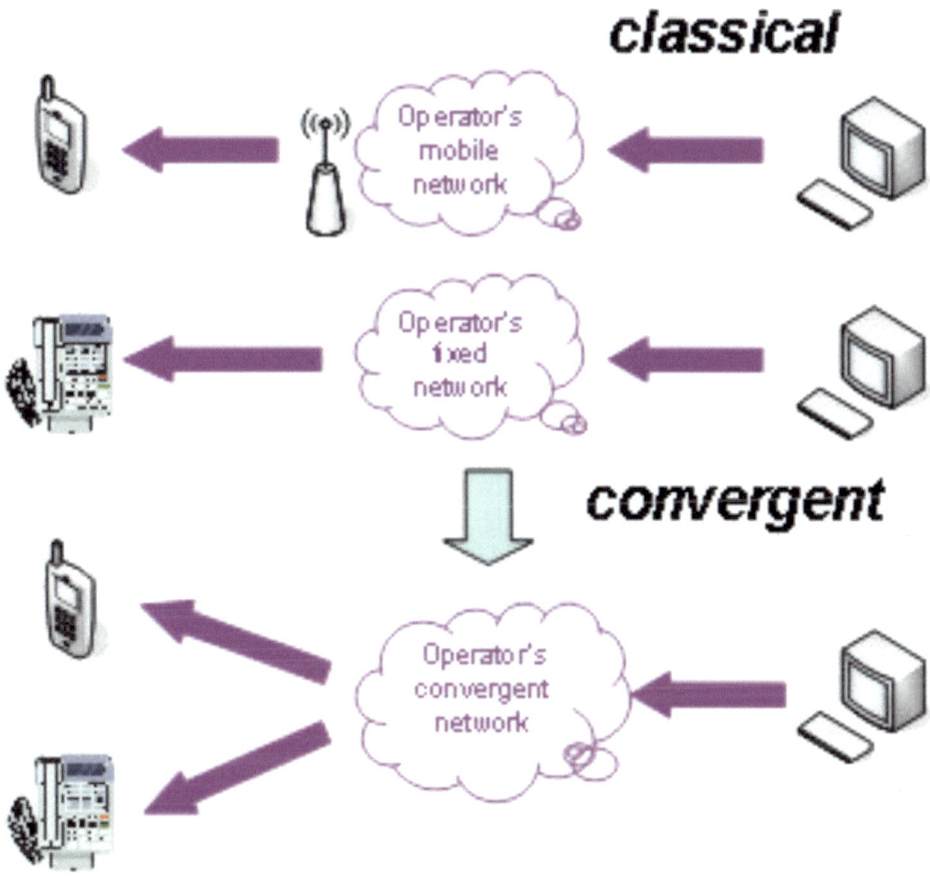

Fig. 7. Intelligent network evolution

In summary, Almira Labs defines itself as a next-generation product and services provider for telecom operators' next-generation converged networks. The main aim of the company is to become one of the key providers within this area at a global level, as its target market is composed of all telephony operators worldwide. Its go-to-market will be undertaken by a direct sales force in both geographically and culturally similar areas such as Iberia, Northern Africa and Latin America, and by a strong partnership with JAIN SLEE network node manufacturers in the rest of the world.

The results of four intense years of R&D effort are condensed in The Five Billion Store Concept (www.fivebillionstore.com). It is an Application Store with services that can be used by any person, from any operator, with any mobile device or fixed line. Useful services at affordable prices. And it is complemented by a developers framework that can be used both by developers (using a JAVA-based library with all the telecom components) and non-technical people who can create complex telecom services using a graphical tool (in prototype stage at the point of writing this book). Developers are offered a typical revenue sharing schema on a site where a single development addresses a huge market of consumers.

This ambitious plan has been developed by a very strong team of professionals with wide experience in this domain. The founders have accumulated a total of more than 25 years experience in working with operators. In addition, they have put together a team of software engineers who have previously cooperated and worked together for many years in the same area.

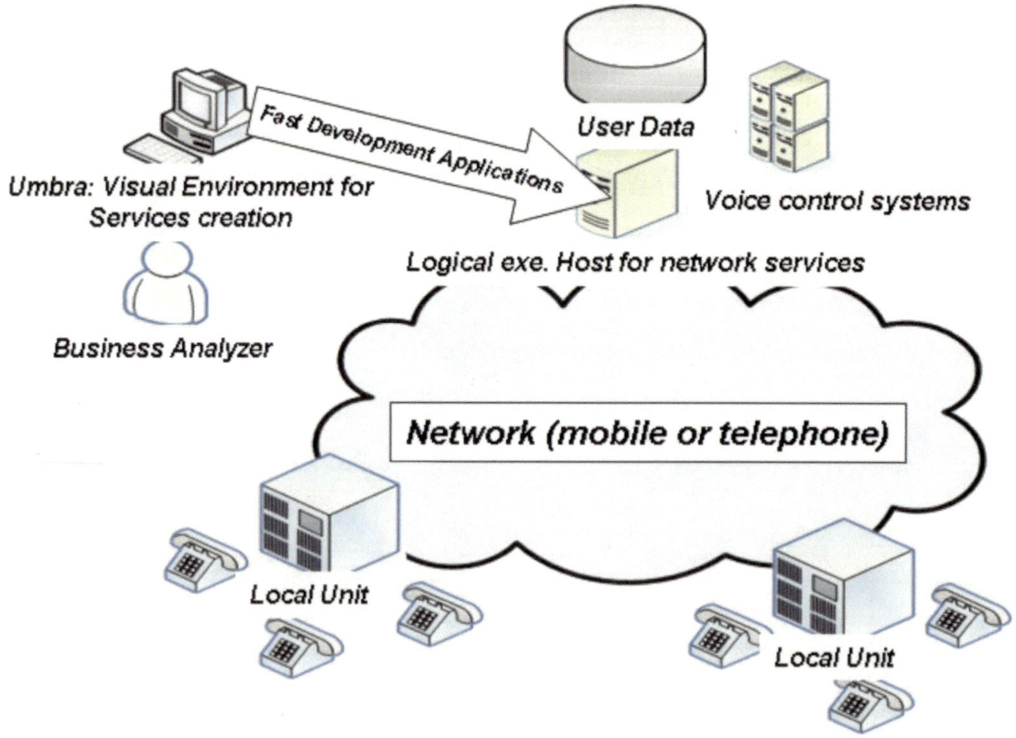

Fig. 8. Almira Labs technology schema

The members of the founding team have a remarkable university involvement. The company's entrance into the Madrid Science Park (MSP) is due to the team's remarkable research profile and its ability to transfer basic university research to an industrial application within a strategic field.

For Almira Labs, the advantages of belonging to MSP are innumerable, but may be summarised in the following points:

- the MSP brand - a prestigious, nationwide-acknowledged brand which raises the Almira Labs project to a higher level, since it ensures the quality of the business project
- associated services, employment service, grant management, marketing management, etc.
- physical space at a reasonable price and with a flexible scheme, which reduces the enterprising risk by controlling fixed expenses
- international services

In brief, the objective is the creation of a product for the development of services for telecommunication operators, which involves a great deal of research in leading fields within the new trends in computing technology; for instance, Modern Driven Architecture (MDA) (Miller and Mujerki, 2003; Kleppe et al., 2003; Meservy and Fenstermacher, 2005), and Design of Object-Oriented Applications or Product Lines. In fact, these fields are currently being studied by the postgraduate members of Almira Labs. Fostering schema from the Madrid Science Park has proved key in allowing such a long research period for current industry standards and Venture Capitalists schemas that prime short-term results.

5.4 Market implications and conclusions

The world is becoming a global market. The globalisation process implies that products can be sold all over the world with the help of new technologies such as the internet, or by using partnership schemas which help a company reach all its potential customers.

However, globalisation also broadens competition. Nowadays every company in the world working in the same domain can sell its products globally, in an efficient way regardless of its geographical location.

This fact involves high risks for Western companies, since an effort in R&D to create a new product or technology can be rapidly copied by companies in emerging countries, especially in Asian countries such as India or China, where labour costs are much cheaper and the workforce is quite large. These countries specialise in adopting technologies and processes and giving them out as services to companies all over the world at a lower price than the Western companies, where those technologies originated.

Almira Labs has identified this risk from the very beginning and has centred its strategy on being a small company with high R&D skills and attempting to always be ahead of competitors in developing countries. In addition, the study of intellectual capital conducted by IADE2 reinforces these ideas and sets the basis for a valuation of Almira Labs' intellectual capital and its market strategy, which is implemented by having a product orientation and keeping a highly innovative profile:

- The product orientation will be the competitive differentiator in our current markets.

Almira Labs is one the first few companies in the world involved in intelligent network services for NGN environments. It is a difficult market to enter, because of the very high-level skills that are needed, but - with time (two years maximum) - more competitors from all over the world will enter the market, thus bringing prices down and improving market conditions. When that happens, Almira Labs products will be ready, so we will be able to compete with new entrants by reducing the costs of our products and, therefore, becoming more competitive.

Keeping a highly innovative profile is a key issue to identify new markets to which Almira Labs could look for the diversification of its activities. Markets evolve very rapidly in the new global economy, especially in the sector of new technologies. When they start maturing and the risks are lower, competition increases - mainly from low labour-cost countries - and, therefore, business profits tend to shrink in a short lapse of time. An occidental technological company like Almira Labs needs to carry out constant research in order to find the next new market or technology with the aim of maintaining its competitive advantage against such competitors.

To implement the market approach described above, Almira Labs is creating partnerships with players in the NGN area, mainly global providers or integrators with an international presence. These partners can sell Almira Labs products all over the world using their already established commercial networks and be in charge of pre- and post-sale processes, including first-level support. This schema alleviates Almira Labs' workload and allows us to concentrate on our core value identified in the study: intellectual capital regarding software creation.

6. Conclusion

The authors have described the current state of the telecommunications market and its shortfalls regarding the generation of a future-proof business model that can give both market and financial sense to the new investments planned for the creation of the Next-generation networks that will drive to 4G and beyond.

The case of Almira Labs has been presented showing how an adequate fostering program can help to develop a long-term R&D plan when the industry is not ready to support it, and that this effort can generate new paradigms for the Telecommunications industry.

A new approach to the VAS market is proposed, putting together the rapid low-cost development of services that are accessible to all kinds of people, regardless of their device, technical skills, geography, age or physical ability. A technology that makes sense from both financial, and social points of view, and which provides innovative services for the population worldwide, at an affordable cost and ease of use.

Operators have the chance of adopting such a consequent technology and to again fight effectively the proprietary Application Store from the device manufacturers, regaining the market share lost and their innovation pace.

7. Acknowledgment

The authors want to thank Madrid Science Park for its continued support to our research activities and acting as facilitators in the fruitful connection between the University (UAM) and the Incubator's SMEs (Almira Labs).

Mr. Miguel Monforte Nicolás wants to thanks his parents for their warmth and support at every step in my life.

Mr. Javier Martín López. I want to thank my wife and kids: Maite, Pablo and Daniel for being so supportive and for not complaining too much about the fact that writing this book stole some precious family time. As well, I want to thank my parents, who made a great effort to get me the best possible education without forgetting about the personal principles.

Mr. Carlos Merino Moreno wants to thank his family for their continuing support.

Acronyms

UAM: University Institute of Innovation and Knowledge Research (Universidad Autónoma de Madrid, Spain).
VAS: Valued Added Service
RCS: Rich Communication Suite
WAC: Wholesale Applications Community
VoLTE: Voice over LTE:
LTE: Long-term Evolution

8. References

Amit, R. and C. Zott (2001). "Value Creation in e-Business." Strategic Management Journal (22)6-7, 493-520.

Applegate, L. M. (2001). "E-business Models: Making sense of the Internet business landscape" in Information Technology and the Future Enterprise: New Models for Managers G. Dickson, W. Gary and G. DeSanctis. Upper Saddle River, N.J.: Prentice Hall.

Barabba, V., C. Huber, et al. (2002). "A multimethod approach for creating new business models: The General Motors OnStar project." Interfaces 32(1): 20-34.

Brancheau, J. C., B. Janz, et al. (1996). "Key issues in information systems management: 1994-95 SIM Delphi results." MIS Quarterly 20(2): 225-242.

Brews, P. J. and C. Tucci (2003). "Building Internet Generation Companies: Lessons from the Front Lines of the Old Economy." Academy of Management Executive 17 (4).

Burgi, P., B. Victor, et al. (2004). "Case study: modeling how their business really works prepares managers for sudden change." Strategy & Leadership 32(2): 28.

Chesbrough, H. and R. S. Rosenbloom (2000). The Role of the Business Model in capturing value from Innovation: Evidence from XEROX Corporation's Technology Spinoff Companies. Boston, Massachusetts, Harvard Business School.

Doumeingts, G. and Y. Ducq (2001). "Enterprise modelling techniques to improve efficiency of enterprises." Production Planning & Control 12: 146-163.

MacInnes, I., J. Moneta, et al. (2002). "Business Models for Mobile Content: The Case of M-Games." Electronic Markets 12(4): 218-227.

Mahadevan, B. (2000). "Business Models for Internet-based e-Commerce: An anatomy." California Management Review 42(4): 55-69.

McKay, A. and Z. Radnor (1998). "A characterization of a business process." International Journal of Operations & Production Management 18: 924.

Osterwalder, A. (2004). The Business Model Ontology - a proposition in a design science approach. Dissertation, University of Lausanne, Switzerland: 173.

Osterwalder, A. and Y. Pigneur (2004). "An ontology for e-business models". In Value Creation from E-Business Models. W. Currie, Butterworth-Heinemann.

Seddon, P. B., G. P. Lewis, et al. (2004). "The Case for Viewing Business Models as Abstraction of Strategy." Communications of the Association for Information Systems 13: 427-442.

Timmers, P. (1998). "Business Models for Electronic Markets." Journal on Electronic Markets 8(2): 3-8.

Weill, P. and M. R. Vitale (2001). Place to space: Migrating to eBusiness Models. Boston: Harvard Business School Press.

Adserá, X. and Viñolas, P. (2003) Principios de Valoración de Empresas, Deusto: Bilbao.

AECA (1996) 'Principios de Valoración de Empresas, Estudio de Aplicabilidad de los Diferentes Métodos de Valoración', Document 5, Madrid.

Belso, J.A. (2004) 'La actuación pública para el fomento de nuevas empresas', Boletín ICE, Vol. 2813, August–September.

Bollinger, L., Hope, K. and Utterback, J. (1983) 'A review of literature and hypotheses on new technology-based firms', Research Policy, Vol. 12, pp.1–4.

Bueno, E. (2003) 'El reto de emprender en la Sociedad del Conocimiento: El capital de emprendizaje como dinamizador del capital intelectual', in E. Genescá, D. Urbano, et al. (Coord.) Creación de Empresas: Entrepreneurship, Barcelona: UAB (Server de Publicacions), pp.251–266.

Butchart, R. (1987) 'A new UK definition of high-technology industries', Economic Trends, Vol. 400, pp.82–88.

Copeland, T., Koller, T. and Murrin, J. (1990) Valuation Measuring and Managing the Value of Companies, 1st ed., New York: John Willey & Sons, Inc.

European Commission (2003) 'Growth paths of technology based companies in life sciences and information technology', EUR, Vol. 17054.

European Commission (2006) 'Putting knowledge into practice. A broad based innovation strategy for the EU', European Innovation (COM(2006) 502 final), Special, November, pp.1–24.

Fariñas, J.C. and López, A. (2006) Las empresas pequeñas de base tecnológica en España: Delimitación, Evolución y Características, DGPYME.

GEM Project (1999) Global Entrepreneurship Monitor, London Business School and Babson College.

Kleppe, A., Warmer, J. and Bast, W. (2003) MDA Explained, Addison-Wesley, ISBN: 0-321-19442-X.

Lamothe, P. and Aragón, R. (2003) Valoración de Empresas Asociadas a la Nueva Economía, Madrid: Pirámide.

March, I. (1999) 'Las claves del éxito en nuevas compañías innovadoras según los propios Emprendedores', CEPADE 21, Revista de Dirección, Organización y Administración de Empresas, Madrid, pp.167–176.

McQuaid, R. (2002) 'Entrepreneurship and ICT industries: support from regional and local policies', Journal of Regional Studies, November, pp.909–919.

Meservy, T. and Fenstermacher, K. (2005) 'Transforming software development: an MDA road map', IEEE Computer, Vol. 38, No. 9, pp.52–58.

Miller, J. and Mujerki, J. (Eds.) (2003) 'MDA Guide Version 1.0.1 Object Management Group', 12 June, www.omg.org/docs/omg/03-06-01.pdf.

Ministry of Trade and Industry (MTI) (2007) High Growth SME Support Initiatives in Nine Countries: Analysis, Categorization and Recommendations, Finnish Ministry of Trade and Industry.

Nevado, D. and López, V. (2002) Capital Intelectual. Valoración y medición, Prentice Hall.

Peña, I. (2002) 'Intellectual capital and business start up success', Journal of Intellectual Capital, Vol. 3, No. 2, pp.180–198.

Proyecto ACREA (2006) Análisis de los Factores de Éxito y Fracaso en el Proceso de Creación de Empresas de Base Tecnológica, Director: Dr. Eduardo Bueno Campos.

Rubiralta, M. (2003) 'El papel de los parques científicos en la incubación de empresas de base tecnológica', Iniciativa Emprendedora, Deusto, Vol. 41.

Shearman, C. and Burrell, G. (1988) 'New technology-based firms and the emergence of new firms: some employment implications', New Technology, Work and Employment, Vol. 3, No. 2, pp.87–99.

Simon, K. (2003) 'Proyecto para la promoción de empresas innovadoras de base tecnológica', in ANCES, La creación de empresas de base tecnológica. Una experiencia práctica, Gobierno Vasco.

Storey, D.J. and Tether, B.S. (1998) 'New technology-based firms in the European Union: an introduction', Research Policy, Vol. 26, pp.933–946.

Veciana, J.M. (2003) 'Creación de empresas como programa de investigación', in J.C. Arnaf (Ed.) Creación de empresa: Los mejores textos, Barcelona: Ariel, pp.19–60.

Veciana, J.M. (2003) 'Creación de empresas como programa de investigación', in J.C. Arnaf (Ed.) Creación de empresa: Los mejores textos, Barcelona: Ariel, pp.19–60.

Ajit Jaokar & Chetan Sharma (2010), Mobile VoIP - approaching the tipping point.

Bertrand, Gilles (2007), The IP Multimedia Subsystem in Next Generation Networks

Fransman, Martin (June 12-14, 2003), Knowledge and industry evolution: the mobile communications industry evolved largely by getting things wrong.

Grøtnes, Endre (2008), Strategies for influencing the standardization process - examples from within, University of Oslo, Department of Informatics

Moconews (2011) *Competition Spikes As More Smartphones Are Sold Than Ever Before* <http://moconews.net/article/419-competition-spikes-as-more-smartphones-are-sold-than-ever-before/>

OnlineMarketing (2011) *Global Mobile Phone Market Share* < http://www.onlinemarketing-trends.com/2011/02/global-mobile-market-share.html>

Celularis (July 2011) *El mercado de smartphones sigue creciendo* <http://www.celularis.com/smartphones/mercado-smartphones.php>

Moore, Geoffrey (1991, revised 1999) *Crossing the Chasm. Marketing and Selling High-Tech Products to Mainstream Customers*, Collins, ISBN 0-06-051712-3

JAINSLEE (2009) *JAIN SLEE*< http://www.jainslee.org/>

Wikipedia (2011). *Dumb Pipe* < http://en.wikipedia.org/wiki/Dumb_pipe>

Telwares (2011). *Tiered-pricing-to-eclipse-flat-rate-data-plans* <http://telwares.wordpress.com/2010/08/26/tiered-pricing-to-eclipse-flat-rate-data-plans/>

Boston (2010). *Spain's Telefonica considers charging Google* <http://www.boston.com/business/technology/articles/2010/02/08/spains_telefonica_considers_charging_google/>

Apple (2011). *Apple's App Store Downloads Top 15 Billion* < http://www.apple.com/pr/library/2011/07/07Apples-App-Store-Downloads-Top-15-Billion.html>

Rajapase (2008). Damith C. Rajapakse. *Fragmentation of Mobile Applications.* <http://www.apple.com/pr/library/2011/07/07Apples-App-Store-Downloads-Top-15-Billion.html>

UniversalDesign (2011). Institute for Human Centered Design. *Universal Design.* <http://www.adaptiveenvironments.org/index.php?option=Content&Itemid=3>

5

Mobile Platforms as Convergent Systems – Analysing Control Points and Tussles with Emergent Socio-Technical Discourses

Silvia Elaluf-Calderwood, Ben Eaton,
Jan Herzhoff and Carsten Sorensen
London School of Economics and Political Science
United Kingdom

1. Introduction

In the field of information systems, mobile platforms as convergent systems represent a new direction for research. To date, platforms have been defined in terms of their composition as physical infrastructure (Gawer, 2009). However, the emergence of new digital and convergent services (e.g. VoIP, IPTV, etc.) as well as overlapping physical mobile telecommunications infrastructures provides the foundations for complex mobile platforms (Herzhoff, 2009e and 2011). Consequently, mobile platforms appear to be more complex than earlier work might indicate.

For the purposes of this chapter, and as an initial step to understand what a mobile platform is, the authors draw on the idea of mobile platforms as defined by Tiwana et al. (2010). They provide a richer definition that includes the complexity of all the contributors to a mobile platform. Together, they form a digital ecosystem (Tiwana et al. 2010) where multiple actors act and interact. A digital ecosystem includes a *platform* that serves as a core on which others can build *modules* that are designed to extend the service possibilities of the platform. It also includes various social actors who build the platform and various modules and a regulatory regime including standards that bind these heterogeneous actors together. In this context, control is a major factor in trying to understand the interactions between the many actors concerned within the ecosystem (Tiwana et al. 2010).

However mobile platforms need to be understood as more than just convergent technical systems, which mix multiple layers of physical and digital infrastructures for the creation and distribution of products and services. Mobile platforms also need to be understood in terms of the socio-technical discourses that play out through control points and tussles between the actors in the platform ecosystem.

Case studies of service convergence (e.g. VoIP, network sharing, mobile application markers, etc.) on platforms provide examples of tussles and control points, around which tussles unfold. Solving these tussles requires a reframing of controls points as socio-technical objects which are driven by the need to share resources and content over networks. In other words, control points as socio-material objects (Orlikowski and Scott, 2008) integrated into a socio-technical system (Herzhoff et al, 2009c). We believe that the technical evolution of mobile platforms will benefit from both a socio-technical and a technical consideration of convergence. This view will enhance the comprehension and ways of providing guidance to network operators when making choices and decisions for the development of future networks associated with mobile service platforms.

Although the notion of convergence has attracted much interest over the past few years in both academic and enterprise spheres, analysis of its effect on platform ecosystems has still to be researched in depth. On one hand, convergence can be interpreted as a meaningless technology fad, but for other observers, convergence is an important factor for the design of new mobile information systems infrastructures and services (Herzhoff, 2010a). In the Information Systems discipline the idea of convergence has been mostly ignored, or applied occasionally in non-technical contexts such as strategic alignment (Herzhoff, 2009b). In recent years a new convergence discourse emerged around next-generation wireless infrastructures and services (Lind 2004). One such manifestation can be seen in the discussion of the mobile Internet and in new converging services connecting mobile telephony networks to the Internet. This new discourse has also evoked interest in the conceptual qualities of the notion of convergence in IS literature. An emergent group of IS researchers suggest that convergence is an important factor to consider in the design of new information infrastructures and services (Lyytinen & Yoo 2002; Yoo et al. 2009; Wareham et al., 2009).

According to Herzhoff (2009b), many different forms of convergence have been developed over the past 30 years, from digital convergence (Yoffie 1996) to cultural and organic convergence (Jenkins 2001). The loose usage of the convergence metaphor in both practice and academia has led to many observers no longer ascribing any meaning to it. In fact, scholars argue, "there seem to be as many definitions of convergence as there are authors discussing the topic" (Appelgren 2004: 246). Therefore, it does not seem to be too farfetched that observers from both practice and academia have begun to label convergence as a buzzword. Another school of thought sees convergence instead as a description of one of the driving forces for technological change (Lyytinen & Yoo 2002). In the meantime, the communications industry is experiencing increasing pressures from users' demands for new applications and services. In order to cope with unforeseen future socio-economic (user/network operator/service provider) demands, network technologies or mobile platforms that promote flexible, agile, dynamic and self-evolving networking are vital.

Mobile platforms, as part of a mobile system of infrastructures, need to be understood in the context of digital infrastructures. The Internet and global mobile telecommunications infrastructures are increasingly converging at different layers. Digital infrastructures are established and operated by a heterogeneous collection of public and private organisations, each governed by its own interests in the collaborative arrangement. The creation and distribution of value is collaborative, yet governed by conflicting interests. Two separate strands of research explore collaboration, conflict and control in digital infrastructure innovation. Research on tussles between participating interests focus on the need to understand the complex relationships between collaboration and conflict. Research on architectural control points emphasises individual organisations' ability to exercise control and generate value. So far these two research strands have not been subjected to a synthesis. Research is necessary in order to understand how the composition of these complex relationships, that both collaborate and conflict, affect the overall value chain of mobile digital infrastructures. An increasing problem faced by telecommunication network operators today is the need to monetise their network assets in the face of diminishing margins on voice and data traffic. There are already many examples where virtual operators provide services using the infrastructure of a physical network provider. The question is then how can networks be shared fairly, between many providers, if providers are unwilling to exchange full information about their subscriber bases? Even if they were to do so, would regulators object on the basis of competition and privacy? Any practical solution must take into account all the stakeholders – users, network, service and application providers, manufacturers and regulators – and their various goals and aspirations. The first step in supplying a viable, long-term solution is to identify the tussles that result from the individual goals and aspirations of the stakeholders or entities (Clark et al., 2005).

In this chapter, we will use these components to present a framework model that integrates these concepts to the analysis of value networks, and in doing so we provide a complementary analysis of mobile platforms as convergent systems: the ecosystem composition, the interactions and the relationships and conflicts that are generated by those interactions.

This chapter aims to present a potential framework for the analysis of the composition of both ecosystems and value networks using tussles and control points. The remainder of this chapter is structured as follows: section 2 presents the definition of tussles and control points using a socio-technical analysis; section 3 introduces control points in the context of mobile digital infrastructures, its relevance and characteristics; section 4 provides a discussion on value networks using the control point analysis, in order to present our framework; finally section 5 concludes this chapter with the final remarks and future research areas to be explored.

2. Tussles

Clark et al. introduced the original idea of tussles in ACM SIGCOMM'02 Conference (2002). The idea was one outcome of a major research project funded by the US Defence Advanced Research Projects Agency (Department of Defence) to explore a Greenfield approach for designing the Internet. The paper points out that one initial assumption of the designers of the Internet was that everyone had aligned interests, which cannot be the case, as explained in the introduction of this paper. Although Clark et al. (2003) mention that the characteristic of tussle might be intrinsic to society, they point out that it also has another side where it slows down innovation and can decrease security on the Internet.

Clark et al. (2005) develop the concept of tussle in the context of network architecture design. They define a tussle as "the on-going contention among parties with conflicting interests" (p. 462). They further specify that in these tussles "different parties adapt a mix of mechanisms to try to achieve their conflicting goals, and others respond by adapting the mechanisms to push back" (p. 462). Although Clark et al. (2003) mention that the characteristic of tussle might be intrinsic to society, they point out that it also has another side where it slows down innovation and can decrease security on the Internet.

Clark et al. (2003) suggest that there are three forms of socio-technical tussles:

- Two or more users with common interests operate in presence of adverse third parties (e.g. government's desire to wiretap).
- Two or more users want to communicate but have conflicting interests. Here a third party might help (e.g. credit card company, to ensure trust).
- One party uses an application but is disturbed by some other intruding party (e.g. e-mail spam).

Clark et al. (2005) develop the ideas further and discuss in their revised journal paper the nature of tussle and in particular the specific role of technology in tussles. The term tussle is defined as an intense disagreement, or dispute, between parties who nonetheless have significant interests in collaborating. This concept can help explain a number of important changes at the core of mobile network innovation. However, the theoretical underpinning of the notion of tussle is still ambiguous. For example, the three types suggested by Clark et al. (2003) seem to be examples rather than a clear taxonomy of tussles. Schmidt and Kochan (1972) argue that a useful and precise conceptualisation of conflict needs to: (1) be devoid of value laden perspectives (2) should be conceptually distinct from both its conditions and consequences, and (3) the concept should be distinct from competition (Herzhoff et al., 2010b). Furthermore, Clark et al. (2005) themselves acknowledge that their multidisciplinary discussion of conflict only scratches the surface. Therefore, this section takes a closer look at some of the state of the art conflict theories within the social sciences, to identify ways to improve the conceptualisation of tussle.

It is necessary to lay down some basic assumptions about the very important distinction between competition and conflict (Economides, 1992 and 2006). Many studies, including the work by Clark et al. (2005), lack the conceptual clarity to differentiate between these two concepts. Others argue that this mainly results from the fact that the precondition of both competition and conflict is goal incompatibility. However, these incompatible goals can also be the result of contested resources, incompatibility of roles or incompatibility of values. Thus, competition is distinct from conflict. There are four different schools of thought on how the distinction between competition and conflict plays out:

- The first makes the distinction based on awareness: in this line of thought conflict is seen as a situation of competition in which parties are aware of their incompatible goals.
- The second school of thought examines how competition is regulated. Hence, competition becomes conflict if it goes beyond the limits of regulatory norms.
- The third school of thought bases the distinction on behaviour: two parties might compete and yet not be in a state of conflict, and will continue to cooperate on a daily basis. The behaviour of each party might be determined by different and incompatible goals, but this is not necessarily the precondition for a conflict to emerge, since this also requires some sort of motivation to interfere. This difference can be described as one of parallel striving (competition) and mutual interference (conflict) (Herzhoff et al, 2009c).
- The fourth school is based on Luhmann's (1995) systems theory. Competition is here seen as a descriptor for the environment of the organisation projected by one party, but direct interaction is not a necessary precondition. However, if direct interactions take place the possibility emerges for one party to communicate a "no" (Luhmann, 1995). It is this negation that may lead to the emergence of a conflict system.

According to Luhmann (1997), conflicts result from a communicated disagreement. Conflicts dissolve the predictability of the acceptance of a communication by taking back the initial complexity reduction and show that more is possible compared to what has been actualised. Therefore, Luhmann sees conflict as an immunisation for society. Conflicts test resistance potential (Luhmann, 1995) and are important for the immunisation of society and for its evolution. However, conflicts tend to become more and more decoupled from the initial disagreement, use more and more resources and attention by the so-called host systems.

Thus, conflicts can become highly integrated social systems in themselves. Luhmann calls this the parasitic character of conflict systems. The conflict system develops a life of its own and feeds itself from the host system, a term that will be discussed further later in this chapter.

Summarised, a systems-theoretical perspective fulfils the criteria of the conceptualisation of conflict put forward by Schmidt and Kochan (1972). It avoids value-laden perspectives and clearly distinguishes between condition and consequences of conflicts as well as between conflict and competition. These theoretical considerations are applied in this body of research, within the context of mobile information infrastructures and services.

Herzhoff et al. (2009a,b, c, and d) study mobile VoIP and mobile network sharing. In this context, network sharing is a cost-effective way of deploying 3G networks, and it has both benefits and drawbacks. Infrastructure sharing for example can be used both in the start-up phase to build coverage quickly and, in the longer term, to build cost effective coverage for areas of low reception. From the point of view of competition, many operators are satisfied with the arrangements established for sharing when it has a vertical distribution in the different telecommunication layers. However, changes in the sources of economic revenue are making it more common for operators to be willing or pushed to share on a horizontal basis — layer to layer - with other mobile operators.

In networking terms, tussles test the strength of control in a value network. If the tussle is too intense or cannot be resolved, then the control is completely overtaken by one player - hence it stops being shared but instead converts itself into a laissez-faire leader - or negotiations might occur, presenting a wide range of possible solutions. The concept of tussle therefore seems to stress more the dynamics of the conflict situation and the different mechanisms the contesting parties put in place. This brings us to the discussion of control and control points in digital infrastructure innovation.

3. Control points and mobile digital infrastructure

This section discusses the possible role of control points as an aid to understanding the complexity of digital infrastructure innovation. There are diverse discussions on the complexities of network architecture and modularity (Voss, 2009; Woodard 2008). With the increasing importance of alliances of participant stakeholders with different, and possibly diverging, interests, the issues of control and the associated process of organising collaboration under conflicting interests are brought to the fore. Hanseth and Lyytinen (2010) propose a high order discussion model on control and form part of a set of organising principles in which the Internet - and the networks linked to it - is composed of multiple layers of distinct information technology capabilities that carry out similar functions at different layers. Tilson et al. (2010) argue that digital infrastructure development is a continual process governed by the paradox of change with reliance on stability and the paradox of control coexisting with generativity. These conflicting interests regarding infrastructure developments can arise from a variety of socio-economic areas. In the case of the Internet, for example, the design is distributed between a large set of architects and developers, user communities and forms of governance (Hanseth & Lyytinen, 2010). The control of different network capabilities is separated and distributed, and the control forms are loosely coupled through architectural network principles. Hence mobile networks usually present one or more actors actively seeking the control of a whole section of a mobile network.

The notion of "control points" has been used in several contexts, for example, to characterise essential architectural design decisions (Woodard, 2008), or to characterise the generation of value (Trossen & Fine, 2005). The concept was developed by the Value Chain Dynamics Working Group at MIT (Trossen & Fine, 2005) in order to understand how commercial benefit is gained from business models emerging in and around the telecommunications industry. Woodard (2008) defines architectural control points as "system components whose decision rights confer architectural control over other components" (p. 361). This effect can be small but also powerful, influencing the whole architectural landscape. Control points can broadly be defined as points at which management can be applied, and any encapsulated functional element of a system can be a control point (Trossen & Fine, 2005).

One starting point in the context of conflicts and the role of control points in the mediation of those points is the question of what are relevant host systems. In mobile telecommunications, Lyytinen and King (2002) studied conflicts in the standardisation arena and distinguish between a marketplace, innovation, and regulatory system. Another approach is to consider functional differentiation. One of the key distinctions for a functional differentiation is between service and infrastructure as suggested by Lyytinen and Yoo (2002). The infrastructure system comprises the network itself, the data pipe, and the transport technology. It can be based on different types of technology (e.g. Wi-Fi, 3G, LTE). The second system is the service system. The service system comprises any type of service, e.g. in the case of mobile VoIP a voice service. Another important function is regulation. This is not strictly limited to a formal governmental regulation authority since many entities can 'regulate'. However, in the context of this paper, the regulatory system comprises all regulatory functions like e.g. spectrum regulation and setting of interconnection charges. Finally, the use system is comprised of all functions in the use domain of a specific service and a specific infrastructure, i.e. the device, the operating system or the user interface. This is a complementary approach to the one proposed by Lessig (2000).

Cyberlaw scholars concerned with the legal regulation of the Internet against abuse provide a complementary view of infrastructure development (Tilson et al, 2010; Herzhoff et al, 2010b; Eaton et al., 2010). Benkler (2005), for example, suggests that appropriate regulatory frameworks in a converged network should orient themselves towards democratic values and he proposes an approach to develop descriptive models based on how laws concentrate or distribute control over production and exchange of information. Lessig (2000) identifies four types of modalities of regulation: (1) laws, (2) social norms, (3) markets and (4) architecture or code. While Lessig applies these modalities within the limited context of regulation, Murray and Scott (2002) argue that the modalities of regulation are not limited to regulation but are part of any form of control system.

Additionally the concept of control points is complemented by the idea of triggers, which considers not only the dynamics but also the interactions between the different systems. Based on Luhmann's Theory of Social Systems (Luhmann, 1995), systems are operationally closed but structurally open. This perspective suggests that outside stimuli or triggers have to be considered. However, the effects of these outside triggers on the system are determined by the internal operations of the system, in this case the mobile digital infrastructure.

A mobile digital infrastructure is formed by an ecology of devices and services aiming to provide a seamless experience to the network users. Enabling technologies within this type of network promotes flexible, agile, dynamic and self-evolving networking capable of coping with unforeseen socioeconomic demands, e.g. user/network operator/service provider, so that the seamless goals can be achieved. There are three components contributing to the definition of a digital infrastructure (Mobile VCE, 2007, 2008):

- Social factors: This component is the voice of the user perception when using services provided by a digital infrastructure. It should be a seamless service, ideally with a featured configuration provided free (or at minimal cost), and requiring little user awareness of changes in formats, protocols or quality of service.
- Economic and business factors: This component is the voice of the network operators. In an operational digital infrastructure it implies the use of an intelligent decision making process. Computational algorithms should provide a working framework to optimise allocation of the resources available within networks. These should be informed by, and configured according to, advanced dynamic service level agreements, discovery service intelligence, digital market oriented application, and regulatory requirements.
- Network factors: A digital infrastructure shall be adaptable when network expansion is required. This adaptability is understood in terms of network capacity and protocol negotiation.

In general, the design of communication infrastructures cannot be considered as an entirely isolated design of each part of the infrastructure without overall insight into the end-to-end delivery of services, since low-level services may prove redundant or ineffective when applied at aggregate levels (Saltzer et al.1984). The increasing number of conflicts caused by the convergence of information and communication technology puts pressure on the existing infrastructure (Clark et al., 2005; Tilson et al., 2010). Parties with conflicting interests get increasingly incentivised to actively engage in interference. These interferences increase the complexity of the infrastructure and may lead to breakdowns in operation. A possible strategy to overcome these problems is the development of digital infrastructures in terms of structural flexibility, e.g. network virtualisation, and control flexibility. This constitutes a dynamic market approach (MVCE, 2008; Irvine, 2002; Bush, 2009). The idea of digital infrastructures faces three main challenges: (1) the role played by heterogeneous systems in terms of transmission power, frequencies, range, quality of service (QoS) requirements, spectral efficiency, and standards (Grøtnes, 2009); (2) the limited or no communication between these systems; and (3) the way systems change rapidly and the way digital infrastructures have to adapt quickly without degradation of service (Herzhoff et al, 2009c).

A digital infrastructure cannot be singled out as a network demand or capacity tussle mediator. The common use of expressions such as "a network capable of coping with unforeseen demands" or "a network able to resolve tussles on demand" represent partial or incomplete views of what a digital infrastructure can do. A digital infrastructure is not able to resolve, using its self-contained resources, all tussles generated internally. A requirement for digital infrastructures is not a justification for an expansion of the network that does not take into account the variations in usage the network might have. A digital infrastructure shall not be the replicator or amplifier of current network hierarchy, or a computational tool to extend current IP networks and protocols (Herzhoff et al, 2009a; Herzhoff, 2009b).

When taken as a whole, these components of tussles, control points, triggers, and digital infrastructure provide a means to address the research question proposed in the introduction of this chapter. This research question concerns how to understand an ecosystem of mobile convergence platforms.

4. Control points and value in networks

Tussles can occur between and within these four different socio-technical systems. The infrastructure system comprises the network itself, the data pipe, and the technology enabling the transport. It can be based on different types of technology (Wifi, 3G, LTE etc.). The service system can be of any type, e.g. in the case of mobile VoIP, a voice service. The regulatory system consists of all regulatory functions such as spectrum and setting of interconnection charges. Finally, the use system consists of all functions in the use domain of a specific service and a specific infrastructure, e.g. the device, the operating system and the user interface. Combining the discussion of regulation with infrastructure, service, and use provides a comprehensive perspective on the aspects relevant to a discussion of control points and tussles in flexible mobile network innovation. The model presented in Figure 1 uses these four elements to explain the relationships between the tussles elements relevant to this analysis.

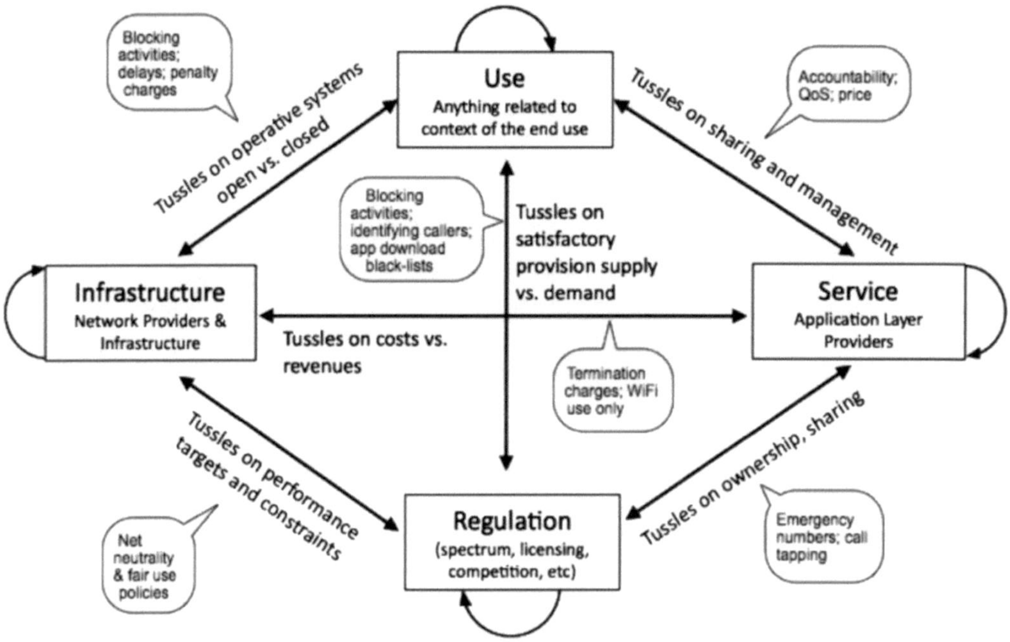

Fig. 1. Tussles and control model of relationships in a mobile platform convergence system (Elaluf-Calderwood et al, 2011)

The figure illustrates the conflicts or tussles that may occur within and between socio-technical entities in terms of the existing infrastructure, the services offered by the various providers, the regulatory system, and market demand (Herzhoff et al., 2009a). Each of the four socio-technical systems presented above has certain functions, which can also be described as control points. These control points can follow different modalities. They can be hierarchical, market-oriented, design-oriented, or community-oriented. In a market environment, control points are defined by the actor(s) interested in the maximum revenue, or stake of control. However they will also expect to limit the scope of usability when subject to regulation. Regulation can exclude certain types of control point (e.g. compulsory provision of emergency services) or determine the limits of power for certain control points (e.g. limitation in charges or service pricing).

Depending on their role in the revenue value-model, an actor could have a set of control points defined based on regulation, which leads one to think that control points are not an off-the-shelf definition but vary depending upon the circumstances in which regulation is applied. As explained, control points enable the controller to exercise power over other players or actors of a socio-technical ecosystem. They represent a socio-technical mechanism expressing the boundaries of areas of economic power in the value networks identified within a telecommunications network.

Trossen & Fine (2005) show how control points can be identified and implemented within communications architectures, and how they can facilitate the construction of potential business models that in turn can be evaluated in terms of viability and sustainability. This manner of use of control points also shows how external triggers, arising from different domains (e.g. changes in technology, the business cycle, industry structure, regulatory policy, customer preference, capital markets and corporate strategy), can lead to control points increasing or decreasing in importance, which in turn affect the strength of business models.

Business models and value chains can be defined in terms of "the way a network of companies intends to create and capture value from the employment of technological opportunities" (Faber et al, 2003). Fine (1998) was one of the first researchers to work on comparative studies using the approach of value chain dynamics, cross-industry comparisons, and the exploration of life-cycles in complex value chains. Fine (1998) proposes a double helix model, which for telecommunications captures this life cycle in four phases – integration, market differentiation, verticalisation and disintegration. It visualises a complex trigger dynamic analysis that leads to the observed integration/disintegration effects. Trossen and Fine (2005) extend this to develop analysis methodologies that allow for segmentation into value chains or value networks. Fine (1998) also discusses the bullwhip effect, whereby a complex value chain can amplify changes in demand, the impact being increased volatility of demand further up the supply chain. While this more traditionally relates to inventory-based value chains, a similar behaviour can be observed in telecommunications (equipment stock) and computer industry (investment in R&D). Mitigating this effect, within the context of future network design, is desirable.

Clark and Blumenthal (2007) apply a socio-economic perspective to network architectural design in a systematic manner and thereby shape the foundation for trust-to-trust principles. Sollins and Trossen (2007) extend the "Design for Tussle" concepts towards a vision for a flexible execution environment that incorporates tussles - and the concerns that drive them - directly into the formation of the dynamic execution environments. As an example of such evaluation, Trossen & Fine (2005) outline the potential application of such an evaluation tool in the area of VoIP, informing decision makers at the regulatory level, in this case the FCC in the US market, on the required speed of regulatory action, a crucial part of an overall design process.

Finding a method to identify the creators of value is a major concern. Eaton et al. (2010) propose the use of control points in the mobile Internet for the determination of value networks in a two-stage model that includes the creation of a map of the various constituent actors within the industry. This map serves to illustrate the businesses that may exist across the industry, and control points are used to examine where and how members of the value network can extract value and the use of triggers in order to understand the sustainability of this economic power given the impact of external factors.

Faber et al. (2003) definition of the business model highlights the networked character of digital infrastructure innovation, the value creation and captures involved in the trade-off, as well as the issues connected with technology design (Ballon, 2009). Value networks are defined as: *"a dynamic network of actors working together to generate customer value and network value by means of a specific service offering, in which tangible and intangible value is exchanged between the actors involved"* (De Reuver, 2009). In doing so, there are three critical dimensions of analysis (Ballon, 2009) mirroring the model in Figure 1; Industry structure and value network; functional and technical architecture; and value creation and capture. Based on the empirical evidence collected by the authors, there are no strong indicators to challenge this description of the fundamentals of tussle creation and management between operators as proposed in figure 1.

The authors encountered a mirrored reflection of the high-end tussles models on the analysis of value networks completed by Ballon (2009): for each component of the proposed model by Herzhoff et al. (2009d), there is a value component in the model proposed by Ballon. If the tussle model proposed an ontology considering the potential relationships between the actors influencing the tussles, then the business model ontology incorporates four different levels of a business model: a strategic, functional, financial, and value configuration level. At the strategic level, a business model is concerned with the value network configuration, i.e. setting up roles and relations between actors, and the physical and virtual flows between them. At the functional level, a business model describes the architecture of a product or service, which is determined by a specific configuration of modules, interfaces and intelligence. At the financial level, a business model describes the cost and revenue sources, as well as the distribution of flows for the actors involved.

Together, these three levels contribute to the fourth and final level of a business model, i.e. the value configuration. We propose that within complex and converging business and digital infrastructures, characterised by value co-creation within a large "industrial architecture", research should not just focus on any clear cut value proposition, but rather on the process of value construction leading to various value configurations. This deals with the way in which actual value is created in the market. While specific design choices also need to be made at this level, the value configuration can also be viewed as the logical outcome of business model design choices made at the previous levels. Figure 2 illustrates the basic, bi-directional relations between the different levels.

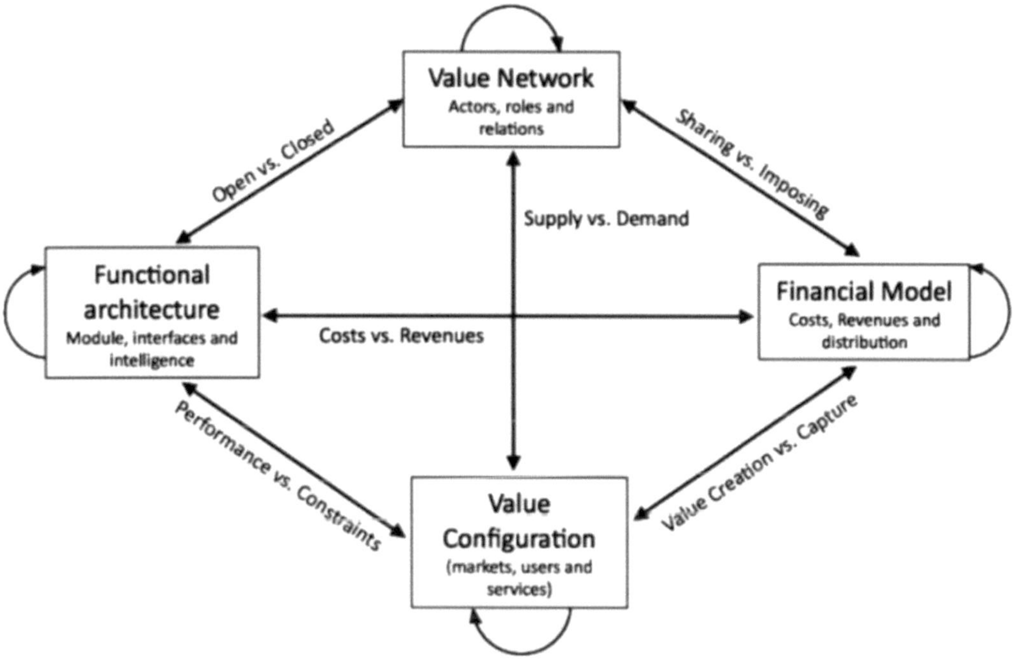

Fig. 2. High-End Tussle model transposed to value networks (Elaluf-Calderwood et al, 2011)

We propose that within complex and converging business and digital infrastructures, characterised by value co-creation within a large "industrial architecture", research should not just focus on any clear cut value proposition, but rather on the process of value construction leading to various value configurations. This deals with the way in which actual value is created in the market. While specific design choices also need to be made at this level, the value configuration can also be viewed as the logical outcome of business model design choices made at the previous levels. Figure 2 illustrates the basic, bi-directional relations between the different levels.

In reality, a range of complex, both direct and indirect, bidirectional relations between the different levels exist. Also, which particular relationship is focused on and in which 'direction' the impact is studied depends upon particular cases and contexts. One of the tasks of a design approach that takes into account contextual contingencies will be to identify the realistic scope for choice available to technology producers and users at the various levels and subsequently work out the impact of the various 'degrees of freedom' among the different levels. However, in order to enhance the clarity of the initial ontology, it is proposed here as a point of departure that the value network is the primary agent, which designs and uses a functional architecture and shares cost and revenues, and that the value configuration is the primary outcome of the business modelling process.

Finally, a value network consists of actors possessing certain resources and capabilities, which interact and together perform value activities or roles, in order both to create value for customers and to realise their own strategies and goals. It is the result of organisational and strategic design, in which control points as an analytical tool provide insightful understanding of the forces in place for the development of business models. The four levels of this framework and their interrelationship need to be detailed, and subsequently the levels or domains need to be extrapolated into a number of parameters, i.e. the crucial configuration parameters that would need to be addressed by any business model aiming for new or improved digital infrastructures products or services.

5. Conclusion

The convergence of mobile telephony networks with the Internet has resulted in an increasing number of conflicts or tussles. These tussles emerge around control points within an infrastructure. These conflicts challenge the key assumptions of information infrastructure design and lead to inefficiencies and system failures. The analysis of tussles based on Luhmann's Theory of Social Systems indicates that tussles can emerge around four different types of system: the infrastructure system, the service system, the regulatory system, and the use system.

These control points can be architectural, hierarchical, community, or market-based. These control points enable a system to produce a communicative disagreement. This communicative disagreement can often be quickly resolved through exercising power (e.g. by the regulator, the dominant actor on the ecosystem) or by putting in play economic factors (e.g. T-Mobile's offer to enable mobile VoIP for higher paying customers, optional IPTv services, etc).

Overall, these conflicts often result in a dynamic self-sustaining autopoetic system - a dynamic set of tussles. In particular, we argue that the notion of control points may be helpful in providing a better understanding of sociotechnical tussles. This may in turn facilitate dialogue between management and information infrastructure engineers. The concept of control points, as a methodological tool for the analysis of the development of network design, has been successfully transposed from its technical origins to become a socio-technical variable. By including the multiple relationships, tussles and ambiguity between stakeholders in an analysis, control points can become a tool that adequately addresses the complexities attached to the development of digital infrastructure design.

Although control points are contributing significantly to the analysis of and planning for tussles, there are some shortcomings to the approach. It is necessary to complement this analysis with a revision of the combination of methods used to exploit value chain dynamics in conjunction with other approaches as part of an analysis of metrics. Furthermore, a conceptual clarification of control points needs to be part of this process, e.g. the role of tussles, granularity of the analysis, value web, etc. Particular stress must be placed on understanding what is controlled, e.g. network behaviour, revenue, resources, functionality, generativity, innovation.

Control points have been used at the network, management and content, and business model layers as a powerful tool to understand challenges brought about by the evolution and fast innovation of the technologies described. This method of analysis can be used to understand the relationships between the different stakeholders in the ecosystem, the roles and functions they bring to the value chain, and the short and long term effects of those relationships.

Further work is required to complement the understanding of the tussle concept. Perhaps the goal ought to be the development of a tussle taxonomy, which clarifies the important distinctions between tussle, conflict, collaboration, and competition. Some tussles will continue to be external to socio-technical approaches and need to be properly identified.

By including a socio-technical metrics definition in this analysis, we have opened a number of research opportunities, which have arisen from the development of socio-economic metrics for network selection algorithms. Metrics that require to be investigated in future might include: 1) profitability metrics, e.g. profitability per byte; 2) trust metrics, for example based on user rating or network strength (e.g. by the inclusion of other people using the network, and who are known to the user); 3) consumer surplus metrics, not only based on network strength but also on other cost profiles converted in utility; and 4) pay-off metrics, e.g. taking into consideration tussles between users and operators. These areas are of relevance to mobile operators in order to identify the optimal value chain and areas of value add that emerge from these new digital infrastructures.

6. Acknowledgement

The authors wish to thank the Flexible Networks area and User interactions programme of the Core 5 Research Programme of the Virtual Centre of Excellence in Mobile & Personal Communications, Mobile VCE (www.mobilevce.com) which has been jointly funded by Mobile VCE's industrial member companies and the UK Government, via the Engineering and Physical Sciences Research Council.

7. References

Appelgren, E. (2004). Convergence and Divergence in Media: Different Perspectives, in Proceedings of the 8th ICCC International Conference on Electronic Publishing, Brasil, June 2004, 237-248.

Ballon, P. (2009) Control and Value in Mobile Communications: A Political Economy of the Reconfiguration of Business Models in the European Mobile Industry. Faculteit Letteren en Wijsbegeerte – Vakgroep Communicatiewetenschappen, Vriej University Brussels: 641.

Benkler, Y. (2005) "The Wealth of Networks; How Social Production Transforms Markets and Freedom", New Haven and London, Yale University Press.

Bush, J. et al (2009) "A Digital Marketplace for Tussle in Next Generation Wireless Networks", VTCFall2009, Tokyo, Japan, September 2009.

Clark, D. and Blumenthal, M. (2007) "End-to-end Arguments in Application Design: The Role of Trust", In Proc. of TPRC.

Clark, D., Sollins, K., Wroclawski, J., Braden, R. (2002) "Tussle in Cyberspace: Defining Tomorrow's Internet," In: Proceedings of the ACM SIGCOMM'02, Computer Communication Review, 32(4).

Clark, D., Sollins, K., Wroclawski, J., Katbi, D., Kulik, J., Yang , X. (2003) "New Arch: Future Generation Internet Architecture", Final Report, retrieved on 21st April 2009 from http://www.isi.edu/newarch.

Clark, D. D., J. Wroclawski, K. R. Sollins, & R. Braden (2005): Tussle in Cyberspace: Defining Tomorrow's Internet. IEEE/ACM Transactions on Networking, vol. 13, no. 3, pp. 462-475.

De Reuver, G. (2009). Governing Mobile Service Innovation in Co-evolving Value Networks. Delft, The Netherlands, Department of Information and Communication Technology — Technische Universiteit Delft.

Eaton, B., S. Elaluf-Calderwood, & C. Sørensen (2010): A Methodology for Analysing Business Model Dynamics for Mobile Services using Control Points and Triggers. In Business Models for Mobile Platforms (BMMP 2010), Berlin

Elaluf-Calderwood, S., J. D. Herzhoff, et al. (2011). Mobile Digital Infrastructure Innovation - Towards a Tussle and Control Framework. European Conference of Information Systems. Helsinki, Finland.

Faber, E et al., "Designing Business Models for Mobile ICT Services," Proc. 16th Bled eCommerce Conference, 2003.

Fine, C. (1998) Clockspeed: Winning Industry Control in the Age of Temporary Advantage, Sloan School of Management, 1998.

Gawer, A. (ed.) (2009) Platforms, Markets and Innovation. Edward Elgar, Cheltenham, UK.

Grøtnes, E (2009) "The Work of an International Standardization Consortia: Paths Towards its Current Structure." International Journal of IT Standards and Standardization Research 7(1): 46-65.

Hanseth, O. & Lyytinen, K. (2010) Design Theory for Dynamic Complexity in Information Infrastructures: The Case of Building Internet. Journal of Information Technology, 25(1): 1-19.

Herzhoff, J. (2011) Unfolding the Convergence Paradox: The Case of Mobile Voice-Over-IP in the UK. Unpublished PhD Thesis, London School of Economics and Political Science.

Herzhoff, J. (2010a). Convergence and Mobility: Just Another Fad or Fashion? A Systems-theoretical Analysis. 18th European Conference on Information Systems (ECIS), South Africa.

Herzhoff, J., S. Elaluf-Calderwood, & C. Sørensen (2010b): Convergence, Conflicts, and Control Points: A Systems-Theoretical Analysis of Mobile VoIP in the UK. In Proceedings of joint 9th International Conference on Mobile Business (ICMB 2010) and 9th Global Mobility Roundtable (GMR 2010), Athens

Herzhoff, J., S. Elaluf-Calderwood, & C. Sørensen (2009a) "Flexible Networks Position paper: The Role of Metrics Within Convergent Networks: Blocks to Sharing Nicely Approaches in the Context of Wireless Information Infrastructures," Internal Report Mobile VCE Core 5 Program, London, UK, LSE

Herzhoff, J.D (2009b) "Design for Tussle - A Case Study on Emerging Conflicts surrounding Mobile VoIP in the UK", Working Paper. LSE

Herzhoff, J., S. Elaluf-Calderwood, & C. Sørensen (2009c) "Flexible Networks Final Phase 1 Deliverable," Internal Report — Mobile VCE Core 5 Research Program, London, UK, June 2009

Herzhoff, J., S. Elaluf-Calderwood, & C. Sørensen (2009d) "Flexible Networks — User and Business Factors: State of the Art Review", Internal Report, Mobile VCE Core 5 Research Programme. London, UK, LSE

Herzhoff, J (2009e) "The ICT Convergence Discourse in the Information Systems Literature - A Second- Order Observation", European Conference of Information Systems (ECIS), Verona, 8. - 10. June 2009.

Irvine, J. (2002) "Adam Smith Goes Mobile: Managing Services Beyond 3G with the Digital Marketplace," European Wireless Conference 2002, Florence, Italy, 22-25 February 2002.

Jenkins, H. (2001). Convergence? I diverge, Technology Review, June, p. 93.

Lessig, L. (2000) Code and Other Laws of Cyberspace, Basic Books

Lind, J. (2004). Convergence: History of term usage and lessons for firm stategists, Center for Information and Communications Research, Stockholm School of Economics, Working Paper.

Luhmann, N. (1995). Social Systems. Stanford, California, Standford University Press.

Lyytinen K., King J. (2002), "Around The Cradle of the Wireless Revolution: The Emergence and Evolution of Cellular Telephony", Telecommunications Policy, 26, pp. 97-100.

Lyytinen, K., Yoo,Y. (2002) "Research commentary: the next wave of nomadic computing", Information Systems Research, Vol. 13, pp. 377– 388.

Mobile VCE (2007) 2020 Vision Mobile VCE Consultation Paper. London, UK. 23p

Mobile VCE (2008) Enabling Flexible Networks — Internal Document. London, UK. 24p.

Murray, A. and C. Scott (2002). "Controlling the New Media: Hybrid Responses to New Forms of Power." The Modern Law Review 65(4): 491-516.

Schmidt, S., Kochan, T. (1972) "Conflict: Toward Conceptual Clarity", Administrative Science Quarterly, September 1972, pp. 359–370.

Sollins, K and Trossen, D (2007) "From Visions to Understanding", Presentation to the Communications Futures Program, MIT, USA.

Tilson, D., Lyytinen, K., & Sørensen, C. (2010). Digital Infrastructures: The Missing IS Research Agenda. Information Systems Research, 20(4): 748-759.

Tiwana, A., B. Konsynsky, et al. (2010). "Research Commentary - Platform Evolution: Coevolution of Platform Architecture, Governance, and Environmental Dynamics." Information Systems Research 21(4): 675-687.

Trossen, D and C. Fine (2005), Value Chain Dynamics in the Communication Industry, MIT Communications Futures Program.

Orlikowski, W.J and S. Scott (2008). Sociomateriality: Challenging the Separation of Technology, Work and Organisation. London, UK, Department of Management. London School of Economics and Political Science: 47.

Wareham, J., Busquets, J., and Austin, R. D. (2009). Creative, convergent, and social: Prospects for mobile computing. Journal of Information Technology, 24 (2), 139-143.

Woodard, J (2008) "Architectural Control Points", Third International Conference on Design Science Research in Information Systems and Technology (DESRIST 2008), Atlanta GA, 7-9 May 2008.

Yoffie, D. (1996). Competing in the Age of Digital Convergence. California Management Review, 38 (4), 31-53.

Yoo, Y., Boland, R. J., Lyytinen, K., Majchrzak, A. (2009). Call for Papers - Special Issue: Organizing for Innovation in the Digitized World, Organization Science, 20 (1), 278-279.

6

Measuring the Return of Quality Investments on Mobile Telecommunications Network

Manuel J. Vilares[1,2] and Pedro S. Coelho[1,3]
[1]*ISEGI, Universidade Nova de Lisboa*
[2]*Banco de Portugal*
[3]*Faculty of Economics, Ljubljana University*
[1,2]*Portugal*
[3]*Slovenia*

1. Introduction

Until recently, methodologies that identified the extent to which various improvements in quality caused financial returns were unavailable (Aaker & Jacobson, 1994; Zeithaml, 2000). Therefore, the benefits of the investment in improving the quality of mobile telecommunications networks have been questioned by many operators. Others have accepted investments in quality as unquestioned generators of returns, with the result that some firms have run into financial difficulties after having incurred heavy investments in quality (Rust, Zahorik & Keiningham, 1995). For instance, some firms faced serious difficulties after winning the Malcom Baldrige National Quality Award in 1990. Also, several surveys conducted during the 90's have indicated a failure of quality implementation approaches (such as Total Quality Management) to increase the economic returns of firms (Ittner & Larcker, 1996; Keiningham, Zahorik &Rust, 1994). In addition, it has been difficult or impossible to choose objectively between different types or levels of quality investment. These negative experiences, plus concerns over cost reductions in many telecommunications operators, caused a real interest in the benefits of quality investments. The identification of perceived quality as a driver of customer satisfaction and loyalty is a well researched field, but firms began to feel the need for methodologies to connect quality investments to the bottom line. In fact, for many quality investments there must be a point above which further investment is unprofitable.

Rust et al. (1995), following the previous work of Rust & Zahorik (1993) presented a very promising approach to this problem named ROQ (Return on Quality). The authors consider a causal chain between quality improvement and profitability, through customer retention and cost reduction in a duopoly context. Danaher & Rust (1996) empirically show that service quality impacts customer attraction and customer usage rates. Bolton & Drew (1991) found a relationship between service change and customer attitudes in the fixed phone industry. Aaker & Jacobson (1994), Anderson, Fornell & Lehmann (1994), Anderson, Fornell & Rust (1997) and Ittner & Larcker (1996, 1998) found significant associations between customer satisfaction (or other related variables) and financial performance data such as Return on Investment (ROI) and accounting returns. Zeithaml (2000) presents a survey of the research for evaluating quality investments. A similar approach is adopted by Rust et al. (2004) in a later survey concerning the measuring of the return of the investments in marketing. Also, Allen & Wilburn (2002) present the state of the art regarding the linkage between customer and employee satisfaction and the financial performance of the firms. More recently Anderson, Fornell & Mazvancheryl (2004) propose the first extensive theoretical and empirical examination of the association between customer satisfaction and shareholder value. Rust, Lemon & Zeithaml (2004) outline a broad approach to estimate return on marketing, focusing on customer equity that enlarges the return on quality approach to different marketing dimensions and Coelho & Vilares (2010) propose an extension of Return on Quality approach to a realistic competitive context applied to the mobile telecommunications industry. In fact, among these works, only Bolton & Drew and Coelho & Vilares work has been developed or applied in the framework of telecommunications (fixed phone and mobile phone industries, respectively).

Notwithstanding the appearance of some literature in recent years, the number of applications of this approach is not only very limited, but some have also adopted unrealistic simplifications that very significantly reduce their operational value. Many of them link quality investments (or quality perceived by customers) to customer's attitudes (and sometimes behavioural intentions), but do not establish a complete connection between the investments and their financial returns. Others find significant associations between customer satisfaction and financial performance, but do not identify the role and importance of the mediating variables. An exception to this is the Kamakura et al. (2002) paper which presents an application encapsulating both the Service-Profit Chain (SPC) framework (Heskett et al., 1994) and the Data-Envelopment Analysis (DEA) framework (Charnes et al., 1994). The authors offer an integrated framework to understand how firms' operational investments are related to customer's perceptions and behaviours and translated into profits.

Building on previous work on the Return on Quality (ROQ) from Coelho & Vilares (2010), Rust et al. (1995), Rust & Zahorik (1993), the present work aims to develop an integrated methodology for estimating the return on quality investments with application in mobile telecommunications industry. More precisely, the aim of this work is to propose a methodology that will enable the identification of profitable quality investments. This includes identifying the profitability of different levels of investments and prioritizing alternative investments, taking into account their return or profitability. The approach uses a chain of causality that assumes that quality investments in the mobile network potentially affect technical quality indicators. This in turn affects customer perceptions and customer satisfaction, thus influencing customer behaviours generators of financial returns to the operator. This methodology will also provide a cost-benefit analysis of the investments on network quality, allowing to estimate several effects in the value chain: 1) the relationship between the quality investments and the improvement of customer perceptions of network quality; 2) the relationship between improvement in the customer perceived quality and improvements in customer satisfaction and loyalty; 3) the relationship between improvement of customer satisfaction and loyalty and improvement of the financial results of the firm. We assume that this relationship is mediated by customer behaviours.

In contrast to the previous work of the same authors where the focus was on the relationship between customer satisfaction and customer behaviours potentially generator of financial returns, this work will focus on the whole causal chain ranging from quality investments, technical network quality indicators and customer perceptions to operator's results originated by changed customer behaviours. We can also be distinguished from some previous approaches to this problem whose aim is to link customer's perceptions and behavioural intentions to profitably. In our approach we explicitly model additional revenues and expenditures generated by quality investments and obtain Return on Investment (ROI) as a result of the predicted future cash-flow.

The remainder of this chapter is divided into three sections. The next section of this chapter describes the methodological approach and identifies the channels through which investments on network quality may improve the financial results of the firm. This section also presents some comments about modelling and data challenges. The third section presents an application of this approach to the mobile telecommunications industry. The final section discusses the main conclusions.

2. Methodological approach

2.1 Structure of the approach

Our modelling chain is shown in Figure 1. First it is assumed that investments in quality improve technical quality of the network. We implicitly assume that technical quality indicators that adequately express the relevant results of quality improvements are available.

The second step represents the relationship between technical quality and perceived quality. An improvement in a technical quality indicator should be at least partially perceived by the customers, resulting in an improvement in perceived quality (Bolton & Drew, 1991; Simester et al., 2000). This perception in turn positively affects customer satisfaction and loyalty. In fact, the existence of a relationship between perceived quality and customer satisfaction is well known and researched (Anderson et al. 1994; Keiningham et al., 1994; Fornell et al., 1996). Note that contrarily to previous proposals (eg. Burton et al., 2003) who assume that this technical quality or actual performance may be a significant direct predictor of customer satisfaction, we propose to consider this effect of technical quality on customer satisfaction as indirect, i.e. mediated by perceived quality.

Next, we consider that perceived quality and customer satisfaction will have a potential positive impact on customer loyalty. The existence of a relationship between customer satisfaction and loyalty is generally accepted and is extensively supported both by theory and empirical results (Anderson et al., 1994; Oliver, 1997a, Oliver, 1997b; Fornell, 1992; Fornell et al. 1996; Hallowell, 1996; Strauss & Neuhaus, 1997).

Furthermore, this improvement in customer satisfaction and loyalty has effects on the financial results of the firm through customer behaviours. In this context we have considered three main channels of customer behaviour: 1) **revenue per customer**. As a result of improved perceived quality and customer satisfaction, it is expected that customers will increase their consumption of the operator services (e.g. usage of the network) and increase the likelihood of buying new products and services from the same operator (Bolton, 1998, Reichheld & Sasser, 1990; Reichheld, 1996; Anderson et al., 1994, Danaher & Rust, 1996; Ittner & Larcker, 1998). In addition, the improvement in satisfaction may render customers more willing to pay higher prices (Zeithaml, Berry & Parasuraman, 1996). Both behaviours contribute to an increase in the average revenue per customer. In our model a possible direct effect from technical quality on customer behaviours and particularly on customer revenue is also considered. In fact, a failure resulting from a low product or service quality (eg. lack of network coverage, call drop) may render the consumption of the product or service impossible during the failure period (impossibility to use the operator network, to maintain a call, etc), and therefore reduce the customer revenue; 2) **customer retention**. The improvement of customer satisfaction and loyalty resulting from the increase in perceived quality is expected to generate a higher probability of staying with the operator, and thus higher customer retention (Reichheld & Sasser, 1990; Rust & Zahorik, 1993; Anderson et al. 1994, Bolton, 1998, Ittner & Larcker, 1998). Customer retention improves market share and may also influence the revenue per customer, which in both cases has a positive impact on the revenues and profits of the firm (Rust & Zahorik, 1993; Rust et al., 1995; Reichheld, 1996); 3) **acquisition of new customers**. More satisfied and loyal customers will recommend the company, potentially increasing the acquisition of new customers (Anderson et al., 1994, Anderson, 1998, Danaher & Rust, 1996). Also, a higher quality of products and services can be advertised, constituting another mechanism through which customer attraction can be obtained. According to Kordupleski, Rust and Zahorik (1993), advertising not accompanied by an effective increase in quality is unlikely to increase market share. Finally, increases in the number of customers through customer acquisition will generally cause an improvement in the returns of the firm.

Finally, additional revenues resulting from the quality investment should be compared to the expenditures in quality improvement in order to obtain the profitability of the investment.

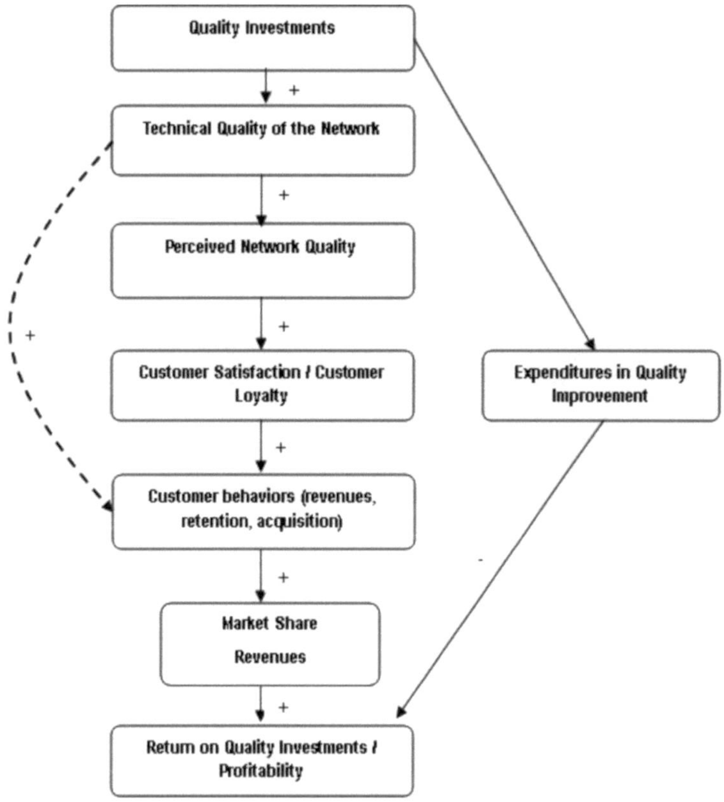

Fig. 1. Modelling chain

In this context, the development of an approach of return on quality from an operational point of view requires an explanation of the relationships between the variables adopted to represent the different stages that mediate between the investment on the quality and the financial performance of the firms.

2.2 Evaluation of the return on network quality investments

In this section, we formalize the relationships implicit to the chain represented in Figure 1, resulting in the calculation of the return on network quality investments.

Assuming that a network quality investment can be expressed through a set of technical quality indicators, *TQ*, then

$$PQ = f_1(TQ) + \varepsilon_1 \tag{1}$$

where *PQ* represents perceived quality and ε a stochastic error term.

The effect of perceived quality on customer satisfaction and loyalty may be modelled as

$$\begin{aligned} S &= f_2(PQ, DS) + \varepsilon_2 \\ L &= f_3(PQ, S, DL) + \varepsilon_3 \end{aligned} \tag{2}$$

where *S* and *L* represent customer satisfaction and loyalty, respectively. *DS* and *DL* represent the other drivers of customer satisfaction and loyalty that are included in the modelling process. These include factors such as company image, customer expectations and perceived price.

Let:

$a_{jt}^{(1)}$: Acquisition rate of firm *j* for competition churners in period *t* (i.e. the probability that a customer leaving one of the competitors moves to firm *j*);

$a_{jt}^{(2)}$: Acquisition rate of firm *j* near new customers in period *t* (i.e., the probability of a new entrant to choose firm *j*);

r_{jt}: Retention rate of firm *j* in period *t* (i.e., the proportion of firm *j* customers that remained in the company)

1- r_{jt} : Proportion of the customers of firm *j* that switched to competition or left the service in period *t*;

v_{jt}: Average revenue per customer of firm *j* in period *t*;

The retention and acquisition rates and the revenue per customer can now be expressed as (for simplicity of notation the indexes representing the firm are omitted):

$$\begin{aligned} r_t &= f_4(S_{t-1}, L_{t-1}, M_{t-1}) + \varepsilon_4 \\ a_t^{(i)} &= f_5(S_{t-1}, L_{t-1}, M_{t-1}) + \varepsilon_5 \\ v_t &= f_6(S_{t-1}, L_{t-1}, M_{t-1}, TQ_t) + \varepsilon_6 \end{aligned} \tag{3}$$

where *M* represents other determinants potentially influencing the levels of customer retention, acquisition and revenue per customer as market characteristics (e.g. market shares) or characteristics of the customer (e.g. socio-demographics, customer profile)[1]. As

[1] By including the lagged market share as an explanatory variable, the number of retained customers in each period plays an accelerating effect. An increase in the retention or acquisition rates will potentially increase the market share that in turn will influence retention and acquisition rates in the next period.

explained previously in equation (3) we also consider a possible direct path from technical quality to revenue per customer.

Let also:

MS_{jt}: Market share of firm j in period t;

N_t: Market size in period t (total number of customers of all competing firms);

N_{jt}: Total number of customers of firm j in period t ($N_{jt} = MS_{jt}.N_t$)

n_t: Market growth rate (measured by the number of customers) in period t;

d_t: Discount rate in period t;

If we assume that customers switch between operators at most once during a period, then the number of customers of any firm j in period t comes from three sources.

1. Number of retained customers, i.e. customers that remain as customers from the previous period. The number of retained customers obviously depends on the customer retention rate of firm j. If we call the number of retained customers $N1$, we will have:

$$N1_{jt} = r_{jt} \; N_{jt-1} \tag{4}$$

2. Number of customers switching from competitors to firm j. This contribution, represented by $N2$, depends on the attraction rate of firm j and the retention rate of the competitors:

$$N2_{jt} = a_{jt}^{(1)} \sum_{i \neq j} N_{it-1}(1 - r_{it}) \tag{5}$$

3. Number of new customers entering the market and attracted by firm j. This contribution depends on the growth of the market in each period and on the attraction rate of new customers for firm j:

$$N3_{jt} = a_{jt}^{(2)} \; n_t \; N_{t-1} \tag{6}$$

If we combine the three sources, then the number of customers of firm j in period t is:

$$\begin{aligned} N_{jt} &= r_{jt} N_{jt-1} + a_{jt}^{(1)} \sum_{i \neq j} N_{it-1}(1 - r_{it}) + a_{jt}^{(2)} n_t N_{t-1} \\ &= \left[r_{jt} MS_{jt-1} + a_{jt}^{(1)} \sum_{i \neq j} MS_{it-1}(1 - r_{it}) + a_{jt}^{(2)} n_t \right] N_{t-1} \quad = r_{jt} \; N_{jt-1} \end{aligned} \tag{7}$$

and the market share of firm j at moment t is given by:

$$MS_{jt} = \left[r_{jt} MS_{jt-1} + a_{jt}^{(1)} \sum_{i \neq j} MS_{it-1}(1 - r_{it}) + a_{jt}^{(2)} n_t \right] \Big/ (1 + n_t) \tag{8}$$

Assuming that the investments in network quality influence the rates r_{jt}, $a_{jt}^{(1)}$, $a_{jt}^{(2)}$, as well as the average revenue per customer v_{jt}, and let us also suppose that the discount rate is equal to d during p periods. Then, the net present value of the incremental returns generated by an investment in quality, $NPVR$, is given by:

$$NPVR = \sum_{k=1}^{p} (1+d)^{-k} \left(v_{jt+k} \Delta N_{jt+k} + \Delta v_{jt+k} N_{jt+k} + \Delta v_{jt+k} \Delta N_{jt+k} \right), \tag{9}$$

with the change on the number of customers given by:

$$\Delta N_{jt} = [\Delta r_{jt}\ MS_{jt-1} + \Delta c_{jt}^{(1)} \sum_{i \neq j} MS_{it-1}(1 - r_{it}) + \Delta c_{jt}^{(2)}\ n_t]\ N_{t-1} \tag{10}$$

where the symbol Δ represents the change of the associated variable following an investment in quality by firm *j*.

In order to obtain the return on a quality investment, let us assume an initial investment of I_0, a level of ongoing expenditure I_1, and a level of expenditure after the investment I (including the cost of maintenance of the new investment), net of any eventual reductions on costs that can be interpreted as negative expenditures.

The present value of the additional expenditure during *p* periods, *NPVAE* is given by:

$$NPVAE = I_0 + (I - I_1) \sum_{k=1}^{p} (1+d)^{-k} \tag{11}$$

So, the return on the investment on quality is given by:

$$ROQ = NPVR / NPVAE \tag{12}$$

The value of *NPVR* for firm *j* depends on: the market share MS_j in each time period for firm *j*, the size of the market , i.e. the number of customers in each period N_t, the market growth rates n_t, the average revenue per customer v_j, the retention and acquisition rates r_{jt}, $a_{jt}^{(1)}$, $a_{jt}^{(2)}$, the discount rate d (assumed to be constant over periods), and the number of periods, p.

Among the whole set of variables that influence the return on quality investments, the average revenue per customer and the customer retention and attraction rates are those variables that firms can influence through quality investments. These variables are at least partially influenced by customer satisfaction and loyalty. In turn, these attitudes are influenced by customer perceptions of the quality of products and services delivered by the firm, which in turn are influenced by the investments in quality.

2.3 Modelling and data challenges

Note that data used in the proposed modelling chain typically result from different sources. Data referring to quality investments and technical quality indicators usually come from technical or quality departments within the operator. Usually it is organized with an aggregation level that doesn't enable the linkage towards individual customers. In fact, it is usually impossible to know the level of network quality (eg. network coverage, quality of the communication, existence of call drops) experienced by each individual customer.

Data relative to customer perceptions and attitudes would be typically obtained from a satisfaction survey. In this framework is extremely important to organize the questionnaire around quality dimensions that can comprehend quality investments. Each quality dimension should be implemented using a set of perceived quality manifests that can be directly associated to technical quality indicators.

Data relative to customer effective behaviours (customer retention, customer acquisition and revenue) may be obtained from marketing files, accounting files or other company sources. It can also be complemented through a longitudinal satisfaction survey where each customer is observed in more than one wave. This allows the collection of customer perceptions and attitudes as well as actual and reported behaviours in the same survey. Also, market sizes and market shares can, in some situations, be obtained from published data regarding the mobile telecommunications industries. In several countries this information is published by telecommunications regulators. In other situations it has to be obtained through customer satisfaction surveys or other specific surveys, including customers not only from the focal operator, but from all relevant operators in the market.

Among the relationships shown in Figure 1, the most difficult to model are the ones that link data of a different type or from a different sources, i.e. the linkage between technical quality and perceived quality (equation 1) and the linkage between customer perceptions and attitudes and effective behaviours (equation 3). These difficulties arise from the fact that these sources frequently use different aggregation levels.

Due to these difficulties, most of the studies published so far have only just used behaviour intentions as a proxy for the effective behaviours. Our experience confirms the results of other authors (see, for instance, Johnson & Gustafsson, 2000, p.7) that state this can generate totally unrealistic results, since repurchase or recommendation intentions are in many situations very far from the true observed behaviours.

3. Application to the mobile telecommunications industry

3.1 Introduction

The following application uses the mobile telecommunications industry in an EU state, which is composed of three operators. The main goal of the application is to evaluate the return of investments in the quality of the telecommunications network. The potential quality investments carried out on the focal firm's network regard the installation of new cell towers and aim to improve three technical quality indicators: the *call drop rate*; the *congestion rate*; and the *coverage rate*. These indicators are those that are most widely used to measure technical quality in mobile telecom and for which there is available data. They are defined as

Call drop rate = number of dropped calls (due to loss of radio frequency or handover failure) / total number of initiated calls

Congestion rate = number of failures in channel attribution / number of attempts of channel attribution

Coverage rate = coverage area / total area

The steps in the process of converting quality investments into financial outcomes are the same as were defined previously. In our application the relationships between technical quality and perceived quality were established using linear regression linking levels of technical quality experienced by customers and their perception regarding the quality of the network measured in a satisfaction survey. The relationships linking perceived quality to customer satisfaction and loyalty were established using structural equation modelling (SEM). Sequentially the modelling of customer retention, acquisition and revenues were based in logistic and linear regression. Finally, the changes in market share and the effects on the financial results following investments in the improvement of the quality of the network will be estimated in the framework of a simulation. In what follows we briefly address each one of these steps.

3.2 Perceived network quality

Data referring to perceived network quality were obtained through the administration of a perceived quality and customer satisfaction survey comprising users of the three mobile phone operators in the market. There were three waves of the survey administered every six months during a period of one and a half years (July 2001, January 2002 and July 2002). The survey is based on a probability sample drawn through random digit dialing of mobile phone numbers from the three operators. The sample size was 1448.

The questionnaire used in the survey queries the overall experience of the respondent with the operator. It includes questions regarding a set of multidimensional constructs: image, expectations, several dimensions of perceived quality (including perceived network quality), customer satisfaction and customer loyalty. To measure perceived network quality we used 4 manifests: perceived network coverage, perceived network availability, perceived network uniformity and perceived communication quality. The first three manifests are the perceptive contra parts of the quality indicators influenced by the planned investments (call drop rate, the congestion rate and the coverage rate). Manifests are measured in an 11 point interval scale (0 to 10) ranging from very negative appreciation to very positive.

Several explanatory models were tested to explain the three target variables (perceived network coverage, perceived network availability, perceived network uniformity), having as potential explanatory variables the three technical indicators (call drop rate, the congestion rate and the coverage rate). As the level of technical indicators experienced by individual customers was not available, we tested the regression models at different regional aggregation levels (municipalities, NUTSIII and NUTSII). Also different time lags were tested in order to account possible lags between changes in the level of network quality and customer perceptions. It was not possible in any case to find relationships with statistical significance between these two sets of variables. This result was not surprising as the estimates for customer perceptions are only representative at NUTSII level. The aggregation of technical indicators at this high regional level hides disparities within each region, making it impossible to relate individual customer perceptions with the actual level of network quality experienced by each customer. The overcome this limitation we have included a new set of questions in customer satisfaction questionnaires querying the customer about his/her perception regarding the experienced level of technical indicators (using the questionnaire applied in the third available wave of the survey). This made it possible to model data at individual level. To measure the experienced level of call drop rate each customer was asked to state the approximate proportion of times that s/he has experienced a call drop during conversation within a 6 month period. To obtain a proxy of congestion rate the customer was asked about the proportion of times that s/he was not able to initiate a call due to network unavailability. Finally, experienced coverage rate was approximated asking the proportion of places where s/he was not able to initiate a call due to lack of coverage within the same time frame.

Three regression models were adjusted (one for each indicator). Using F tests for the significance of the three regressions, the null hypothesis that all model coefficients are zero was always rejected ($p<.001$), showing the relevance of the three models. The effects of perceived call drop rate, congestion rate and coverage rate on perceived network coverage, perceived network availability, perceived network uniformity are presented in Figure 2 and were always found significant using *t* tests ($p<.001$). For each percentage point of reduction in the call drop rate perceived network uniformity will grow .25 points (in the 11 point scale), for each percentage point drop in the congestion rate perceived network availability will grow .26 points and for each percentage point of increase in coverage rate the perceived network coverage grows .19 points.

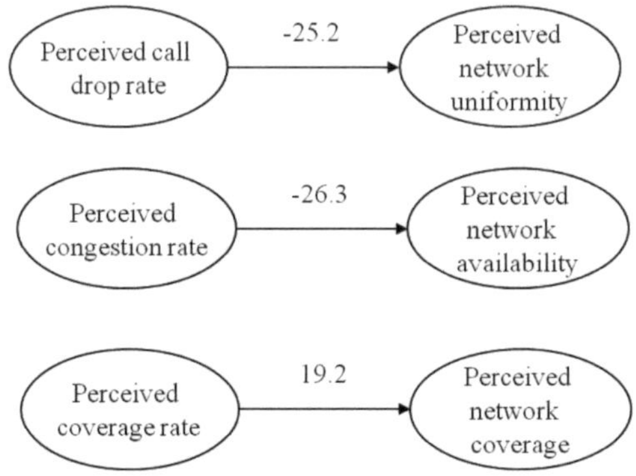

Fig. 2. Modelling of network perceived quality

3.3 Customer satisfaction and loyalty

Customer satisfaction and loyalty are modelled using data from the customer satisfaction survey also used to model perceived network quality. Structural equation modelling (SEM) was used to model customer satisfaction and loyalty. Construct indexes were produced in a 0 to 100 scale. The model is constituted by 12 multi-item constructs and is summarized in Figure 3. The model was estimated using partial least squares (PLS) and the parameters significance was accessed using bootstrapping with 1000 replicates. Final results shown good explanatory power both for customer satisfaction (R^2=0.72) and loyalty (R^2=0.64). The

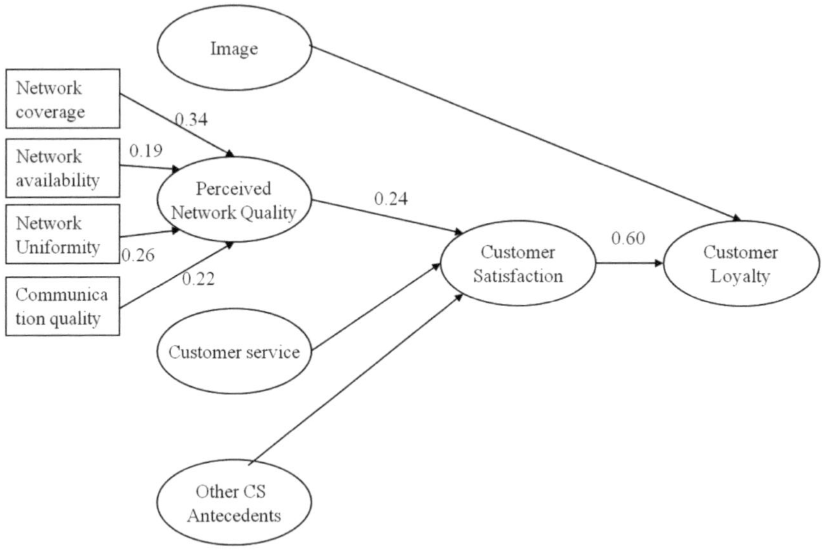

Fig. 3. Customer satisfaction model

perceived network quality construct had significant impact of 0.24 ($p<.001$) on customer satisfaction. The impact of this construct on customer loyalty is indirect (mediated by customer satisfaction) and equal to 0.14. The weights of perceived network quality manifests are all significant ($p<.001$) and the respective estimates are shown in Figure 3.

3.4 Customer retention

Customer retention is modelled using panel data from the same customer satisfaction survey used to estimate the relationship between perceived network quality and customer loyalty.

The determinants of customer retention that have been examined as potentially relevant were the variables intervening in the customer satisfaction/loyalty models with particular emphasis on customer satisfaction and loyalty indexes. We also considered other possible explanatory variables related to the customer profile (such as type of contract, time as a customer, etc), socio-demographic variables, as well as the market share of each operator. In fact, empirical evidence from previous studies performed by the authors showed that a key variable in the process of leaving or choosing a future operator is whether or not an operator is used by the persons we regularly contact (due to the pricing policies for calls within and between networks). A way of including this effect is through the market shares that reflects the different sizes of the customer bases of the mobile phone operators and that appears as an explanation of customer retention.

A weighted logistic regression was used to model the retention of each customer. The weights of the logistic regression were chosen in order to reproduce the actual retention rates known for each operator obtained from a market size survey. This is a parallel survey whose main goal is to estimate market size, market shares and churn and acquisition rates for each operator.

After testing different specifications, a logistic regression was retained, having as explanatory variables customer loyalty and market share lagged one period (cg Figure 4).

The Chi-squared test for the model significance always rejected the null hypothesis that all model coefficients are zero ($p<.001$). The estimate of loyalty impact on the logit of retention probability is 0.014 ($p<.005$) and the impact of market share 1.67 ($p<.005$). Results confirm that in addition to customer loyalty, the relative size of operators strongly influences churning decisions. As an example, a ten point increase in customer loyalty index for a medium operator with a 30% market share and an average loyalty of 70 (on a 100 point scale) is associated with a 1.2 percentage point increase in retention rate in the period of one semester.

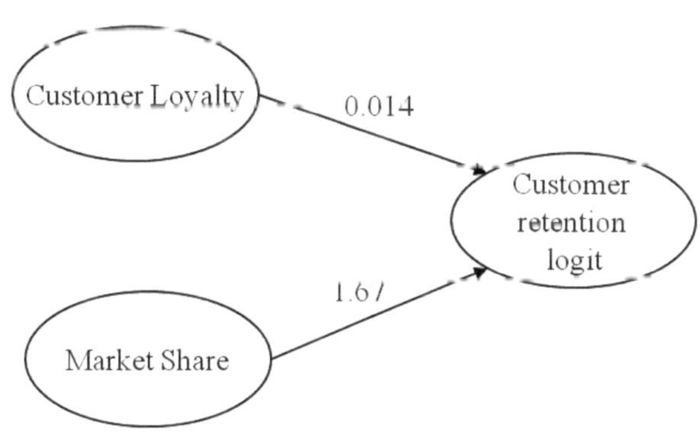

Fig. 4. Modelling of customer retention

3.5 Customer acquisition

Customer acquisition is modelled using data from two waves of a quarterly market size survey referring to January and July 2002. The survey is also based on a probability sample drawn through random digit dialing of fixed phone and mobile phone numbers. The questionnaire queries if the respondent is a mobile phone user, who the operator is and other characteristics of the service usage. In each wave, market size, market share, and the number of customers entering and leaving each operator is obtained. As customer acquisition could not be modeled at the individual level, we have used regionally aggregated data (seven regions). Using these data we modelled both customer acquisition rates: the acquisition rate of new customers ($a^{(1)}$) and the acquisition of customers leaving the competition, ($a^{(2)}$).

The determinants of customer acquisition that were examined as potentially relevant were the intervening variables in the customer satisfaction/loyalty model, as well as the market share of each operator. Linear regression was used to model the regional acquisition rates. After testing different specifications, the selected models explain customer acquisition in both cases through the regional market share of the firms. In fact, the parameter estimates for customer satisfaction and loyalty did not show statistical significance whenever the market share was present in the model. The underlying reason for the importance of market share as an explanatory variable for acquisition rates was already presented in the context of modelling customer retention. It may be possible that the relative size of the operator is so important in the choice of another operator that it completely cancels out the recommendation effect on acquisition. Nevertheless, this result may also be an artefact of the data used for modelling as only two data points were available. However, this does not mean that in the estimated models loyalty does not influence acquisition rates, but only that this relationship may be indirect and lagged. In fact, higher customer loyalty will, in our model, cause higher customer retention, which in turn will cause a higher market share and acquisition rates in the future. When simulating the market for several periods, this chain of events may be explicitly included. In our final model both acquisition rates are explained by market share lagged one period (cf. Figure 5).

Using F tests for the significance of both regressions, the null hypothesis that all model coefficients are zero is rejected in both cases showing the relevance of the model. The regional market share of each operator has positive and significant effects on both customer acquisition rates being equal to 0.82 (p<001) and 0.71 ($p<0.001$), respectively. The determination coefficients are 0.68 when explaining the acquisition of new customers and 0.29 for the acquisition rate from competitors.

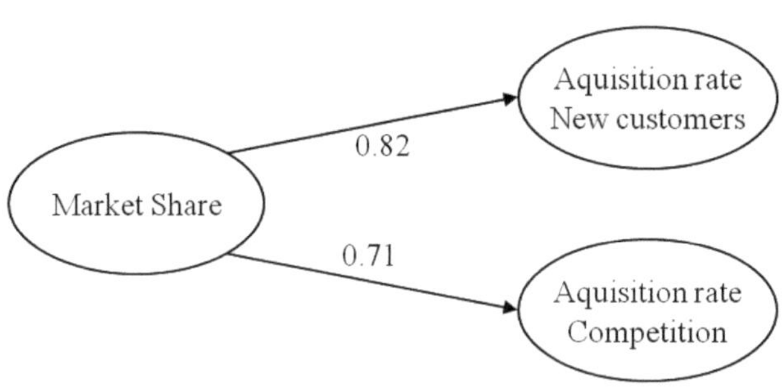

Fig. 5. Modelling of customer acquisition

3.6 Average revenue per customer

The average revenue per customer is modelled using data from the most recent wave of a customer satisfaction survey (also used to model customer retention). The determinants of customer acquisition that have been examined as potentially relevant to explaining revenues per customer were the variables intervening in the customer satisfaction/loyalty models, variables related with customer profile (such as the type of contract), socio-demographic, the market share of each operator and the three technical quality indicators (call drop rate, congestion rate and coverage rate). The reason for testing the possible direct effects from the technical quality indicators is related to the fact that a better performance of the network (and consequently fewer call drops, congestion rates and/or higher coverage rates) may result in an increase in usage/traffic and a consequent increase in the revenue per customer. Modeling on individual data was abandoned due to the difficulty to identify the levels of technical indicators effectively experienced by each customer. Therefore, we have used regionally aggregated data (seven regions). Models were estimated with a sample size of 20 observations (one observation was missed since the samples size for one operator in one of the seven regions was too small to estimate the average revenue per customer). Linear regression was used to model the regional revenue per customer. After testing different specifications, the selected model explains customer average expenditure with the type of customer contract and the coverage rate (Figure 6). Note that among the three technical indicators that have been tested (call drop rate, coverage rate and congestion of the network) only the coverage rate appeared to have a direct influence on the average revenue. A possible justification for this finding is that the calls not initiated (due to network congestion) or not concluded (due to call drop) tend to be successfully completed afterwards (by a new call). But, calls that are not completed (or initiated) due to a lack of coverage are not (at least partially) substituted and customer revenue drops. Also, among customer profile variables, only the type of contract contributes significantly to explaining average expenditure. This result was to be expected since it was already well known that customers with post-paid contracts have higher expenditures then the ones with pre-paid plans. Finally, it should be stressed that it was not possible to obtain evidence to support the idea that customer satisfaction/loyalty influences the level of use of the service and therefore the average expenditure per customer.

The *F* test for the significance of the regression rejected the null hypothesis that all model coefficients are zero ($p<.01$) showing the relevance of the model. The determination coefficient is 0.41. Both coverage rate (72.8; $p<.07$) and type of contract (52.8; $p<.01$) were found to have positive and significant impacts on average customer expenditure.

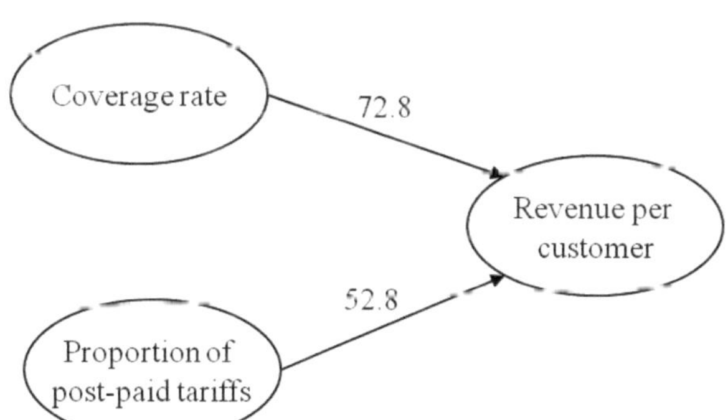

Fig. 6. Modelling of revenue per customer

3.7 Simulation

Equations (7) to (12) may be used to obtain the net present value of additional revenues resulting from the planned investment in the quality of the network. The application of the general framework proposed for evaluating the return of quality investments for these mobile telecommunications market is represented in Figure 7. The simulation starts by considering two alternative investments in network cells as planned by the focal operator (Investment A and Investment B). The level of investment A is 22 million Euros and for investment B 29 million Euros, allowing the focal company to reduce congestion rate and call drop rate in a specific area. The investment will also contribute to an improvement in the coverage rate. The improvement in technical quality indicators experienced by the customers will improve perceived network quality and customer loyalty. Initially the improvement in customer loyalty and coverage rate will improve customer retention and revenues per customers. In each period the market share will be improved through the gains in customer retention and acquisition rates obtained in previous periods, which in turn will promote in the new period, customer retention and acquisition. The increase in the number of customers along with the increase in revenues per customer will give rise to additional revenues that can be accounted for. Running the model within a time frame, it is possible to obtain the increase in revenues for each period and its present value using an appropriate discount rate. The period considered for the simulation is 10 half-years (the period for which the focal firm evaluates the return of other type of investments) and we have considered a discount rate of 5% typically used by the focal company.

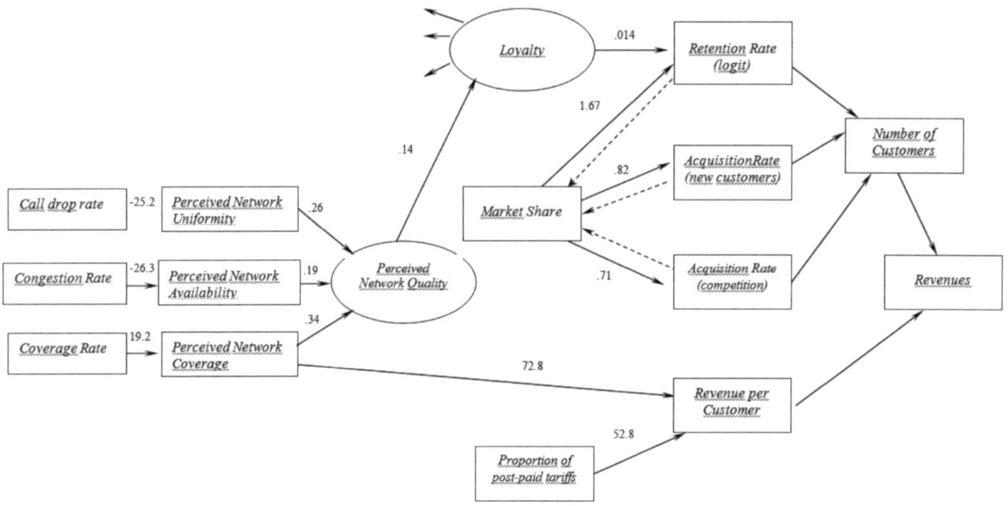

Fig. 7. Complete model used for simulation

Table 7 presents the net present values of additional revenues at each of the 10 periods included in the simulation. The planned investment A will originate additional present revenue of about 23.2 million Euros. The alternative investment B originates present revenue of about 26.6 million Euros.

The profitability of the investment may be evaluated by comparing the additional revenues with the value of additional expenditures associated with the investment. The return on investment (ROI) is 5.3% for investment A and -8.4% for investment B. Note that the ROI depends on the considered discount rate. The internal rate of return (IRR) is 5.8% for alternative A and 3.7% for B. The result is that the latter alternative is not profitable for the level of discount rate used by the company.

These results show that the present methodology can be used to simulate the results of different types of investment in network quality, allowing the configuration of investments most adequate for a certain purpose to be determined (to maximize profitability, to maximize net profit, etc).

	Investment A		Investment B	
Period	*Aditional profits*	*NPV for aditional profits*	*Aditional profits*	*NPV for aditional profits*
1	487.975 €	464.738 €	559.024 €	532.403 €
2	1.011.413 €	917.381 €	1.158.703 €	1.050.977 €
3	1.570.409 €	1.356.578 €	1.799.143 €	1.554.168 €
4	2.164.879 €	1.781.051 €	2.480.247 €	2.040.505 €
5	2.794.680 €	2.189.705 €	3.201.846 €	2.508.730 €
6	3.459.596 €	2.581.604 €	3.963.690 €	2.957.767 €
7	4.159.324 €	2.955.954 €	4.765.425 €	3.386.699 €
8	4.893.453 €	3.312.082 €	5.606.575 €	3.794.751 €
9	5.661.451 €	3.649.422 €	6.486.523 €	4.181.270 €
10	6.462.642 €	3.967.502 €	7.404.486 €	4.545.712 €
NPV		**23.176.017 €**		**26.552.982 €**
NPVAE		**22.000.000 €**		**29.000.000 €**
ROI		**5,3%**		**-8,4%**
IRR		**5,8%**		**3,7%**

Table 1. Simulation results

4. Conclusions

This work proposes and applies a methodology that enables a cost-benefit analysis of quality investments in the context of mobile telecommunications industry.

The adopted approach aims to estimate additional revenues generated by an investment in a certain time frame along with the additional expenditures associated with the investment. Combining the estimates of additional revenues with the values of additional expenditure one may calculate the profit and profitability associated with the investments. Therefore, these results may be used to choose between alternative investments by comparing their potential profitability, allowing a rational allocation of available resources, and allowing mobile telecommunications managers to approach the investment in quality as any other investment in a competitive environment. In our application it was clear that one of the two competing investments in network cells was not profitable for the used discount rate.

Moreover it becomes possible to simulate the results of different levels and configurations in quality investments, allowing the establishment of priorities for resource allocation in order to maximize profits or profitability. Also, with this approach it is clear that the increase in expenditure on quality alone does not induce an increase in the profits of the firm. Where to spend and how much to spend needs to be identified. Moreover, it becomes clear that the benefits of quality investment are not all immediate and should be evaluated in the long term.

The development of the proposed approach from an operational point of view requires the estimation of the relationships between variables that describe the different stages of the causal chain that mediates between quality investments and the financial outcomes that result from such investments.

We have shown that it is possible to estimate financial returns from each type of quality investment, given sufficient knowledge and measurement of the critical paths in the value chain. These paths involved in the chain will include (1) variables that measure technical quality indicators, (2) constructs such as customer perceptions of quality, customer satisfaction, and customer loyalty, (3) variables that represent customer behaviours as retention, acquisition and level of expenditure and (4) financial results. In addition, there will be predictive variables unique to particular industries including market and customer characteristics. For example, in our application to the mobile telecommunications industry, market share, geographical network coverage of the network and proportion of customers using each type of billing/charging plan appeared as relevant predictors of customer behaviours.

Some limitations of this work should be stressed. Firstly, we consider that the improved level of quality perceptions resulting from an investment in quality will maintain constant over the simulation time frame. We do not take into account that the competitive advantages earned by an operator through the investments in improving service quality may be cancelled out if consumers' expectations rise or if other competitors follow suit. Nevertheless, this does not mean that the model does not consider a competitive framework, since it is possible to explicitly consider the actions of competition.

The different steps in the model chain involve different populations, different levels of aggregations (geographic and temporal) and different sources of data. Consequently, in many cases, estimation was done at a high level of aggregation and with a limited number of observations. More work should be done in order to try to establish relationships at individual level or at least at lower levels of aggregation.

Also, most of the models involved in our chain were estimated using data from one to three periods of observation. The application presented should be revisited using a longer series of observations in order to investigate non detected lags in the relationships established.

Some results, such as not having found a direct effect from customer loyalty on customer acquisition rates or on average expenditure per customer, may be an artefact of this specific market or company. This approach should be tested in other countries and with companies with different market shares. Also, the links between customer loyalty and behavioural variables as revenue per customer may be mediated by other factors as consumption rate, products and services consumed and price sensitivity. Further research explicitly considering these mediating factors may yield a greater insight on these processes.

Finally, the specifications that have been adopted in the models were in general linear. Yet the research in quality and customer satisfaction and loyalty has emphasized the potential for nonlinear relationships (see, in particular, Kano et al., 1996 and Jones & Sasser, 1995). We have tested some non-linear specifications and the quality of the results was not significantly improved. Nevertheless, with more observation periods or within the application in other countries, non-linear specifications may become relevant.

5. References

Aaker, D., Jacobson, R. (1994). The financial information content of perceived quality. *Journal of Marketing, 58*, 191-202.

Allen, D.R., Wilburn, M. (2002) *Linking Customer and Employee Satisfaction to the Bottom Line*, ASQ Quality Press.

Anderson, E. (1998). Customer Satisfaction and the Word-of-Mouth, *Journal of Service Research*, *1*(1), 1-14.

Anderson, E., Fornell, C. & Lehmann, D. (1994). Customer Satisfaction, Market Share and Profitability. *Journal of Marketing*, *58*, 53-66.

Anderson, E., Fornell, C. & Mazvancheryl, S. (2004). Customer Satisfaction, and Shareholder Value. *Journal of Marketing*, *68*, 172-185.

Anderson, E., Fornell, C. & Rust, R. (1997). Customer Satisfaction, Productivity and Profitability: Differences between Goods and Services. *Marketing Science*, *16*(2), 129-145.

Bolton, R. (1998). A Dynamic Model of the Duration of the Customer's Relationship with Continuous Service Provider: The Role of Satisfaction. *Marketing Science*, *17*(1), 45-65.

Bolton, R., Drew, J. (1991). A Longitudinal Analysis of the Impact of Service Changes on Customer Attitudes. *Journal of Marketing*, *55*, 1-9.

Burton, S., Sheather, S. & Roberts, J. (2003). Reality or Perception? The Effects of Actual and Perceived Performance on Satisfaction and Behaviour Intention. *Journal of Service Research*, *5*(4), 292-302.

Coelho, P.S., Vilares, M. (2010). Measuring the return on quality investments. *Total Quality Management and Business Excellence*, *21*(1), 21-42.

Charnes, A., Cooper, W., Lewin, A. & Seiford, L. (1994). *Data Envelopment Analysis: Theory, Methodology and Applications*, Kluwer Academic Publishers, Norwell, MA.

Danaher, P., Rust, R. (1996). Indirect Financial Benefits from Service Quality, *Quality Management Journal*, *3*(2), 63-75.

Fornell, C. (1992). A National Customer Satisfaction Barometer: The Swedish Experience. *Journal of Marketing*, *56*, 6-21.

Fornell, C. (2003). Boost Stock Performance, Nation's Economy. *Quality Progress*, *36*(2), 25-31.

Fornell, C, Johnson, M., Anderson, E., Cha, J. & Bryant, B. (1996). The American Customer satisfaction Index: Nature, Purpose and Findings. *Journal of Marketing*, *60*, 7-18.

Hallowell, R. (1996), "The Relationship of Customer Satisfaction, Customer Loyalty, and Profitability: An Empirical Study", *International Journal of Services Industry Management*, *7*(4), 27-42.

Hart, C., Johnson, M. (1999), "Growing the trust relationship", *Journal of Marketing Management*, *8*(1), 8-19.

Heskett, J., Jones, T., Loveman, G., Sasser, W. and Schlesinger, L. (1994). Putting the Service Profit Chain to Work. *Harvard Business Review*, *72*(2), 164-174.

Hromi, J. (ed.) *The Best on Quality, International Academy for Quality*, Vol 7. Milwaukee, WI: The quality press.

Ittner, C., Larcker, D. (1996). Measuring the Impact of Quality Iniciatives on Firm Finantial Performance. In Soumeh Ghosh and Donald Fedor (Eds.) *Advances in the Management of Organizational Quality*, Vol. *1* (pp. 1-37). JAI.

Ittner, C., Larcker, D. (1998). Are Non-financial Measures Leading Indicators of Financial Performance? An analysis of Customer Satisfaction. *Journal of Accounting Research*, *36*(3), 1-35.

Jacobs, F., Latham, C. & Lee, C. (1998). The Relationship of Customer Satisfaction to Strategic Decisions. *Journal of Managerial Issues*, *10*(2), 165-182.

Johnson, M. D., Gustafsson, A. (2000). *Improving Customer Satisfaction, Loyalty and Profit*, San Francisco: Jossey-Bass.

Jones, T. O., Sasser JR, E. (1995). Why Satisfied Customers defect. *Harvard Business Review, 73*, Nov-Dec.

Kamakura, W.A., Mittal, V., Rosa, F., Mazon, J. (2002). Assessing the Service-Profit Chain. *Marketing Science, 21*(3), 294-317.

Kano, N., Seraku, N., Takahashi, F., Tsuji, S. (1996). Attractive Quality and Must Be Quality, In Hromi, J.D. (Ed.), *The Best of Quality*, Vol. *7*, Milwaukee, WI: ASQC Quality Press.

Keiningham, T., Zahorik, A. & Rust, R. (1994). Getting Return on Quality. *Journal of Retail Banking Services, 16*(40), 7-12.

Kordupleski, R. E., Rust, R. & Zahorik, A. J. (1993). Why Improving Quality Doesn't Improve Quality (Or Whatever Happened to Marketing?). *California Management Review, 35*(3), 82-95.

Oliver, R. (1997a), *Satisfaction: A Behavioural Perspective on the Consumer*. Boston: Irwin/McGraw-Hill.

Oliver, R. (1997b). New directions in the study of the consumer satisfaction response: anticipated evaluation, intend cognitive-affective process and trust influences on loyalty. In J.W. (Ed.), *Advances in Consumer Research*, ACR, Provo, UT.

Oliver, R. (1999). Whence Consumer Loyalty?. *Journal of Marketing, 63* (special issue), 33-44.

Reichheld, F. (1996). *The Loyalty Effect*. Boston: Harvard Bussiness School Press.

Reichheld, F. & Sasser, L. (1990). Zero Defections: Quality Comes to Services. *Harvard Business Review, 68*, 105-111.

Rust , R. T.; Ambler, T.; Carpenter, G. S, Kumar, V. and Srivastava, S. (2004). Measuring Marketing Productivity: Current Knowledge and Future Direction. *Journal of Marketing, 68*, 76-89.

Rust, R.T.; Lemon, K. & Zeithaml, V. (2004). Return on Marketing: Using Customer Equity to Focus Marketing Strategy. *Journal of Marketing, 68*, 109-127.

Rust ,R.T., Zahorik, A.J. (1993). Customer Satisfaction, Customer Retention and Market Share. *Journal of Retailing, 69*, 193-215.

Rust, R.T., Zahorik, A.J. & Keiningham, T.L. (1995). Return on Quality (ROQ): Making Service Quality Financially Accountable. *Journal of Marketing, 59*, 58-70.

Simester, D., Hauser, J., Wernerfelt, B. & Rust, R. (2000). Implementing Quality Improvement Programs Design to Enhance Customer Satisfaction: Quasi-Experiments in the United States and Spain, *Journal of Marketing Research, XXXVII*, 102-112.

Strauss, B., Neuhaus P. (1997). The Qualitative Satisfaction Model. *International Journal of Services Industry Management, 9*(2), 69-88.

Zeithaml, V.A. (2000). Service Quality, Profitability and the economic Worth of Customers: What We Know and What We Need to Learn. *Journal of the Academy of Marketing Science, 28*(1), 68-75.

Zeithaml, V.A, Berry, L.L. & Parasuraman, B.A. (1996). The Behavioral Consequences of Service Quality. *Journal of Marketing, 60* (April), 31-46.

7

Network Effects in the Mobile Communications Industry: An Overview

Juan Pablo Maicas[1] and F. Javier Sese[2]
[1]*Departamento de Dirección y Organización de Empresas, Facultad de Economía y Empresa, Universidad de Zaragoza*
[2]*Departamento de Dirección de Marketing e Investigación de Mercados, Facultad de Economía y Empresa, Universidad de Zaragoza*
Spain

1. Introduction

Information technology (IT) markets, in general, and mobile telecommunications, in particular, represent a large and growing portion of today's economy. They are one of the main sources of economic growth in modern economies (Greenspan 2000) and provide the basis for the development of a knowledge-centric world. The rapid development of these markets and the central role that they play in the economy have attracted the attention of a large number of researchers who have shifted their emphasis from traditional markets to the so-called New Economy. When dealing with these markets, however, it is important to understand that they are governed by a unique set of characteristics that may render well-established principles and managerial practices invalid (Shapiro and Varian 1998).

One of the most significant distinctive features of these markets is the central role played by the network of users. In IT markets, customers derive utility not only from the product/service itself, but also from the networks surrounding these products (Frels, Shervani, and Srivastava 2003). This is because the installed base of users offers benefits to existing and potential customers in the form of reduced uncertainty, compatibility, the transfer of technical and non-technical information between members of the network and the increased availability and quality of complements, among others (Farrell and Klemperer 2007). This feature explains why IT markets are frequently referred to as network markets (Shankar and Bayus, 2003; Tanriverdi and Lee, 2008). Mobile communications, video games and software are just three examples of businesses where network effects drive market competition and consumer behavior. The network of users becomes a central strategic asset for assessing the firm's current and future competitive position (Shankar and Bayus 2003; McIntyre and Subramaniam 2009) and, as a result, the understanding of network effects has become a top priority for researchers and practitioners alike.

The previous discussion can help explain why network effects have emerged as a trendy topic in recent years in economics and management literatures. This research stream focuses on understanding how network effects alter the way in which firms compete. This increasing interest is due to the evidence that network industries seem to challenge much of the thinking derived from previous models and findings (Shapiro and Varian, 1998; Suárez, 2005). We can observe a trend toward examining value creation in IT markets on the basis of the interdependencies that exist within a market. This trend has intensified in recent years with the proliferation of social networks and other new media that have resulted in an increasingly networking society where individuals can easily interact with each other. In this new environment, a firm's success and survival critically depend on a proper understanding of how these networks operate.

In this chapter, our primary objective is to provide an overview of the current state of the networks effects literature highlighting their role in the mobile communications industry in order to gain a better understanding of one of the key forces that underlie the development of these markets. To do so, we will review the relevant works in the field of network effects and network markets, offer empirical support for the conceptual arguments and identify how network effects operate in mobile communications.

The second section of the chapter elaborates on the concept of network effects as well as on the different types of network effects that the literature has identified. Networks can be conceived as a whole - which we refer to as *direct* or *pure network effects* - in which each individual in the network contributes equally to creating value for the others. This implies that adding one customer to the market equally increases the utility of all users who are already in the network. Networks effects can also be *indirect* or *market-mediated*. In this case, the utility of each user is not directly influenced by other users consuming the good at the same time, but for the growth in the other side of the market. Recently, researchers have introduced a new category of network effects, *personal network effects*, which acknowledges that network effects may be localized (Birke and Swann, 2006; Farrell and Klemperer, 2007; Maicas, Polo and Sese, 2009a). For instance, a mobile phone user gains more when her friends join the mobile network than when a stranger does. In other words, there are different densities in the network (Ahuja, 2000) with different consequences on customer utility.

The third section of the chapter focuses on the telecommunications industry and discusses the different sources of network effects that can be identified therein. We distinguish between direct and localized network effects. Direct network effects emerge as a result of pricing strategies implemented by mobile phone carriers and based on the origin and destination of the calls (price discrimination between on-net and off-net calls). Personal network effects refer to those instances where mobile service providers offer special tariffs for calls to members of the social network (family, friends…).

Finally, we will discuss the managerial implications of network effects in mobile communications. The significant contribution of networks to the utility derived by individuals is important for firms and they have begun to strategically manage this market-based asset. For example, operators try, as mentioned, to increase the contribution of direct network effects to utility by charging different prices depending on whether the calls made by the user are directed to members of the same operator (low or on-net tariff) or to members outside the operator (high or off-net tariff). In this same context, firms aim to promote personal network effects by offering special tariffs to a group or social subset of users (e.g. customers can select five contacts to whom they can call with a reduced tariff). Overall, these strategies influence the value that users derive from the networks surrounding the product in an attempt to ultimately improve a firm's competitive position and gain competitive advantage.

2. Network effects: Definition and types

2.1 Network effects: Concept and implications for markets

In recent years, network effects have received a great deal of attention in various disciplines including economics, management and marketing (Birke and Swann 2006; Farrell and Klemperer 2007; Srinivasan, Lilien, and Rangaswamy 2004; Wang, Chen, and Xie 2010). Drawing upon these recent research developments, network effects can be defined as follows. A good exhibits network effects when the utility of a user increases with the number of other users consuming the good (Katz and Shapiro, 1985; Farrell and Klemperer, 2007). That is, this definition acknowledges that the utility of the user is driven, not only by the product itself, but, even more importantly, by the network that surrounds that product so that the larger the network, the higher the utility derived from consuming the product.

It is clear from the definition offered that network effects have significant implications for the way markets operate and firms compete. For example, when the network of users is large and consumers can derive a high utility from it, customers' willingness to pay increases, with the subsequent potential impact on firm competitive performance (Shapiro and Varian, 1998; Shankar and Bayus, 2003). Network effects can also affect the distribution of the market (consumers) across the available alternatives (companies). By having a positive effect on customers' utility functions, firms or technologies with a high market share are able to obtain a higher level of profitability (Katz and Shapiro, 1985, 1994; Farrell and Saloner, 1985; Farrell and Klemperer, 2007). Taken to an extreme, network effects may create *winner-takes-all markets*, in which one company emerges as dominant and the other firms, which may have superior products or technologies, must abandon the market (Shapiro and Varian, 1998; Liebowitz, 2002).

The key implications of network effects for competition and strategy have resulted in an increasing interest among academics and practitioners in measuring and quantifying network effects in a variety of business. However, there is a clear mismatch between the development of the theoretical and empirical work in the network effects literature (Farrell and Klemeperer, 2007). From a theoretical point of view, research on network effects has mainly aimed at analysing their impact on market competition using analytic and game-theory models. Overall, these studies provide support for the notion that, when network effects are present, the firm's installed customer base can be considered a key asset to gain abnormal returns (Shankar and Bayus, 2003; Shapiro and Varian, 1998). Moreover, as noted previously, network effects may create *winner-takes-all* outcomes (Arthur, 1996).

From an empirical perspective, most studies provide support for the theoretical postulates that (1) network effects are very important in information technology markets and (2) they significantly influence market competition. For instance, Gandal (1994) analyzes the spreadsheet market and concludes that network effects are very important and that there is a prize for compatibility with Lotus. In the same market, Brynjolfsson and Kemerer (1996) find that the price of the market leader in the late 1980s was much influenced by the size of its installed customer base. More recently, Dranove and Gandal (2003) study the battle between DVD and DIVX and show that the preannouncement of a competing technology (DIVX) delayed the adoption of DVD by many consumers. Srinivasan, Lilien and Rangaswamy (2004), using a sample of 63 different office products and consumer durables, show that, in technologically intense products, increases in network externalities are associated with increased survival duration. Overall, these studies offer empirical evidence of the significant effect of network effects on individual user behavior which, aggregated across the entire population of the market, determine the level of competition in the market.

2.2 Types of network effects

The attention devoted to network effects in the literature has also resulted in an increasing interest in identifying different types of network effects as a prior step to understanding their different consequences on the marketplace. The literature has traditionally classified network effects into two types: (1) direct or pure network effects and (2) indirect or market-mediated network effects.

Direct network effects are present when "adoption by different users is complementary, so that each user's adoption payoff, and his incentive to adopt, increases as more others adopt" (Farrell and Klemperer, 2007: 1974). This type of network effects is easily understood when we think of examples such as the e-mail, fax or telecommunications. Here, the technology has no value in isolation. It can only produce utility to consumers when other users adopt the technology. More formally, each user's utility function increases with the number of additional users of the technology/product, so the larger the installed user network, the higher the utility derived from the product (Katz and Shapiro, 1985). This is because a large installed base allows the firm to offer more benefits to potential customers compared to companies with smaller customer bases (Farrell and Klemperer, 2007). These benefits can take the form of reduced uncertainty, compatibility, the transfer of technical and non-technical information between members of the network and the increased availability and quality of complements. The presence of these benefits encourages consumer adoption of the product (Gatignon and Robertson, 1985; Valente, 1995) and increases its utility over and above its stand-alone product performance.

Indirect network effects arise "through improved opportunities to trade with the other side of a market" (Farrell and Klemperer, 2007: 1974). They imply that customer utility from the primary product (i.e. the hardware) increases as more complements become available (i.e. software). In turn, this availability of complementary products depends on the installed user network of the primary product (Stremersch et al. 2007). Prior research has typically referred to the primary product, such as a television set, a mobile handset or a DVD player, as "hardware", and to the product that complements the primary product, such as television programs, mobile phone applications and music or movies, as "software". In the presence of indirect network effects, we can observe the "chicken-and-egg" paradox (Katz and Shapiro, 1994). This happens when consumers wait to adopt the primary product until enough complements are available. At the same time, manufacturers of complements delay releasing new complements until enough consumers have adopted the product. This effect can be dangerous for the diffusion of the technology as it may delay consumer adoption and reduce the interest of manufacturers in designing and releasing new complementary products. Stremersch et al. (2007) have recently carried out an empirical application in nine indirect network effects markets with the aim of resolving the paradox and they found that hardware sales lead to software availability, and not the other way around.

An important characteristic of this first classification of network effects into direct and indirect network effects is that it gives the same weight (importance) to all the customers in the utility function. This implies that adding one customer to the market equally increases the utility of all the users who are already in the network. However, researchers have recently recognized that network effects may be localized (Birke and Swann, 2006; Farrell and Klemperer, 2007; Maicas, Polo and Sese, 2009a). For instance, a mobile phone user gains more when her friends adopt than when a stranger does. In other words, there are different densities in the network (Ahuja, 2000) which should be taken into account to analyze user behavior. Personal or local network effects, microexternalities and strong ties are the terms used by prior research to refer to this type of network effects (Swann, 2002; Suárez, 2005; Birke and Swann, 2006).

Personal network effects explicitly take into account the differences that exist in the contribution of each network member to the utility function (Birke and Swann 2006). They refer to the utility that an individual obtains from the adoption by a given individual. This utility may be positive, neutral or negative, depending on the person that joins the network. For example, a mobile user may derive a high utility when her boyfriend or her brother joins the network, while this utility becomes zero when a stranger does. Thus, network benefits are not homogeneous (Ahuja, 2000); consumers find more benefits from interacting with their social subset than with the rest of the installed base. This quotation from the Economist in 2007 perfectly illustrates this notion:

"Although mobile phones make it easier to keep in regular touch with a wide group of friends, for example, it turns out that a typical user spends 80% of his or her time communicating with just four other people"

Finally, recent research has introduced a new concept, marginal network effects (Farrell and Klemperer, 2007), which refers to the increase in the incentives of potential users to adopt the technology as network size grows. It means that a firm with a larger network will also have a better market position, not only because of its current market share, but also because the probability of future dominance is higher. Current users will have incentives to stay within the firm's network and the incentives of potential users will also increase.

3. Network effects in the mobile telecommunications industry

Mobile communications is a paradigmatic example of an industry where network effects drive market competition. Srinivasan, Lilien and Rangaswamy (2004) suggest that mobile communications present a high degree of network effects: they rate among the highest in a list of 45 goods/services that are believed to be intensive in network effects. Previous empirical evidence in mobile communications (Kim and Kwon 2003; Birke and Swann 2006) shows clear signs of network effects, even when the networks are perfectly compatible. This is the case of the European context in which the technological standard is the same for all the operators (Gandal 2002). Birke and Swann (2006) explore the role of network effects in users' choice of mobile service providers in the UK. They find that (1) there are direct network effects even in the absence of price differences between on-net and off net calls and that (2) individual choice is heavily influenced by the choices made in the individual's social network. Doganoglu and Grzybowski (2007) analyze demand for mobile telecommunications in Germany and their results suggest that network effects play a significant role in the diffusion of mobile services. Grajek (2010) acknowledges that, in spite of being crucial to the understanding of mobile communications, the empirical literature has almost completely ignored the impact of network effects on the level of industry competition. The author estimates price elasticities in the Polish market and concludes that, if we do not consider network effects, the elasticity of demand could be substantially overestimated. Fuentelsaz, Maicas and Polo (2010) study market competition in the European mobile communications industry and find that the level of network effects shows significant differences among countries with respect to the level of competition. Maicas, Polo and Sese (2009a) investigate the role of local network effects in explaining customer choice in the Spanish mobile telecommunications industry. Their results reveal that local or personal network effects play a key role in determining mobile users' choice of supplier. In other words, the probability that a user selects a mobile service provider increases with the number of members of his social network already subscribed to that provider.

Following the classification of network effects suggested in the prior section, we can identify different types of network effects in mobile communications.

Direct network effects are present in the mobile communications industry even when the networks are perfectly compatible (Kim and Kwon, 2003; Birke and Swann, 2006). In this market, pure network effects mainly arise due to the differences between on-net and off-net tariffs (Laffont, Rey and Tirole, 1998). Operators charge different prices depending on whether the call made by the user is directed to a member of the same operator (low or on-net tariff) or is made to a member outside the operator (high or off-net tariff). These strategies aim to encourage existing customers to stay with the provider and to attract new customers because the larger the user network, the lower the average cost of the calls for their customers. At the same time, these strategies increase consumer switching costs (Maicas and Sese 2008) because leaving the company implies losing the benefits derived from making cheaper calls to the members of the same network, which reduces competition in the market and confers market power to the firms.

Indirect network effects are associated with the increase in utility derived from the availability of complements to the primary product or service (Stremersch et al. 2007). In mobile communications, as more and more individuals are interested in the technology, software manufacturers will also have higher incentives to design and release new applications, features or devices that will increase the utility of using the primary product. For example, as more and more individuals become mobile users, handset manufacturers were encouraged to introduce new devices with more complex features and with a wider range of options (front camera, GPS technology, Bluetooth technology) to increase the utility of the mobile experience.

Similar to pure network effects, in the mobile telecommunications industry, personal network effects also come mainly from pricing strategies implemented by mobile operators. In this case, the benefit for the user is not associated with the calls directed to the members of the same operator, but with the calls made to a particular group of users that constitutes her *social subset* (family, friends,...). The underlying logic is that, although all users in a network are a potential source of network effects, some users matter more than others (Birke and Swann, 2006). Thus, network benefits are not homogeneous (Ahuja, 2000) but consumers find more benefits from interacting with their *social subset* than with the rest of the installed base. Consequently, customer choice behavior will be more influenced by her *social subset* than by the rest of the users in the network. Maicas, Polo and Sese (2009a) studied the importance of personal network effects in customer choice of mobile carrier. Their results reveal that personal network effects play a key role in determining mobile users' choice: the probability that a customer selects a mobile phone company increases with the number of members of her social network already subscribed to that firm. In addition, they also acknowledge that the influence of personal network effects on choice of mobile supplier differs among consumers and, therefore, they investigate the drivers of each customer's sensitivity to personal network effects. They find that relationship characteristics (e.g. relationship duration, service usage, cross-buying behavior) are important drivers of local network effects. Users with an intense service usage, who purchase few services from the firm and who have recently joined the company are more sensitive to local network effects. The results also show that more sophisticated users have a higher valuation of local network effects.

4. Conclusion and managerial implications

In mobile communications, the user network is a critical strategic asset for assessing the firm's current and future competitive position. This is because there are strong interdependencies within the market and customers derive utility, not only from the product or service itself, but, even more importantly, from the network of users surrounding these products (Frels, Shervani, and Srivastava 2003). In today's competitive environment, one of the greatest challenges that practitioners face in mobile communications is to manage the user network optimally in order to maximize current and future profitability. In this section, based on prior research and the observation of best firm practices, we discuss some strategies that may help companies in leveraging the effect of network effects to increase firm profitability and value.

First, as we already noted, firms can implement strategies that both make existing consumers more willing to stay in the network and new customers more predisposed to join the network. This can be done, for example, by price discriminating against customers who call to other networks (off-net calls) as these calls are significantly more expensive than those made to members of the same network. Alternatively, firms can use additional incentives (e.g. promotions, rewards) to encourage customers to join the network. This increase in the size of the network will increase the probability that other users join the network in the future (marginal network effects).

In addition to increasing pure network effects, firms should pay attention to the social networks, as customers' choice has been found to be mainly driven by the behavior of their social group. Thus, in addition to implementing general strategies to promote pure network effects, firms should also develop strategies to increase personal network effects in an attempt to build a big network that is the result of many smaller social networks. To achieve this goal, the firm may, for example, offer price discounts to groups of friends and family if they all belong to the same network. The importance of these strategies that promote the building of social networks inside the company's installed base comes from empirical evidence showing that, according to the Pareto principle (or the 80-20 rule), 80% of the calls are directed to less than 20% of the people that belong to the social network.

But firms should not only rely on their strategies to manage network effects in a way that increases firm profitability. Prior research has shown that customers are very heterogeneous and that they have different sensibilities to network effects (Maicas, Polo and Sese 2009a). Thus, firms should first understand customer behavior with respect to network effects before they can design an optimal strategy to leverage the impact of network effects on firm value. For example, customer relationship characteristics, such as the mobile plan selected – prepaid vs. postpaid, minutes of use, length of the relationship, purchase of additional services or products and mobile phone bills, are stored in firms' databases and can be used to identify the profile of customers who are more sensitive to network effects. Customer demographics or attitudinal variables can also help the company better understand the responses of customers to network effects.

In addition to firm strategies and customer behavior, the degree of competition in a market is another force that can moderate the impact of network effects on firm performance. When competition is intense, gaining market share is critical for survival because the presence of network effects increases the probability that individuals will join the firm with the largest user network. This is why we see price wars and huge investments by firms directed at expanding the customer base. Once a firm has built a large market share, it becomes one of the most critical assets of the firm. In addition to the direct benefits that the company can obtain from the large number of users of its products, the company benefits from higher acquisition rates, higher barriers to entry (it is very difficult for competitors to enter a market with network effects dominated by companies with large market shares), lower price sensitivity and increased market power.

Finally, as a result of the consequences of network effects on marketing competition, we should also consider the role played by regulatory authorities. As we have already acknowledged, the presence of network effects confers market power on firms with high market shares, allowing them to charge higher prices, reduce product or service quality, create entry barriers and, as a consequence of all this, obtain abnormal returns (Klemperer, 1987, 1995; Farrell and Klemperer, 2007). In a context of low competition in mobile telecommunications, some national regulatory authorities have introduced measures directed at promoting competition, including Mobile Number Portability (Maicas, Polo and Sese 2009b). Although these policies are also motivated by the high level of switching costs in the market, regulatory authorities can alter the degree of network effects and their effectiveness in increasing firm performance.

5. Acknowledgments

The authors appreciate the financial support received from the Spanish Ministry of Education and Science and FEDER (ECO2008-04129; SEC2008-04704) and the Regional Government of Aragón (PI 138/08; S09-PM062). F. Javier Sese gratefully acknowledges financial aid from the Spanish government (Ministry of Education) through the program: "Estancias de movilidad en el extranjero "José Castillejo" para jóvenes doctores" (JC2010-0305) and the hospitality of the Institute of Marketing, University of Muenster (Germany). This is a fully collaborative work where both authors have contributed equally. Errors remain our sole responsibility.

6. References

Ahuja, G. (2000). "Collaboration Networks, Structural Holes, and Innovation: A Longitudinal Study". *Administrative Science Quarterly*, 45 (3), 425–455.

Birke, D. and Swann, P. (2006). "Network Effects and the Choice of Mobile Phone Operator". *Journal of Evolutionary Economics*, 16 (1-2), 65–84.

Brynjolfsson, E. and Kemerer, C. (1996). "Network Externalities in Microcomputer Software: An Econometric Analysis of the Spreadsheet Market". *Management Science*, 42 (12), 1627–2647.

Doganoglu, T. and Gryzbowski, L. (2007). "Estimating Network Effects in the Mobile Telephony in Germany". *Information Economics and Policy*, 19 (1), 65–79.

Dranove, D. and Gandal, N. (2003). "The DVD-vs.-DIVX Standard War: Empirical Evidence of Network Effects and Preannouncement Effects". *Journal of Economics and Management Strategy*, 12 (3), 363–386.

Farrell, J. and Klemperer, P. (2007). *Coordination and Lock-In: Competition with Switching Costs and Network Effects*. In M. Armstrong, and R. Porter ed., *Handbook of Industrial Organization*, Volume 3, Elsevier.

Farrell, J. and Saloner, G. (1985). "Standardization, Compatibility, and Innovation". *RAND Journal of Economics*, 16 (1), 70–83.

Frels, J.K., Shervani, T. and Srivastava, R.K. (2003), "The Integrated Networks Model: Explaining Resource Allocations in Network Markets". *Journal of Marketing*, 67 (January), 29–45.

Fuentelsaz, L., Maicas, J.P. and Polo, Y. (2010). "Switching Costs, Network Effects, and Competition in the European Mobile Telecommunications Industry". *Information Systems Research*, forthcoming.

Gandal, N. (1994). "Hedonic Price Indexes for Spreadsheets and an Empirical Test of Network Externalities". *RAND Journal of Economics*, 25 (1), 160–170.

Gatignon, H. and Robertson, T.S. (1985). "A Propositional Inventory for New Diffusion Research". *Journal of Consumer Research*, 11 (4), 849–867.

Grajek, M. (2010). "Estimating Network Effects and Compatibility: Evidence from the Polish mobile market". Information Economics and Policy, 22(2), 130-143.

Greenspan, Alan (2000), "The Revolution in Information Technology," Conference on the New Economy, Boston College, Boston, Massachusetts.

Katz, M. and Shapiro, C. (1994). "Systems Competition and Network Effects". *Journal of Economic Perspectives*, 8 (2), 93–115.

Katz, M. and Shapiro, C. (1985). "Network Externalities, Competition, and Compatibility". *The American Economic Review*, 75 (3), 424–440.

Kim, H. and Kwon, N. (2003). "The Advantage of Network Size in Acquiring New Subscribers: A Condicional Logit Análisis of the Korean Telephony Market". *Information Economics and Policy*, 15 (1), 17–33.

Klemperer, P. (1987). "Markets with Consumer Switching Costs". *The Quarterly Journal of Economics*, 102 (2), 375-394.

Klemperer, P. (1995). "Competition when Consumers Have Switching Costs: An Overview with Applications to Industrial Organization, Macroeconomics, and International Trade". *Review of Economic Studies*, 62 (4), 515–539.

Laffont, J.J., Rey, P. and Tirole, J. (1998). "Network Competition: II. Price Discrimination". *RAND Journal of Economics*, 29 (1), 38–56.

Liebowitz, S.J. (2002). *Re-Thinking the Network Economy*. Amacom.

Maicas, J.P., Polo, Y. and Sese, F.J. (2009a). "The Role of (Personal) Network Effects and Switching Costs in Determining Mobile Users' Choice". *Journal of Information Technology*, 24 (2), 160–171.

Maicas, J.P., Polo, Y. and Sese, F.J. (2009b). "Reducing the Level of Switching Costs in Mobile Communications: The Case of Mobile Number Portability". *Telecommunications Policy*, 33 (9), 544–554.

Maicas, J.P. and Sese, F.J. (2008). "Análisis de la Intensidad de los Costes de Cambio en la Industria de la Telefonía Móvil". *Cuadernos de Economía y Dirección de la Empresa*, 35, 27–56.

McIntyre, D.P. and Subramaniam, M. (2009). "Strategy in Network Industries: A Review and Research Agenda". *Journal of Management*, 35 (6), 1494–1517.

Shankar, V. and Bayus, B.L. (2003). "Network Effects and Competition: An Empirical Analysis of the Home Video Game Industry". *Strategic Management Journal*, 24 (4), 375–384.

Shapiro, C. and Varian, H.R. (1998). *Information Rules: A Strategic Guide to the Network Economy*. Boston, M. A.: Harvard Business School Press.

Srinivasan, R., Lilien, G.L and Rangaswamy, A. (2004). "First In, First Out? The Effects of Network Externalities on Pioneer Survival". *Journal of Marketing*, 68 (January), 41–58.

Stremersch, S., Tellis, G.J., Franses, P.H. and Binken, J.L.G. (2007). "Indirect Network Effects in New Product Growth". *Journal of Marketing*, 71 (July), 52–74.

Suárez, F. (2005). "Network Effects Revisited: The Role of Strong Ties in Technology Selection". *Academy of Management Journal*, 48 (4), 710–720.

Swann, P. (2002). "The Functional Form of Network Effects". *Information Economics and Policy*, 14 (3), 417–429.

Tanriverdi, H., and Lee, C.-H. (2008). "Within-Industry Diversification and Firm Performance in the Presence of Network Externalities: Evidence from the Software Industry". *The Academy of Management Journal*, 51 (2), 381–397.

Valente, T.W. (1995). *Network Models of the Diffusion of Innovations*. Cresskill, NJ: Hampton Press.

Wang, Q., Chen, Y. and Xie, J. (2010). "Survival in Markets with Network Effects: Product Compatibility and Order-of-Entry Effects". *Journal of Marketing*, 74 (July), 1–14.

Part 2

Technical Oriented

8

Multiband and Wideband Antennas for Mobile Communication Systems

Mustafa Secmen
Electrical and Electronics Engineering Department, Yasar University
Turkey

1. Introduction

The popularity of mobile communication systems has increased remarkably during the last decade and the market demand still continues to increase. As a fundamental part of these systems, antenna is one of the most important design issues in modern mobile communication units. Although there are several similar definitions, an antenna can be mainly described as a device, which transforms the electromagnetic waves in an antenna to radiating waves in an unbounded medium such as air in transmitting mode and vice versa in receiving mode. Because antennas are dependent on frequency, they are designed to operate for certain frequency bands.

The rapid growth of mobile communication systems has forced to the use of novel antennas for base and mobile station applications (mobile phone, notebook computer, personal digital assistants (PDA), etc.). Earlier, mobile systems were designed to operate for one of the frequency bands of 2G (second generation) systems, which are Digital Cellular System (DCS), Personal Communications Service (PCS) and Global System for Mobile Communications (GSM) networks. Currently, many mobile communication systems use several frequency bands such as GSM 900/1800/1900 bands (890-960 MHz and 1710-1990 MHz); Universal Mobile Telecommunication Systems (UMTS) and UMTS 3G expansion bands (1900-2200 MHz and 2500-2700 MHz); and Wi-Fi (Wireless Fidelity)/Wireless Local Area Networks (WLAN) bands (2400-2500 MHz and 5100-5800 MHz) where the list of frequently used frequency bands is given in Table 1 (Best, 2008).

Conventionally, because a single antenna can not operate at all of these frequency bands of mobile communication, multiple different antennas covering these bands separately should be used. However, usage of many antennas is usually limited by the volume and cost constraints of the applications. Therefore, multiband and wideband antennas are essential to provide multifunctional operations for mobile communication. A multiband antenna in a mobile communication system can be defined as the antenna operating at distinct frequency bands, but not at the intermediate frequencies between bands. For example, a triple band antenna for GSM 900/1800/1900 bands can cover the frequency bands 890-960 MHz and 1710-1990 MHz (Ali et al., 2003); however, it does not operate properly at the frequencies such as 1200 MHz or 2500 MHz. On the other hand, a wideband antenna operates at every frequency points within a given frequency band. For example, a wideband antenna covering UMTS, extended UMTS and WLAN 2400 bands functions at every frequency points between 1900 and 2700 MHz (Secmen & Hizal, 2010; Caso et al., 2010). At this point, the readers may wonder what "the antenna operates at this frequency properly" means. This chapter follows in the brief explanation of this question by describing crucial antenna parameters for mobile communication systems. Afterwards, this chapter provides types and examples of multiband antennas used in mobile communication. Finally, wideband antennas are investigated in the last section of this chapter.

Wireless Application	Alternate Description(s)	Frequency Band (MHz)
GSM 850	AMPS (Advanced Mobile Phone System)	824-894
GSM 900		890-960
GPS (Global Positioning System)		1565-1585
GSM 1800	DCS 1800	1710-1885
GSM 1900	PCS 1900; CDMA 1900 (Code Division Multiple Access)	1850-1990
UMTS	W-CDMA (Wideband Code Division Multiple Access); IMT 2000 (International Mobile Telecommunication)	1885-2200
Extended UMTS	LTE 2600 (Long Term Evolution); WiMAX (Worldwide Interoperability for Microwave Access) 2500	2500-2690
Wi-Fi/WLAN (IEEE 802.11 b/g/n)	ISM 2450 (Industrial, Scientific and Medical)	2400-2484
Wi-Fi/WLAN (IEEE 802.11 y)		3650-3700
Wi-Fi/WLAN (IEEE 802.11 a/h/j)	HIPERLAN (High Performance Radio Local Area Network); U-NII (Unlicensed National Information Infrastructure)	5150-5825

Table 1. The frequency bands for mobile communication applications

2. Main antenna parameters for mobile communication systems

In antenna terminology, the frequency bandwidth of an antenna is generally characterized either with the lower and upper limits of frequency band (f_L and f_u) or the percentage (%) bandwidth for a center frequency, which is given as:

$$\%\ bandwidth = \frac{f_u - f_L}{f_c} \times 100 \tag{1}$$

where f_c is the center frequency of the band as the arithmetic mean of lower and upper frequency limits. The bandwidth of an antenna is defined as the frequency range which the performance of antenna satisfies specified standards of some antenna parameters (Balanis, 2005). Therefore, in order to operate properly at the specified frequency bandwidth, the antenna should meet the given standards of these parameters for all frequencies within the frequency bandwidth. Although there are many parameters for different antenna applications, only the important ones regarding to the performance standards for mobile communication systems are mentioned briefly here.

2.1 Input impedance

Depending on the impedance of the antenna and the line feeding the antenna, a certain fraction of transmitted power to the antenna reflects from antenna without radiation. This power fraction is usually described as the return loss (RL) (or sometimes called as mismatch loss) in decibel scale as

$$RL(dB) = -20\log|\Gamma| \tag{2}$$

where Γ is the reflection coefficient which is given by

$$\Gamma = \frac{Z_{ANT} - Z_0}{Z_{ANT} + Z_0} \tag{3}$$

where Z_{ANT} is the complex input impedance of the antenna and Z_0 is the impedance of feeding line. As the alternative way of describing the reflected power from the antenna, the term of voltage standing wave ratio (VSWR) is also used with a formal definition given by

$$VSWR = \frac{1+|\Gamma|}{1-|\Gamma|} \tag{4}$$

VSWR provides a more quantitative indication about mismatch between the antenna and feeding line impedances that *VSWR*= 1 indicates perfect matching.

Because the complex impedance of antenna is a function of frequency, both return loss and VSWR depend on the operating frequency. Thus, if the antenna operates at a given frequency bandwidth, the impedance of the antenna should satisfy application-specific criterion such as VSWR≤ 2 or equivalently *RL* ≥ 10 dB at all frequencies within the bandwidth. In base station systems, the constraint of VSWR≤ 2 (or sometimes shown as VSWR 2:1) is usually sufficient, which corresponds to about 10% reflected power from the antenna. On the other hand, mobile station antennas such as handheld antennas are typically designed to have VSWR≤ 3 for multiband systems due to very tight volume constraints (Rahmat-Samii et al., 2008).

2.2 Radiation pattern and beamwidth

An antenna radiation pattern is defined as a graphical representation of power distribution or field strength of the antenna as a function of space coordinates. These coordinates are usually selected as elevation (θ) and azimuth (ϕ) angles of spherical coordinate system. There are many types of representation of radiation pattern of an antenna. One of them is the three dimensional (3D) graph whose examples can be found in many antenna books (Balanis, 2005). However, the drawing of 3D graph is usually difficult and unnecessary due to the symmetry of antenna radiation pattern. Therefore, instead of 3D radiation pattern, a more comprehensive representation of radiation pattern called as polar plot is used. Polar plot is actually a planar cut from 3D radiation pattern as shown in Fig. 1(a). Same pattern can be presented in the rectangular plot, as shown in Fig. 1(b). Both patterns are normalized to the pattern's peak, which is pointed to $\theta = 0$ in this case and given in decibel scale.

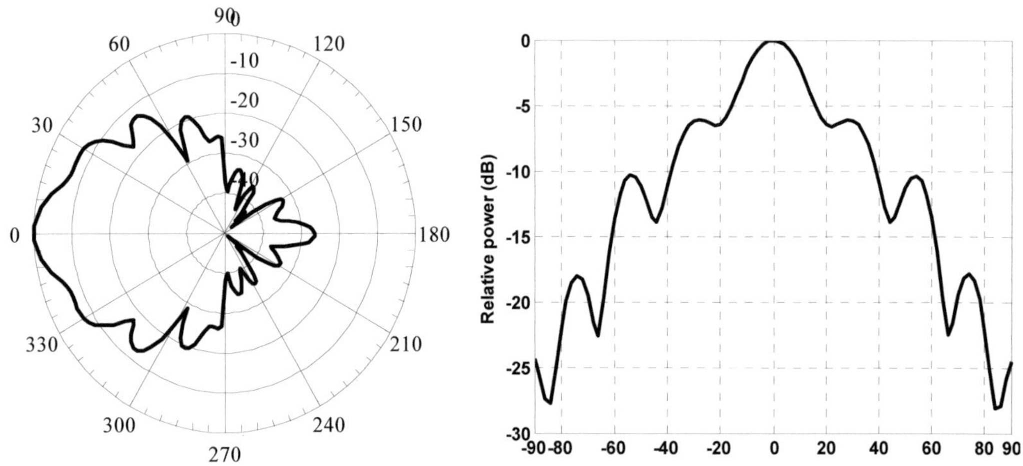

(a) (b)

Fig. 1. (a) Polar plot and (b) Rectangular plot representation of radiation pattern

In antenna terminology, planar cuts from 3D pattern are considered for two main planes, which are E-plane and H-plane for linearly polarized antennas. The E-plane is defined as the plane containing the electric field vector and the direction of maximum radiation; and H-plane is the plane containing the magnetic field vector and the direction of maximum radiation. Therefore, by representing plots of an antenna in both planes, which are orthogonal, power distribution of the antenna in whole space can be comphrehended well without drawing 3D pattern.

The beamwidth of the antenna is defined as the angular distance (width) between two half power points in the radiation patterns, where half power level is 3 dB below than maximum radiation power. The beamwidth parameter is usually expressed as "3 dB beamwidth" in the antenna applications for both E plane (elevation beamwidth) and H plane (azimuth beamwidth). This parameter can be also considered as effective angular width of the antenna that important portion of radiated antenna power is focused within this angular beamwidth. Theoretically, omnidirectional (equal radiation at all directions) pattern in azimuth plane and wide beamwidth in elevation plane are desired for mobile units. Practically, mobile handset antennas may have very wide beamwidth such as 180° in both planes. In indoor or outdoor base station applications, antennas having wide 3 dB beamwidth (90° or 120°) are preferred to provide sufficient angle coverage in azimuth plane; whereas, the elevation beamwidth of these antennas varies typically between 10° or 70° within the frequency bandwidth of the antenna. GSM systems with three-sector configuration typically use antennas having 3 dB beamwidth of 65° (Collins, 2009).

When radiation pattern of an antenna is handled, the front-to-back (F/B) ratio of antenna is also an important parameter in mobile communication applications. This parameter is roughly defined as the ratio of maximum radiated field in forward (mainlobe) direction (0°

in Fig. 1(a)) to the radiated field in the opposite (backlobe) direction (180° in Fig. 1(a)). This ratio is generally desired to be about 30 dB in outdoor base station applications in order to minimize the interference between back-to-back oriented antennas. On the other hand, the required F/B ratio for indoor applications can be low (Secmen & Hizal, 2010). In mobile phone antennas, the backlobe radiation is usually directly oriented to the head of a human body; therefore, this radiation level is desired to be as low as possible corresponding to high F/B ratio. In notebook computer antennas, the desired radiation pattern is omnidirectional; consequently, F/B ratio should be low that the antennas with F/B ratio of 0.5 dB can be employed by using symmetric patch antenna structures (Guterman et al., 2006).

2.3 Gain

The gain of an antenna is defined as the ratio of the power intensity radiated by the antenna in a given direction (usually in spherical coordinate angles θ and ϕ) divided by the intensity radiated by a lossless isotropic antenna, which radiates the power at all angles equally. In a mathematical form, it can be formulated as

$$gain = G(\theta,\phi) = 4\pi \frac{U(\theta,\phi)}{P_{in}} \tag{5}$$

where $U(\theta, \phi)$ is the radiation (power) intensity and P_{in} is total input (accepted) power of the antenna. In antenna applications, gain is usually considered as maximum gain taken in the direction of maximum radiation. Therefore, gain drops at most 3 dB below maximum gain within the beamwidths of the antenna. Gain requirements may vary according to different applications of mobile communication. For example, in outdoor base station applications, the standard gain requirement is generally between 10 and 20 dBi (dBi: gain in dB scale relative to isotropic antenna) within frequency bandwidth, which is usually achieved with array structures (Arai, 2002). For indoor mobile communication, moderate gain (5-7 dBi) is usually sufficient (Serra et al., 2007; Secmen & Hizal, 2010). However, the gain of the antenna may decrease even to 1 dBi within the designated frequency band for handset applications (Rahmat-Samii et al., 2008).

2.4 Polarization

The polarization of the antenna is roughly defined as the orientation of electric field vector of the radiated wave of the antenna with time. While the electric field in linearly polarized wave oscillates in either horizontal or vertical directions, it circulates around direction of propagation vector in circularly polarized wave. In order to transfer maximum power between transmitter and receiver antennas, both antennas should have same polarization. However, in general, the polarization of receiver antenna is not the same as the polarization of the incident wave radiated by transmitter antenna. Consequently, power transfer is reduced, which is called as polarization loss factor (*PLF*). Mathematically, this loss is expressed in decibel scale as (Balanis, 2005).

$$PLF(dB) = 20\log(\vec{\rho}_r \bullet \vec{\rho}_t) \tag{6}$$

where $\vec{\rho}_r$ and $\vec{\rho}_t$ are unit (polarization) vectors of receiver and transmitter antenna, respectively. Accordingly, when the case, where linearly polarized transmitter and receiver antennas are orthogonally oriented, is considered; no power is transferred theoretically between antennas. Therefore, a single linearly polarized antenna can not be used directly in mobile communication systems such as base station application that another linearly polarized receiver antenna, i.e. a mobile phone antenna, can be hold in any tilted position even orthogonal to base station antenna and this case results in zero transferred power. On the other hand, in circular polarization case, there exists no complete power loss (mismatch) that some portion of transmitted power is always transferred to linearly polarized receiver antenna for any spatial orientation. For this purpose, circular polarization is frequently used in mobile communication systems in order to prevent complete mismatch (Haapala et al., 1996; Wong et al., 2002). However, achieving circular polarization within wide frequency bandwidth is difficult; therefore, as compared to linearly polarized antennas, circularly polarized antennas in mobile communication systems have relatively narrow frequency bandwidth. Consequently, in order to optimize polarization mismatch and frequency bandwidth, dual-polarized antenna systems, which include either two orthogonal linearly polarized antennas (Secmen & Hizal, 2010) or an antenna excited by two orthogonal feeds (Guo et al., 2002), are commonly used in base station applications. Moreover, dual-polarized antennas can provide space-saving polarization diversity at the base station point to increase the performance of mobile systems that ±45° dual-polarized (slant-polarized) antennas are currently in almost universal use for base station systems (Caso et al., 2010).

2.5 Mutual coupling

When identical antenna elements are placed in an array or multiple different antennas are used, they interact with each other. This interaction between elements due to their close proximity is called mutual coupling, which affects the input impedance as well as the radiation pattern. It is noted previously that in base station applications, more than one similar antenna can be implemented to either acquire higher gain with array structures or at least provide dual-polarization with two antenna elements or feeds. Furthermore, in mobile station applications, even multiple different antennas can be used in a limited available space to provide multiband operation (Boyle & Massey, 2006). For these antenna systems, the mutual coupling is simply defined as the interference value between two antenna elements or feeds, which is desired to be as low as possible. Mathematically, in N element antenna system, the mutual coupling S_{ij} in between ith and jth antenna elements can be evaluated in decibel scale as

$$S_{ij}(dB) = 20\log\frac{b_i}{a_j}\bigg|_{a_k=0 \;\, for\;\, k\neq j} \tag{7}$$

where a_j is the amplitude of transmitted wave from jth antenna and b_i is the amplitude of received wave from ith antenna that transmitted waves on all other antennas except jth antenna are set to zero. In base station systems, the specification for mutual coupling between antenna elements is typically -20 dB (or 20 dB isolation) within the frequency bandwidth. The mutual coupling effect in these systems using polarization diversity (one antenna with two orthogonal feeds) is usually higher than the systems using spatial diversity (different antennas). As for mobile station applications such as mobile phone or notebook computer antennas, the mutual coupling requirement may increase up to -10 dB (Rahmat-Samii et al., 2008).

2.6 Cross polar discrimination

Most dual polarized antenna systems employed for polarization diversity purpose are demanded that each antenna port receives signals only from its designated linear polarization (co-polarization). However, unfortunately practical antennas also receive unwanted signals from orthogonal polarization called as cross polarization (X-polarization). Cross polar discrimination is the ratio of received co-polar signal level to cross polar signal level. In order to show the cross polar discrimination, radiation patterns (co-polar and cross polar) of an indoor mobile communication antenna are given in Fig. 2 for both principal planes (Secmen & Hizal, 2010). According to this figure, the cross polar discrimination values are approximately 30 dB in the boresight direction (90° in Fig. 2). However, as shown in the patterns, providing constant cross polarization discrimination within beamwidth is difficult that this value falls to 20 dB for 60° degrees in principal H-plane. Nevertheless, cross polar discrimination needed to provide polarization diversity is not large that typical cross polar discrimination requirement for the mobile communication systems is around 25 dB in the boresight direction and 10 dB at the edges of beamwidth (Collins, 2009).

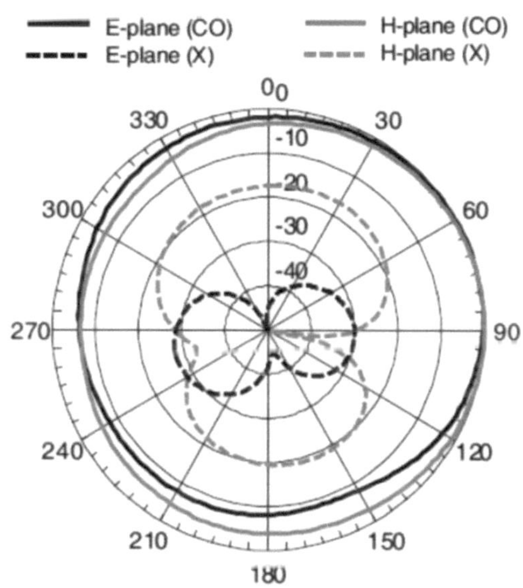

Fig. 2. The radiation patterns in both planes for an indoor base station antenna system where CO and X indicate co-polarization and cross-polarization (Secmen & Hizal, 2010)

2.7 Intermodulation

When the signals with multiple frequencies ($f_1, f_2, \ldots, f_n$) are received by a nonlinear device, intermodulation frequency terms (f_1-f_2, f_1+f_2, $2f_1$-f_2,...) are generated. Although an antenna is actually a linear device, it may slightly deviate from linearity when sufficiently high power is transmitted or received by the antenna. This nonlinearity is usually formed due to mechanical joints or nonlinear materials used in the antenna. The intermodulation level is crucial especially in base station applications that the intermodulation frequencies can degrade the performance of the communication system. The intermodulation frequency terms may easily fall inside the frequency band of interest. For example, two transmitted frequencies (f_1= 935 MHz and f_2 = 955 MHz) in frequency band of GSM 900 can generate 3rd order intermodulation term at the frequency, $2f_1$-f_2 = 915 MHz, which again falls into GSM 900 band. Therefore, the intermodulation levels are desired to be as low as possible that typical signal level for base station applications is between -180 dBc and -120 dBc (dBc: power in dB scale relative to carrier power). On other hand, when mobile station systems such as mobile phone or notebook computer are considered, the intermodulation issue is not so serious that the power handled in these systems is not as high as generating remarkable intermodulation frequency terms. Therefore, intermodulation terms are usually ignored in these applications.

2.8 Specific Absorption Rate (SAR)

For a mobile phone or notebook computer antenna located to the position, which is nearby to a human body, some portion of transmitted power is absorbed by the human body. The specific absorption rate (SAR) is basically defined as the absorbed power density at a particular point of the human body. SAR can be quantitatively expressed as (Huang & Boyle, 2008)

$$SAR = \frac{dP_{abs}}{\rho dV} = \frac{\sigma |E|^2}{2\rho} \tag{8}$$

where dP_{abs} is absorbed power within an infinitesimal volume of dV; E is the peak electric field strength within dV; ρ and σ are mass density and conductivity of the human body. SAR is important that certain regulations about SAR, which are based on the biological effects of thermal heating due to radiation, should be satisfied. The IEEE standard about SAR indicates that maximum allowed 1-g averaged maximum SAR is 1.6 W/kg and whole-body averaged peak SAR is 0.08 W/kg. 10-g averaged maximum SAR value is commonly used as 2 W/kg in Europe countries.

3. Multiband antennas for mobile communication

In order to realize multiband operation, a wide variety of antenna types, which uses different multiband techniques, is used. Fundamental multiband techniques will be explained in the following part of this section. Next, basic multiband antenna types designed for mobile communication systems will be given.

3.1 Multiband techniques

3.1.1 Higher order resonances

One of the basic ways of getting multiband operation is to utilize from higher order resonances. This principle is explained in Fig. 3 that a monopole antenna is often used with a length of $\lambda/4$ (Fig. 3(a)). For this case, the antenna resonates at f_o with electric field minimum at the feed. However, a similar condition of minimum electric field at the feed also exists when same antenna's length corresponds to $3\lambda/4$ (Fig. 3(b)). Therefore, the monopole antenna can also resonate at $3f_o$. Other higher resonances also exist at higher frequencies such as $5f_o$. Higher order resonances are used in many types of antennas such as dipoles, helices, patches and slots. In (Huang & Boyle, 2008), a normal mode helical antenna mounted on a typical mobile phone is given. According to the results, the antenna has the resonances at frequencies f_o and $2.6f_o$ that higher order resonances principle almost holds for this case.

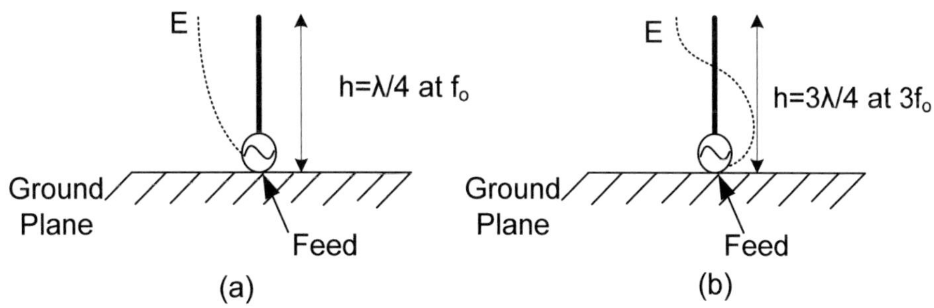

Fig. 3. (a) A monopole antenna resonating at f_o (b) Same antenna resonating at $3f_o$ (E is the electric field magnitude)

3.1.2 Multiple resonant structures

The most popular technique for obtaining multiband antenna system is the usage of multiple resonant structures. Here, two or more resonant structures, which are closely located in space or even co-located with a single feed, are used. This is illustrated in Fig. 4 for dual-band applications that the antennas in both cases have operation center frequencies f_1 and f_2. They are typical examples for corporate feed that two resonant structures are excited simultaneously. On the other hand, sometimes multiple resonant structures can be fed in series way as shown in Fig. 7(b) that the second resonant structure can be excited after the first structure is excited.

The multiple resonant structure technique is also frequently used in mobile communication systems to achieve multiband mobile antennas. For example, in (Haapala et al., 1996), dual frequency antenna systems for handsets are proposed. The designed structures are the combination of monopole and helical antennas as shown in Fig. 4(b) that multiple resonances at two different frequencies are acquired for dual-band operation at GSM 900 and 1800 bands.

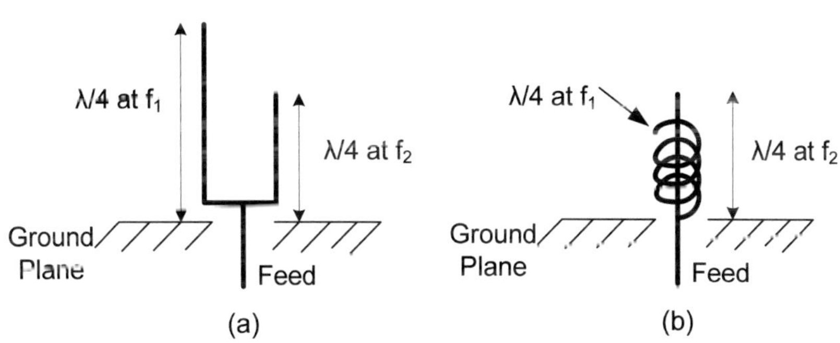

Fig. 4. (a) Two monopole antennas for dual-band operation (b) A helical antenna resonating at the frequency f_1 and a monopole antenna resonating at the frequency f_2 for dual-band operation

3.1.3 Parasitic resonators

Another method to obtain multiband characteristics is the implementation of parasitic resonators to the antenna system. In this technique, an extra parasitic element is added to the fed antenna for the operation at different frequency, but this element is not directly fed as in Yagi-Uda antenna (Balanis, 2005). It is parasitically coupled from near field of the antenna and resonates at another frequency. An example for this technique is given in Fig. 5 for a triple band application (Manteuffel et al., 2001). In this study, the antenna initially operates at GSM 900 and 1800 frequency bands without parasitic element. However, with the addition of the parasitic element, a triple band antenna for GSM 900, 1800 and 1900 frequency bands is realized.

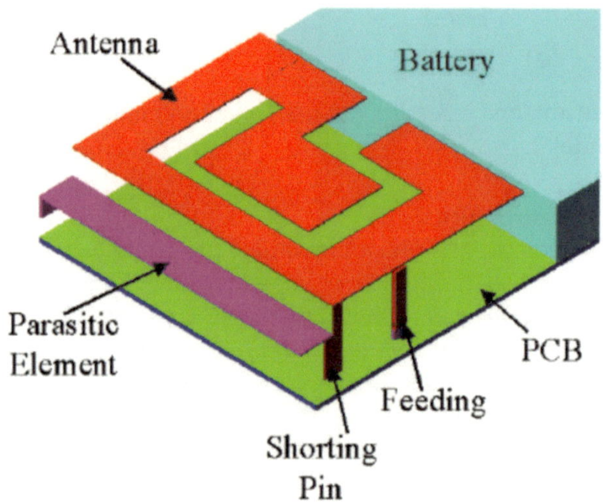

Fig. 5. A folded patch antenna with parasitic element for a triple band application (Manteuffel et al., 2001)

3.2 Monopole (whip) and helical antennas

One of the extensively used antennas in the earlier mobile communication systems is the monopole antenna and it is still used in applications such as United States CDMA networks. Monopole antennas have a very simple form containing a whip with height $\lambda/4$ above a ground plane, two of which are shown in Fig. 4(a) for possible dual-band operation. It has linear polarization characteristics and omnidirectional radiation pattern in H plane making this antenna an attractive choice especially for mobile unit applications. Several different forms of monopoles are given in Fig. 6 for a mobile handset system. However, since the size of ground plane greatly influences the radiation characteristics, it should be large in order to obtain ideal omnidirectional pattern. As a solution to this problem, sleeve dipole in Fig. 6(e) is an interesting antenna that it actually behaves as asymmetrically fed half-wave dipole with monopole like radiation. This antenna is used in private mobile handset systems such as emergency services. A dual-band sleeve dipole antenna operating at AMPS and GSM 1900 frequency bands can be found in (Ali et al., 1999) for a notebook computer application.
These forms of monopoles in Fig. 6 have generally large heights for mobile communication systems. In order to reduce the height of the monopoles, several different wire type antennas such as helical, wound coil or folded loop antennas are used for multiband

operations (Katsibas et al., 1998; Lee et al., 2000). Among these antennas, helical antenna, which is given in Fig. 4(b) in conjunction with a whip for dual-band operation, is the most popular. While axial mode helical antenna provides endfire radiation (parallel to the axis of the helix) pattern and circular polarization, normal-mode helical antenna gives linear polarization and similar radiation pattern with monopole antenna. Some of dual-band helical antennas used in mobile station systems are given in Fig. 7, where the first design uses two helical antennas with different radii and the second design uses antennas with different pitches (Wong, 2003). As another application of helical antenna in mobile communication systems, an intelligent quadrifilar helical antenna for satellite mobile communications is presented in (Leach, 2000).

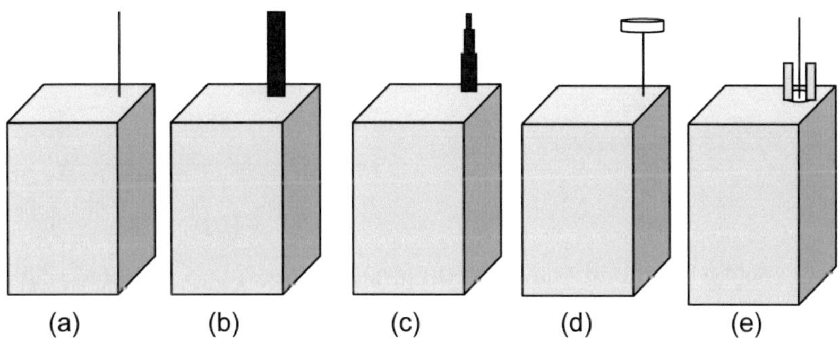

Fig. 6. (a) Wire monopole (b) strip monopole (c) retractable monopole (d) capacitive loaded monopole (e) sleeve dipole

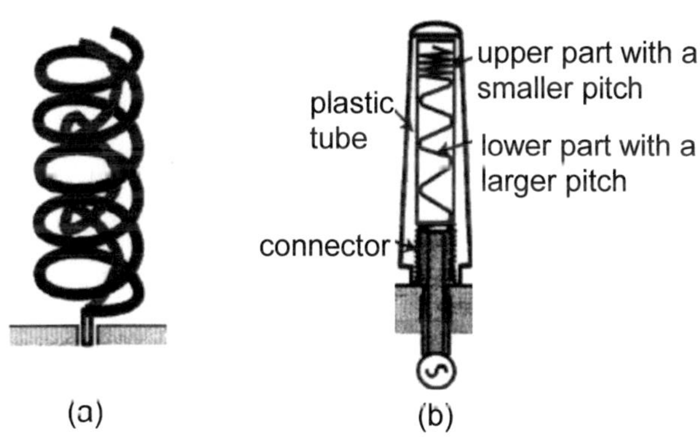

Fig. 7. (a) Two helical antennas with different radii (b) two helical antennas with different pitches (Wong, 2003)

In spite of their simple structures, all these monopole and helical antennas have still high dimensions especially for mobile station systems. Besides, these antennas can be considered as external antennas since they are usually mounted outside the mobile systems such as mobile handset, and external antennas are more sensitive to the position of nearby objects, for instance, head of a human (Rahmat-Samii et al., 2008). For these reasons, internal printed monopole antennas supplying lower profile and higher bandwidth for multiband operations are generally preferred. Some typical examples of internal printed monopole antenna for dual-band operation are given in Fig. 8 (Chen et al., 2001; Chen & Chen, 2004).

Both antennas in these studies provide return loss higher than 10 dB for GSM 1800 and WLAN 2400 bands.

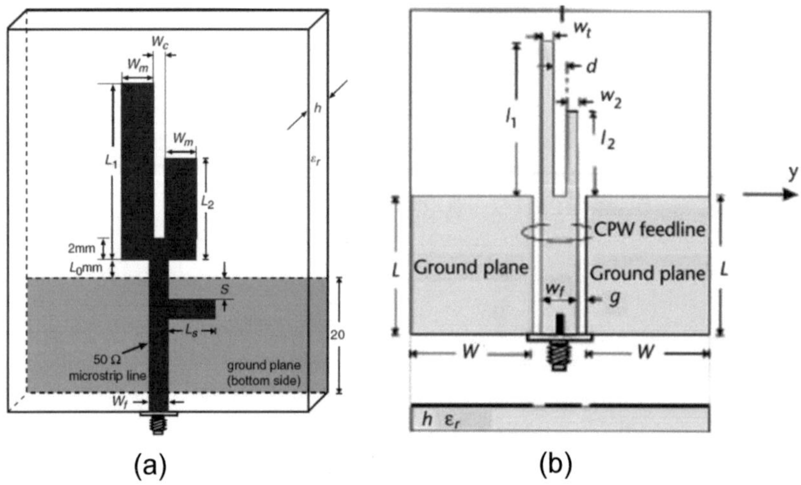

Fig. 8. (a) A microstrip fed dual-band printed monopole antenna (Chen et al., 2001) (b) A coplanar waveguide (CPW) fed dual-band printed monopole antenna (Chen & Chen, 2004)

3.3 Inverted F Antennas (IFA)

The classical monopole type antennas commonly require very large ground plane in order to have maximum radiation of the antenna parallel to the ground plane for principal E-plane. One possible solution for this problem can be to employ an antenna having maximum radiation towards normal to the ground plane; then, ground plane can be one side of the terminal. For this purpose, a quarter-wave monopole is first folded to form an inverted L antenna (ILA), and then it is modified to commonly known inverted F antenna (IFA) that the modification steps are given in Fig. 9 (Huang & Boyle, 2008).

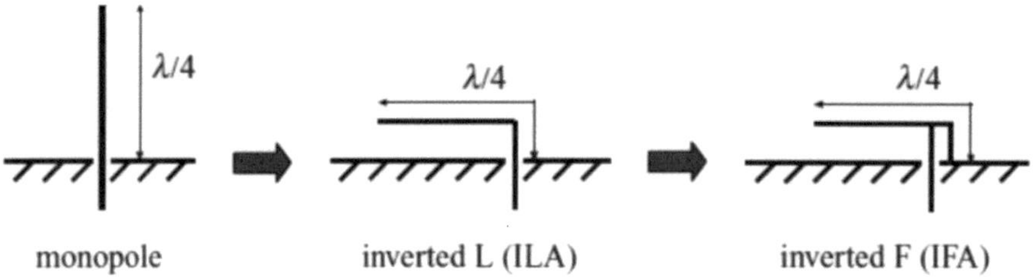

Fig. 9. Modification steps of IFA from monopole antenna (Huang & Boyle, 2008)

When IFA in this figure is investigated, with its image, the antenna appears as a two wire transmission line with a short circuit at the end. The IFA is widely used as an internal antenna especially in mobile handset and notebook computer applications. Many modifications have been made to IFA that IFAs operating at dual WLAN bands (2.4 and 5 GHz) have been proposed (Yeo et al., 2004). The printed forms of inverted L or F antennas are also very popular and widely used for multiband operations in mobile communication systems (Wong et al., 2003; Wang et al., 2007).

3.4 Planar Inverted F Antennas (PIFA)

In terms of mechanical reliability and elegancy, internal antennas are preferred in mobile units. The planar inverted F antenna (PIFA) is the most typical internal antenna especially for mobile handset applications that most of antennas in current mobile units are small, multiband and modified PIFAs. As shown in Fig. 10, a planar inverted F antenna is achieved by short circuiting radiating patch to the antenna's ground plane with a shorting pin or plate. Although PIFA seems to be modified from IFA by just replacing radiating wire in IFA with radiating patch, both antennas have different radiation mechanisms. PIFA can be actually considered as a modification of half-wavelength long microstrip patch antenna.

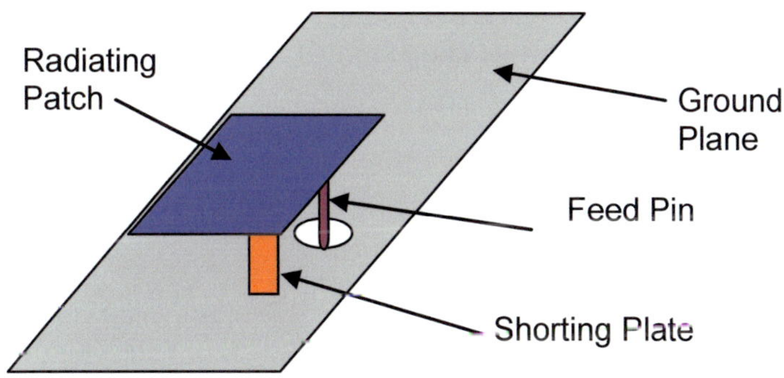

Fig. 10. Configuration of a typical planar inverted F antenna

Compared to the conventional external monopole antennas, PIFAs are less easily broken off. In addition, the ground plane in PIFA reduces the possible backward radiation, for instance, towards the head of a human, leading to lower SAR values. PIFA can resonate at a much smaller antenna size, which is desired and an attractive feature for mobile station applications. Furthermore, by cutting slots in the radiating patch, the resonance path can be modified; therefore, the antenna size can be further reduced. Besides, an intelligent design about the shape of the patch and the positions of the feed and shorting pins results in the existence of multiple resonance paths, causing multiband operations. A sample PIFA for dual-band operation is given in Fig. 11 (Boyle, 2008).

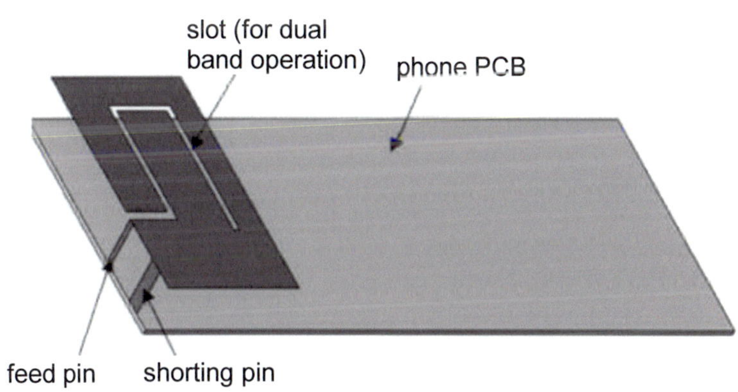

Fig. 11. A dual-band PIFA structure (Boyle, 2008)

The theory of this structure is investigated in detail in (Boyle, 2008). For the antenna in Fig. 11, it can be roughly explained that the inner part of this structure (slot) provides high frequency component of dual-band, whereas the outer part provides a low frequency component. Several PIFA antennas and their extended versions are reported for multiband operations including triple band (Manteghi & Rahmat-Samii, 2006), quad band (Ciais et al., 2004) and even six-band (Guo & Tan, 2004) for mobile communication systems.
In (Manteghi & Rahmat-Samii, 2006), a compact triple band PIFA operating in WLAN 2400 (2.4-2.5 GHz) band and two different UNII bands (5.15-5.35 GHz and 5.7-5.85 GHz) is presented. As shown in Fig. 12(a), three different resonance frequencies are generated by adding J-shaped slot and a quarter wavelength slot on the radiating patch. The fabricated two element antenna array is also given in Fig. 12(b) that total size for the antenna part is approximately 50 mm x 13 mm x 4 mm. The proposed antenna provides return loss higher than 10 dB for the mentioned bands.

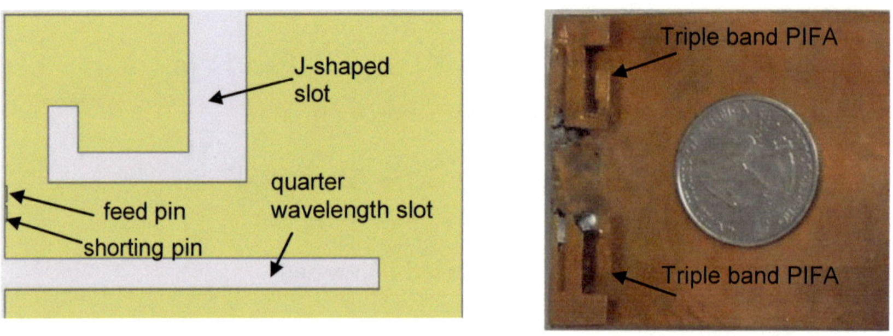

(a) (b)

Fig. 12. (a) A triple band PIFA (b) Array of two elements of triple band PIFA (Manteghi & Rahmat-Samii, 2006)

The paper presented in (Ciais et al., 2004) uses several multiband techniques such as multiple resonant structures (cutting slots) and parasitic resonators in order to implement a quad band PIFA. This antenna covers GSM 900 band by providing VSWR less than 2.5 and GSM 1800, 1900 and UMTS bands by providing VSWR less than 2. The antenna in (Guo & Tan, 2004) proposes a compact PIFA with a parasitic plate and folded stub for mobile handsets. This antenna covers GSM 900, 1800, 1900; GPS, UMTS and ISM2450 bands with return loss better than 6 dB and it occupies only 36 x 17 x 8 mm^3 total volume. There exist many different types of PIFA for mobile communication systems, which can be found in (Wong, 2003) for the readers interested in this antenna type.

3.5 Low profile antennas

The profile of a monopole (printed or planar antenna) or PIFA can be further reduced by some miniaturization techniques such as folded or meandered structures. The folded structures are mainly associated with bending, wrapping or folding of the monopole antennas into more complicated configurations such as S-shaped (Lui et al., 2004) or T-shaped structures (Chen et al., 2006). On other hand, a typical example for a single meandered structure is given for a printed monopole in Fig. 13 that meandered structures can be also combined with other configurations such as an inverted L-element in order to obtain multiband operation.

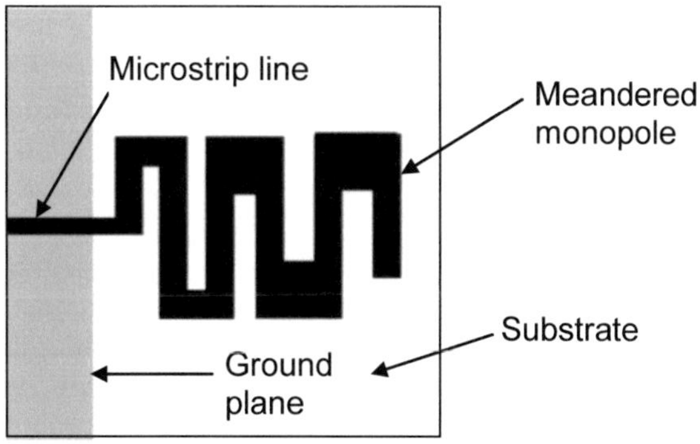

Fig. 13. A meandered printed monopole antenna

Low profile antennas have great importance due to its reduced size that for instance, this kind of low profile monopole in Fig. 13 is very suitable for integration within mobile phone applications as a built-in antenna. As the application of meandered type antenna, the antenna in (Ali et al., 2003) uses a driven meandered line element in addition to two parasitic structures for a triple band application of mobile phone handset. The antenna can be tuned to operate either in GSM 850, 900 and 1900 bands or GSM 850, 900 and 1800 bands by providing VSWR≤2.5 within the given frequency bands. In another realized antenna (Teng & Wong, 2002), a structure consisting of three meandered lines and wrapped into a compact rectangular box is presented for GSM 900, 1800 and 1900 frequency bands. The proposed antenna covers the required bandwidths of GSM 900, 1800 and 1900 by having VSWR less than 2.5 and gain ranging from 1.4 to 3.6 dBi. In a relatively recent study (Jing et al., 2006), a compact multiband meandered printed antenna is represented. The mentioned antenna, whose geometry is given in Fig. 14, has actually three meandered monopoles, which can be considered as three radiating elements or branches. The first (through the path a-b-c-d) and second (through the path a-b-c-e) branches provide resonances at GSM 900 band. The third branch (through the path a-b-f) and additional branch (g-g′) provide resonances at 2 GHz and WLAN 2400 band, respectively. According to the results, this antenna is found to operate in five different bands of GSM 900, 1800, 1900; UMTS 2000 and WLAN 2400 by giving VSWR less than 2.5 and gain between about 1 and 3.2 dBi.

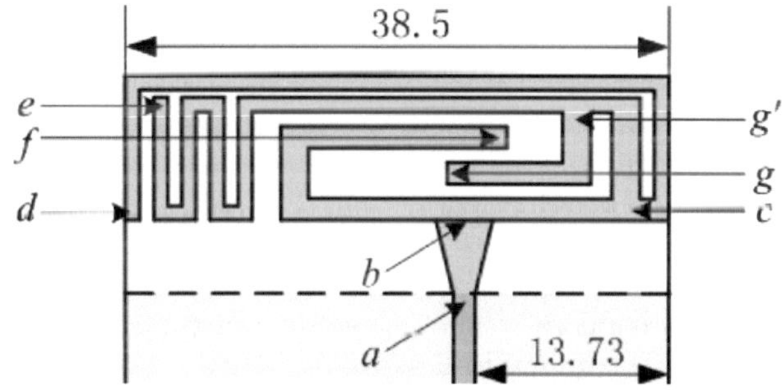

Fig. 14. A compact multiband meandered printed antenna (Jing et al., 2006)

As being another type for low profile antenna, folded structures have been reported in the literature. In the study in (Di Nallo & Faraone, 2005), a novel antenna structure, which can also be called as folded inverted conformal antenna (FICA), has significantly higher bandwidth than a dual-band PIFA operating in GSM 900 and 1800 bands. Besides, it provides resonance at the third band around 2 GHz, which is suitable for UMTS applications. A special design of folded planar monopole is presented in (Lin, 2004) that the proposed antenna can cover GSM 900, 1800 and 1900; UMTS and ISM 2450 frequency bands with constraint of VSWR≤2.

Chip antennas, which can be also included in very low profile antennas, are frequently used in mobile station units such as mobile handsets. The chip antenna is a compact surface mountable device consisting of a high permittivity substrate (such as ceramic) and conducting patterns printed or embedded on it. Low temperature cofired ceramic (LTCC) technology is usually used that the substrate is composed of multilayered thin sheets, and the conducting strips are printed and connected on these sheets via metal posts. The metallic path can take different forms of helix, meander or spiral (Wong, 2003). There are two major types of chip antennas. The first one has a ground plane printed on the bottom of the substrate; however, it has generally narrow bandwidth and low radiation efficiency. For this purpose, in the most of today's chip designs, the chip antenna does not have an underlying ground plane as shown in Fig. 15 (Moon & Park, 2003). The chip part of the presented antenna has total volume of 48 mm³ and operates at dual ISM bands (2.4 and 5.8 GHz) by providing VSWR≤2 within these frequency bands.

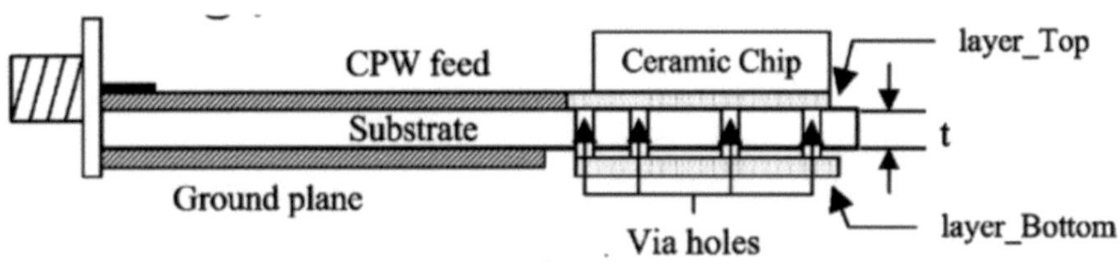

Fig. 15. The configuration of a dual-band chip antenna (Moon & Park, 2003)

4. Wideband antennas

In order to increase the bandwidth of an antenna, several methods such as using thick and low permittivity substrates, stacked and suspended structures, aperture or L-probe coupling, parasitic resonators and planar designs with different shapes (circular, triangular, etc.) can be considered. Wideband antennas normally occupy larger space than multiband antennas in the applications and the profile can be even higher with possible array configurations to obtain higher gain. Therefore, wideband antennas are mostly preferred in indoor or outdoor base station applications rather than mobile handset or notebook computer applications. Besides, while satisfying only VSWR (or return loss) requirement within the desired frequency bands is usually sufficient for mobile unit applications, additional criteria such as high gain and high isolation between the antenna elements should be satisfied for wideband antennas in base station applications. The commonly used wideband antennas in mobile communication systems are described as follows.

4.1 Microstrip patch antennas

Microstrip patch antenna is a well-known printed resonant structure consisting of a conducting patch, a substrate and a ground plane as shown in Fig. 16. Microstrip antenna's patch shape can be any continuous shape such as square, rectangular, circular, ring and elliptical, where rectangular patch is the most common.

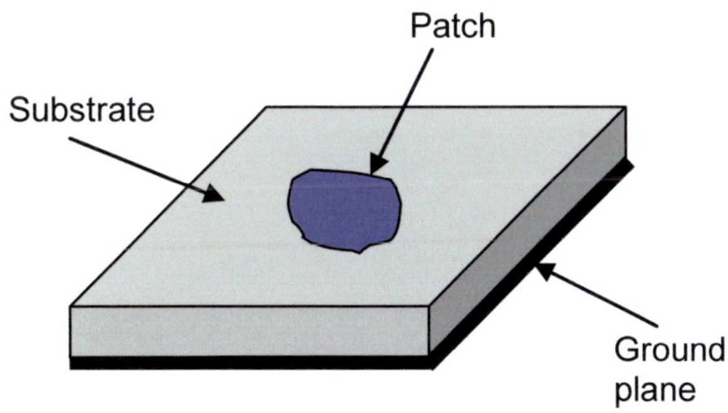

Fig. 16. Microstrip patch antenna configuration

This antenna is heavily preferred due to its low profile, lightweight, easy fabrication and being conformable to planar and nonplanar surfaces. With its original configuration, the antenna has narrow bandwidth, which is more suitable for multiband operations that some multiband patch antenna designs have been developed in literature (Chiou & Wong, 2003). However, by applying techniques such as using thick and low permittivity substrates, aperture coupling, stacked patched or cutting different shaped slots in the patch, its bandwidth can be widened, which makes them more convenient for base station applications. Wideband dual-polarized patch antennas have especially attracted much attention due to their ability of eliminating multipath fading. For example, the antenna in (Caso et al., 2010) proposes a dual-polarized microstrip antenna using both aperture coupling and stacked patch as wideband techniques. The geometry and fabricated view of the antenna are given in Fig. 17 that it operates between 1700 MHz and 2700 MHz (45 percent bandwidth), which includes GSM 1800, 1900; UMTS and extended UMTS; ISM frequency bands. Within the given bandwidth, the antenna provides return loss higher than 10 dB, isolation between ports higher than 22 dB and cross polar isolation higher than 20 dB. For a 2x1 array structure, the antenna gain is measured between 8 and 11 dBi in the entire band of interest, which is sufficient for most of the base station applications.

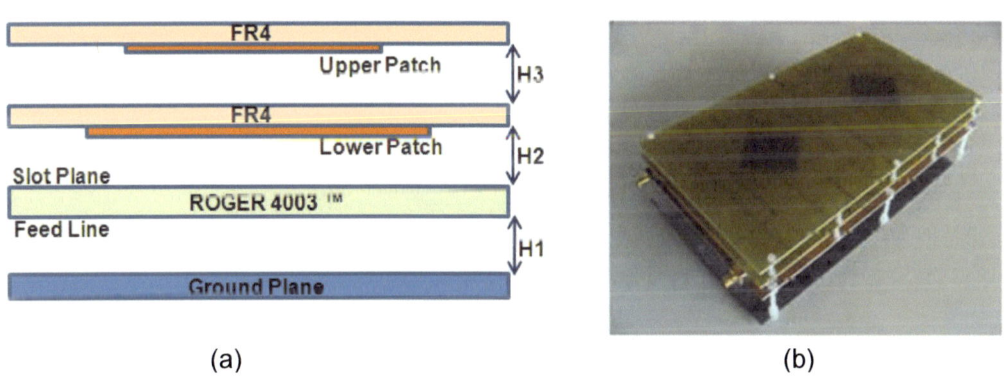

Fig. 17. (a) Stack-up view geometry of the single antenna element (b) Fabricated 2 x 1 prototype of the antenna (Caso et al., 2010)

4.2 Suspended plate antennas

A suspended plate antenna comprises from a thin plate conductor (patch) placed above a grounded low permittivity dielectric substrate (usually air) as shown in Fig. 18. It is usually fed by L or T shaped probes or planar strips in order to increase the bandwidth. These antennas have common advantages of easy fabrication, low cost and large bandwidth.

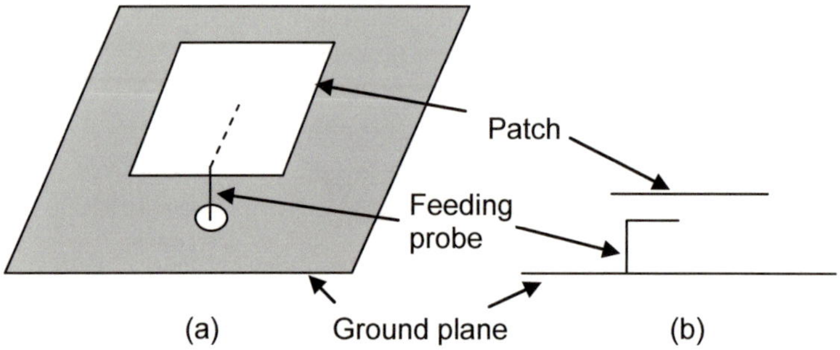

Fig. 18. (a) Isometric and (b) side views of the suspended plate antenna

There are many suspended plate antennas available for mobile communication systems. In (Secmen & Hizal, 2010), an inverted L-shape fed suspended plate antenna is designed for wideband indoor base station applications. The simulation and manufactured views of the proposed dual-polarized antenna are shown in Fig. 19. The antenna is initially fed with a

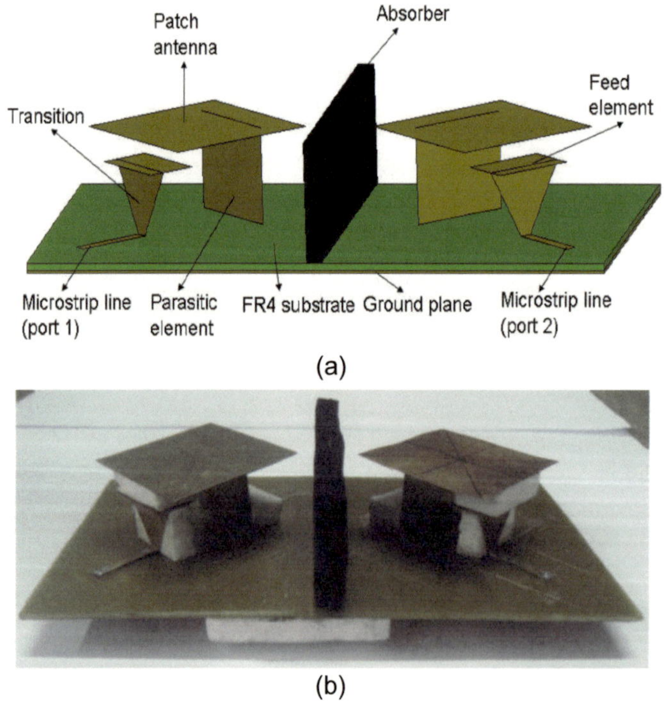

Fig. 19. (a) Simulation and (b) manufactured views of the suspended plate antenna in (Secmen & Hizal, 2010)

microstrip line instead of a probe, then with a bowtie transition, the incident power is transmitted to the suspended patch antenna via coupling from planar strip feed element. The antenna operates within the frequency bandwidth of 1900-2700 MHz (about 34 percent bandwidth) by performing return loss higher than 15 dB, isolation higher than 22 dB and cross polar discrimination in the boresight higher than 25 dB. Besides, the antenna has sufficiently wide 3-dB beamwidth values in both principle planes (minimum 66 degrees for E-plane and 125 degrees for H-plane); therefore, the proposed antenna can be used for indoor mobile communication applications.

4.3 Dielectric resonators

A dielectric resonator antenna (DRA) is mainly composed of a block of dielectric material on a conducting ground plane as shown in Fig. 20, where different geometrical shapes like hemisphere and rectangular instead of circular cylinder are available for DRA. DRA has some superiority over microstrip and printed antennas such as low profile, lightweight and small size. Besides, since there exists no radiating metal patch on the antenna, there is no conduction loss and this brings relatively lower loss compared with the microstrip

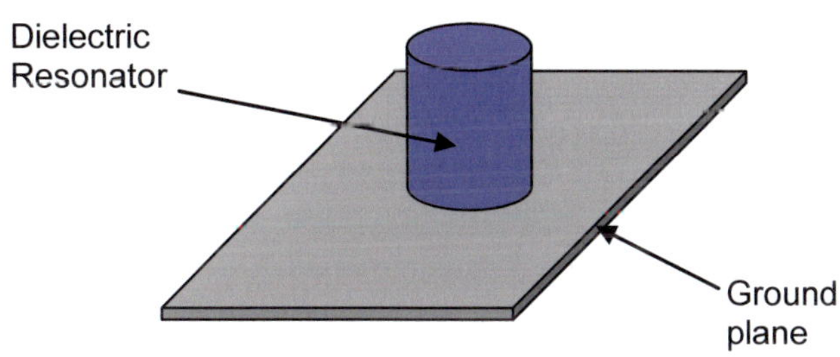

Fig. 20. A circular cylindrical dielectric resonator antenna on a ground plane

antenna especially for higher millimeter wave frequencies. Therefore, in mobile communication systems, it is usually used in WLAN applications, which have relatively higher frequencies (2400, 3600 or 5100 MHz) than GSM frequency bands. DRA also has the advantage of easy, simple and flexible excitation through the use of a coaxial probe, a microstrip line, an aperture coupling. For these reasons, DRA is increasingly popular and attractive to the researchers studying on mobile communication antennas. The resonance frequencies of a DRA are predominantly determined by its size and shape, and dielectric constant of the material (ε_r) that the dimensions can be significantly reduced by selecting materials with high dielectric constant. However, in order to maintain thermal stability, materials with dielectric constants lower than 30 are selected (i.e., ceramic with ε_r= 9.2). But, since the dimensions of the antenna can be still large at mobile communication frequency bands, many advanced designs have been developed in order to reduce the dimension with small ε_r values (Lan et al., 2003). DRAs are commonly used in wideband WLAN applications that many recent studies are available in the literature. For example, in (Mahender et al., 2010), a wideband U-shaped dielectric resonator antenna for WLAN application is given, which performs return loss higher than 10 dB and gain higher than 6.2 dBi for the frequency bandwidth 5.1-6 GHz including two different bands (5.15-5.35 GHz and 5.725-5.825 GHz) of a WLAN system. In a newly reported study (Brar & Sharma, 2011); a wideband aperture coupled pentagon shape DRA is presented for WiMAX (Worldwide Interoperability for Microwave Access) applications as shown in Fig. 21. The antenna operates from 2.55 GHz to

3.9 GHz (42 percent bandwidth) covering almost two WiMAX (2.5-2.7 GHz and 3.3-3.8 GHz) frequency bands. The antenna has return loss higher than 10 dB; gain higher than 3 dBi and moderation cross polarization levels within the given bandwidth.

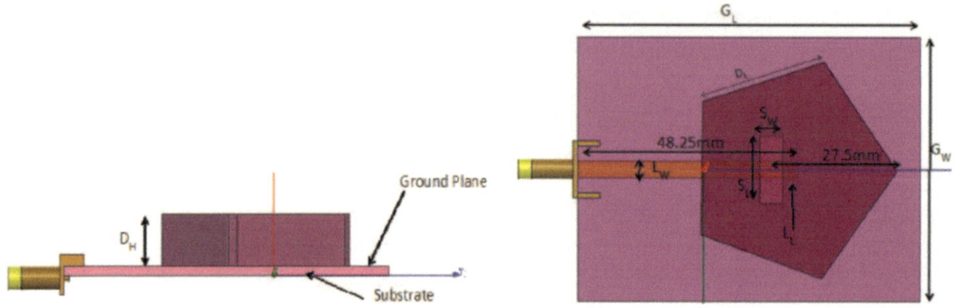

(a) (b)

Fig. 21. (a) Side view and (b) top view of the antenna presented in (Brar & Sharma, 2011)

4.4 Planar monopoles

One of the basic approaches to making an electrically small antenna wideband is to make it plump. Therefore, in order to increase the bandwidth of a simple whip type monopole antenna, the radiating wire element should be replaced by planar elements in order to be more convenient for wideband applications. These planar elements can be square, rectangular, trapezoidal, cross-plate or conical shapes. For example, in (Wong et al., 2005), a square planar monopole with three-branch feeding strip is introduced with a bandwidth of about 10 GHz (about 1.4-11.4 GHz) that these antennas are usually called as ultra-wideband (UWB) antennas. Although these planar monopoles are comparably larger than the other wideband antennas described above, they are mostly preferred in mobile communication systems due to its very wideband characteristics. As an example, a wideband dual-sleeve monopole antenna with cone shape is presented in (Zhang et al., 2011) for indoor base station applications. The structure of the antenna is shown in Fig. 22 that by a top-loading circular patch shorted to the ground plane through four shorting probes, a significant size reduction is achieved. The antenna's impedance bandwidth for VSWR≤ 2 is calculated to be from 730 to 3880 MHz, which covers GSM 900, 1800, 1900; UMTS and extended UMTS, WLAN 2400 and 3600 bands. Because the antenna's gain is considerably low (from 2.5 to 6.7 dBi within the bandwith), it is more suitable for indoor applications rather than outdoor applications, which needs higher gain.

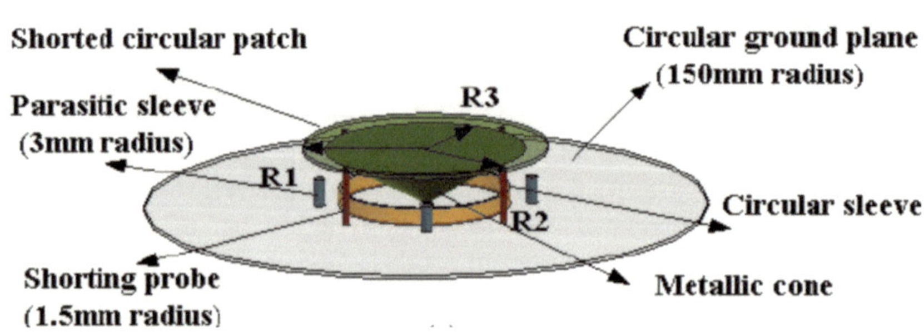

Fig. 22. The structure of the proposed antenna in (Zhang et al., 2011)

5. Conclusions

The explosive demand for mobile communication and information transfer using personal devices such as mobile phone or notebook computer has caused the need for major advancements of antenna design. With the development of 3G and even 4G technologies, multiband and wideband antennas operating at additional frequency bands such as UMTS and LTE are required. In this chapter, it is initially presented the fundamental parameters of the antenna to be taken into account while designing an antenna and determining the operating frequency bands. Afterwards, types of multiband antennas, which are used especially in mobile units, are described. Here, the techniques to make an antenna convenient for multiband operations are given; then, different antennas such as monopoles, PIFAs are examined with several examples in the literature. In the last part, the types of wideband antennas (microstrip patch antenna, DRA or planar) used in mobile communication, which are more appropriate for base station or access point applications, are presented. In conclusion, the engineers interested in mobile communication acquire an initial comprehension about fundamentals and characteristics of multiband and wideband antennas used in mobile communication systems. The readers can utilize from the given references for more detail.

6. References

Ali, M.; Okoniewski, M.; Stuchly, M. A. & Stuchly, M. M. (1999). Dual-Frequency Strip-Sleeve Monopole for Laptop Computers, *IEEE Transactions on Antennas and Propagation*, Vol. 47, No. 2, (February 1999), pp. 317-323, ISSN 0018-926X.

Ali, M.; Hayes, G. J.; Hwang, H.-S. & Sadler, R. A. (2003). Design of a Multiband Internal Antenna for Third Generation Mobile Phone Handsets, *IEEE Transactions on Antennas and Propagation*, Vol. 51, No. 7, (July 2003), pp. 1452-1461, ISSN 0018-926X.

Arai, H. (2002). Outdoor and Indoor Cellular/Personal Handy Phone System Base Station Antenna in Japan, In: *Handbook of Antennas in Wireless Communications*, L. C. Godara (ed.), 345-370, CRC Press, ISBN 0-8493-0124 6, New York, USA.

Balanis, C. A. (2005). *Antenna theory: Analysis and Design*, John Wiley & Sons, Inc., ISBN 0-471-66782-X, New York, USA.

Best, S. R. (2008). Electrically Small Multiband Antennas, In: *Multiband Integrated Antennas for 4G Terminals*, D. A. Sanchez-Hernandez (ed.), 1-32, Artech House, ISBN 978-1-59693-331-6, Boston, USA.

Boyle, K. R. & Massey, P. J. (2006). Nine-Band Antenna System for Mobile Phones, *Electronics Letters*, Vol. 42, No. 5, (March 2006), pp. 265-266, ISSN 0013-5194.

Boyle, K. R. (2008). Multiband Multisystem Antennas in Handsets, In: *Multiband Integrated Antennas for 4G Terminals*, D. A. Sanchez-Hernandez (ed.), 33-52, Artech House, ISBN 978-1-59693 331-6, Boston, USA.

Brar, M. K. & Sharma, S. K. (2011). A Wideband Aperture-Coupled Pentagon Shape Dielectric Resonator Antenna (DRA) for Wireless Communication Applications, *IEEE International Symposium on Antennas and Propagation Society*, pp. 1674-1677, ISBN 978-1-4244-9561-0, Spokane, Washington D. C., USA, July 3-8, 2011.

Caso, R.; Serra, A. A.; Rodriguez-Pino, M.; Nepa, P. & Manara, G. (2010). A Wideband Slot-Coupled Stacked-Patch Array for Wireless Communications, *IEEE Antennas and Wireless Propagation Letters*, Vol. 9, (October 2010), pp. 986-989, ISSN 1536-1225.

Chen, H.-D & Chen, H.-T. (2004). A CPW-Fed Dual-Frequency Monopole Antenna, *IEEE Transactions on Antennas and Propagation*, Vol. 52, No. 4, (April 2004), pp. 978-982, ISSN 0018-926X.

Chen, H.-M.; Lin, Y.-F.; Kuo, C.-C. & Huang, K.-C. (2001). A Compact Dual-Band Microstrip-Fed Monopole Antenna, *IEEE International Symposium on Antennas and Propagation Society*, pp. 124-127, ISBN 0-7803-7070-8, Boston, USA, July 8-13, 2001.

Chen, S.-B.; Jiao, Y.-C.; Wang, W. & Zhang, F.-S. (2006). Modified T-Shaped Planar Monopole Antennas for Multiband Operation, *IEEE Transactions on Microwave Theory and Techniques*, Vol. 54, No. 8, (August 2006), pp. 3267-3270, ISSN 0018-9480

Chiou, T.-W. & Wong, K.-L. (2003). A Compact Dual-Band Dual-Polarized Patch Antennas for 900/1800-MHz Cellular Systems, *IEEE Transactions on Antennas and Propagation*, Vol. 51, No. 8, (August 2003), pp. 1936-1940, ISSN 0018-926X.

Ciais, P.; Staraj, R.; Kossiavas, G. & Luxey, C. (2004). Design of an Internal Quad-Band Antenna for Mobile Phones, *IEEE Microwave and Wireless Component Letters*, Vol. 14, No. 4, (April 2004), pp. 148-150, ISSN 1531-1309

Collins, R. (2009). Base Station Antennas for Mobile Radio Systems, In: *Antennas for Base Stations in Wireless Communications*, Z. N. Chen & K.-M. Luk (eds.), 31-93, Mc-Graw Hill, ISBN 978-0-07-161289-0, New York, USA.

Di Nallo, C. & Faraone, A. (2005). Multiband internal antenna for mobile phones, *Electronics Letters*, Vol. 41, No. 9, (April 2005), pp. 514-515, ISSN 0013-5194.

Guo, Y.-X.; Luk, K.-M. & Lee, K.-F. (2002). Broadband Dual Polarization Patch Element for Cellular-Phone Base Stations, *IEEE Transactions on Antennas and Propagation*, Vol. 50, No. 2, (February 2002), pp. 251-253, ISSN 0018-926X.

Guo, Y.-X. & Tan, H.-S. (2004). New Compact Six-Band Internal Antenna, *IEEE Antennas and Wireless Propagation Letters*, Vol. 3, (December 2004), pp. 295-297, ISSN 1536-1225.

Guterman, J.; Moreira, A. A. & Peixerio, C. (2006). *IEEE Antennas and Wireless Propagation Letters*, Vol. 5, No.1, (December 2006), pp. 141-144, ISSN 1536-1225.

Haapala, P. ; Vainikainen, P. & Eratuuli, P. (1996). Dual Frequency Helical Antennas for Handsets, *IEEE Vehicle Technology Conference*, pp. 336-338, ISBN 0-7803-3157-5, Atlanta, GA, USA, April 28-May 1, 1996.

Huang, Yi. & Boyle, K. (2008). *Antennas: From Theory to Practice*, John Wiley & Sons, Inc., ISBN 978-0-470-51028-5, London, UK.

Jing, X.; Du, Z. & Gong, K. (2006). A Compact Multi-Band Planar Antenna for Mobile Handsets, *IEEE Antennas and Wireless Propagation Letters*, Vol. 5, No.1, (December 2006), pp. 343-345, ISSN 1536-1225.

Katsibas, K. D.; Balanis, C. A., Tirkas, P. A. & Birtcher, C. R. (1998). Folded Loop Antenna for Mobile Hand-Held Units, *IEEE Transactions on Antennas and Propagation*, Vol. 46, No. 2, (February 1998), pp. 260-266, ISSN 0018-926X.

Lan, K. ; Chaudhuri, S. K. & Safavi-Naeini, S. (2003). *IEEE International Symposium on Antennas and Propagation Society*, pp. 926-929, ISBN 0-7803-7846-6, Ontario, Canada, June 22-27, 2003.

Leach, S. M.; Agius, A. A. & Saunders, S. R. (2000). Intelligent Quadrifilar Helix Antenna, *IEE Proceedings- Microwaves, Antennas and Propagation*, Vol. 147, No. 3, (June 2000), pp. 219-223, ISSN 1350-2417.

Lee, E.; Hall, P. S. & Gertner, P. (2000). Dual-band Folded Monopole/Loop Antenna for Terrestrial Communication Systems, *Electronics Letters*, Vol. 36, No. 23, (November 2000), pp. 1990-1991, ISSN 0013-5194.

Lin, S.-Y. (2004). Multiband Folded Planar Monopole Antenna for Mobile Handset, *IEEE Transactions on Antennas and Propagation*, Vol. 52, No. 7, (July 2004), pp. 1790-1794, ISSN 0018-926X.

Liu, W.-C.; Chen M.-C. & Chung, S.-J. (2004). Printed Double S-Shaped Monopole Antenna for Wideband and Multi-Band Operation of Wireless Communication, *IEE Proceedings-Microwaves, Antennas and Propagation*, Vol. 151, No. 6, (December 2004), pp. 473-476, ISSN 1350-2417.

Mahender, P.; Natarajamani, S. & Behera, S. K. (2010). Inverted U-Shaped Dielectric Resonator Antenna for WLAN, *IEEE International Conference on Communication, Control and Computing Technologies*, pp. 9-11, ISBN 978-1-4244-7769-2, Ramanathapuram, India, October 7-9, 2010.

Manteghi, M. & Rahmat-Samii, Y. (2006). Novel Compact Tri-band Two-element and Four-element MIMO Antenna Designs, *IEEE International Symposium on Antennas and Propagation Society*, pp. 4443-4446, ISBN 1-4244-0123-2, Albequerque, USA, July 9-14, 2006.

Manteuffel, D.; Bahr, A.; Heberling, D. & Wolff, I. (2001). Design Considerations for Integrated Mobile Phone Antennas, *Eleventh International Conference on Antennas and Propagation*, pp. 252-256, ISBN 0-85296-733-0, Manchester, UK, April 17-20, 2001.

Moon, J.-I. & Park, S.-O. (2003). Small Chip Antenna for 2.4/5.8-GHz Dual ISM-Band Applications, *IEEE Antennas and Wireless Propagation Letters*, Vol. 2, No.1, (December 2003), pp. 313-315, ISSN 1536-1225.

Rahmat-Samii, Y.; Guterman, J.; Moreira, A. A. & Peixeiro C. (2008). Integrated Antennas for Wireless Personal Communications, In: *Modern Antenna Handbook*, C. A. Balanis (ed.), 1079-1142, John Wiley & Sons, ISBN 978-0-470-03634-1, New York, USA

Secmen, M. & Hizal, A. (2010). A Dual-Polarized Wide-Band Patch Antenna for Indoor Mobile Communication Applications, *Progress In Electromagnetics Research*, Vol. 100, (2010), pp. 189-200, ISSN 1559-8985.

Serra A. A.; Nepa, P.; Manara, G. & Tribellini, G. & Cioci, S. (2007). A Wide-Band Dual-Polarized Stacked Patch Antenna, *IEEE Antennas and Wireless Propagation Letters*, Vol. 6, No.1, (December 2007), pp. 141-143, ISSN 1536-1225.

Teng, P. L. & Wong, K. L. (2002). Planar monopole folded into a compact structure for very-low profile multi-band mobile phone antenna, *Microwave and Optical Technology Letters*, Vol. 33, No. 1, (April 2002), pp. 22-25, ISSN 0895-2477.

Wang, Y.-S.; Lee, M.-C. & Chung, S.-J. (2007). Two PIFA-Related Miniaturized Dual-Band Antennas, *IEEE Transactions on Antennas and Propagation*, Vol. 55, No. 3, (March 2007), pp. 805-811, ISSN 0018-926X.

Wong, K.-L.; Chang, F.-S. & Chiu, T.-W. (2002). Low-Cost Broadband Circularly Polarized Probe-Fed Patch Antenna for WLAN Base Station, *IEEE International Symposium on Antennas and Propagation Society*, pp. 526-529, ISBN 0-7803-7070-8, Boston, USA, June 16-21, 2002.

Wong, K.-L.; Lee, G.-Y. & Chiou, T.-W. (2003). A Low-Profile Planar Monopole Antenna for Multi-Band Operation of Mobile Handsets, *IEEE Transactions on Antennas and Propagation,* Vol. 51, No. 1, (January 2003), pp. 121-125, ISSN 0018-926X.

Wong, K.-L. (2003). *Planar Antennas for Wireless Communications,* John Wiley & Sons, Inc., ISBN 0-471-26611-6, New Jersey, USA.

Wong, K.-L.; Wu, C.-H. & Su, S.-W. (2005). Ultrawide-band square planar metal-plate monopole antenna with a trident-shaped feeding strip, *IEEE Transactions on Antennas and Propagation,* Vol. 53, No. 4, (April 2005), pp. 1262-1269, ISSN 0018-926X.

Yeo, J.; Lee, Y. J. & Mittra, R. (2004). A novel dual-band WLAN antenna for notebook platforms, *IEEE International Symposium on Antennas and Propagation Society,* pp. 1439-1442, ISBN 0-7803-8302-8, Monterey, USA, June 19-25, 2004.

Zhang, F. G.; Wu, W.; Lei, G. & Gong, S. (2011). A Wideband Dual-Sleeve Monopole Antenna for Indoor Base Station Application, *IEEE Antennas and Wireless Propagation Letters,* Vol. 3, (January 2011), pp. 45-48, ISSN 1536-1225.

9

Comparative Analysis of IEEE 802.11p and IEEE 802.16-2004 Technologies in a Vehicular Scenario

Raúl Aquino-Santos, Antonio Guerrero-Ibáñez
and Arthur Edwards-Block
Faculty of Telematics, University of Colima,
Av. Universidad 333, Colima
México

1. Introduction

A major challenge of the automobile industry and safety authorities is how to improve the way cars can communicate either among themselves or with infrastructure designed to assist drivers. Sichitiu and Kihl in [1] describe a taxonomy based on the way nodes (in this case, cars) exchange data. Their work involves two forms of vehicular communication: vehicle-to-vehicle communication (IVC) and vehicle to roadside communication (RVC). IVC can employ either a one hop strategy between two cars (SICV) or multi-hop strategy between many cars (MIVC). It is important to note that multi-hop strategies begin with one car but use several other cars to relay the information to the car requiring the information. Furthermore, the communication strategy can also be either ubiquitous (URVC) or scarce (SRVC). Because of the highly dynamic nature and multiple demands inherent in Vehicular Communication Networks (VCN), these networks have their own very unique requirements:

- The radio transceiver technology must provide omni-directional coverage.
- Rapid vehicle-to-vehicle communications must keep track of dynamic topology changes.
- Highly efficient routing algorithms need to fully exploit network bandwidth.

The increased interest in vehicle-to-vehicle (IVC) and vehicle-to roadside communication (RVC) is due, in part, to the need to expand the amount of information relayed to vehicles. As previously mentioned, the information relayed today is no longer limited to cellular telephone service. As the need to transmit more information grows, so must the technology used to carry that information from car to car or from communications tower to tower. Some applications are more suitable for vehicle-to-roadside communications in applications that involve automatic payment, route guidance, cooperative driving and parking management, just to name just a few. However, there are other applications that are more appropriate for vehicle-to-vehicle communications, including intelligent cruise control, intelligent maneuvering control, lane access and emergency warning, among others. Basically, there are three main categories of applications that have been targeted: (i) road safety applications, (ii) traffic efficiency applications, and (iii) value-added applications. Each application or service possesses different requirements in terms of coverage area, message delay and throughput (Table 1).

Application category	Latency tolerance	Range	Delay requirements
Road safety	Low latency	Local range	Pre-crash sensing/warning (50 ms). Collision risk warning (100ms).
Traffic efficiency	Some latency is acceptable	Medium range	Traffic information - Recommended itinerary (500 ms).
Value-added services	Long latency is accepted	Medium range	Map downloads update - Point of interest notification (500ms).

Table 1. Application categories: examples and requirements.

In RVC and IVC, vehicles require on-board computers and wireless networks to allow them to contact other similarly-equipped vehicles in their vicinity or to roadside access points which, in simpler terms, is a specific point where the information is first introduced into the network. By exchanging information, in the near future, RVC and IVC will be able to provide information about the local traffic situation, real-time vehicle diagnostics and a variety of value-added services to improve comfort and safety.

Future developments in automobile manufacturing will also include new and expanded communication technologies in the area of entertainment, context awareness information, and remote diagnostics of both the vehicle and its occupants. The major goals of IVC and RVC are to provide increased automotive safety, to achieve smooth traffic flow, and to improve passenger convenience by offering passengers both information and entertainment.

Vehicle-to-vehicle communication (IVC) systems based on wireless ad-hoc (dynamically self organizing) networks represent a promising solution for future mobile communication because they minimize communication costs (licensed frequency spectrum and mobile communications based on VoIP, etc.) and guarantee the low delays required to exchange safety-related data between cars. IVC will soon allow vehicles to self organize themselves locally in ad-hoc networks without any pre-installed infrastructure. Having cars carry all of the communications infrastructure in them will require fewer unsightly communications towers and permit each vehicle to be a tower in itself, in the sense that they will be able to transmit, relay and receive information without any visible infrastructure. Communications in future IVC systems will not be restricted to neighbored vehicles travelling within a limited radio transmission range, as is currently the case in wireless scenarios. Future IVC systems will provide multi-hop communication capabilities to relay data and information by employing intermediate vehicles located between any specific sender and receiver as relay nodes.

One of the most obvious issues relating to IVC or RVC is the velocity of the mobile devices. One of the effects of velocity is to make signal strength extremely variable. Chu and Stark show in [2] that fading (interference) is a direct function of velocity. Based on simulations, they observed that signal strength is best maintained between vehicles that travel at lower velocities because the signal strength varies more slowly, thus creating much less

interference between any two consecutive transmissions. Of course, with less fading, there is a smaller chance of the signal suffering communications breakdown.
Another important aspect of IVC or RVC communication is the environment in which the vehicles are moving. It is known that different physical environments lead to different performances. For example, in an urban scenario, vehicles will suffer more multi-path interference (when the signal splits into many signal that come from different directions) than in a freeway scenario. Multi-path interference is very important because if the signal is split and arrives at a specific location from different angles, they will not arrive at the same time, and this will make understanding the signal sent difficult or impossible. Multi-path interference is primarily caused by the presence of buildings and other obstacles (trees, communication towers, billboards, etc.) in urban environments cause diffraction and scattering. Moreover, researchers must also consider the different velocities implicit in different scenarios. Generally speaking, drivers require more time and a greater distance to come to a complete and safe stop. Vehicles will travel at a higher velocity and be more widely spaced in a freeway scenario because drivers require greater reaction times than in urban settings. Distance and relative velocity, therefore, are very important because they significantly influence communications. For example, in urban scenarios, inter-vehicular distances are very small for prolonged periods of time because reduced spacing due to merging and frequent stops. Consequently, closely spaced vehicles can exchange more data than in freeway scenarios, where the distances and velocities between vehicles are substantially greater. It is important to recall that in peer-to-peer communication, the distance between peers must be small enough for the entire duration of the communication. Therefore, vehicles that predictable maintain lower speeds and spacing, along with many predictable stops, can transmit greater uninterrupted information streams. The speed and spacing factors lead us to consider the dynamics of vehicular movements, particularly inter-vehicular distance and their relative velocity and position as they move along to streets or roadways. Consequently, different models must be developed to predict vehicular movement in highly dynamic and varied real-world scenarios.
Another issue that can affect IVC or RVC communication is the technology employed, each technology prioritizes different features, such as frequency, bandwidth, and transmission power.
This work analyzes two emerging wireless technologies that can be employed in RVC communication, IEEE 802.1p and IEEE 802.16-2004, in an urban scenario.
The remainder of this chapter is organized as follows: Section 2 describes research related to IEEE 802.11p and IEEE 802.16-2004 technologies and Section 3 describes the simulated scenario simulated and results we obtained. Finally, Section 4 provides a summary of our work and offers suggestions for future research.

2. Related work in IEEE 802.11p and IEEE 802.16-2004 technologies

Numerous researchers have worked to overcome issues related to IVC and RVC communication (e.g. [3-10]). In 2004, the IEEE group created the IEEE 802.11p (Wireless Access in Vehicular Environments, (WAVE)) task force [11]. The workforce established a new standard that essentially employs the same PHY layer of the IEEE 802.11a standard, but uses the 10 MHz bandwidth channel instead of the 20 MHz bandwidth of IEEE 802.11a. With respect to the MAC layer, WAVE is based on a contention method (i.e. CSMA/CA), similar to other standards in this group. The purpose of this standard is to provide the minimum set of specifications required to ensure interoperability between wireless devices attempting to communicate in potentially rapidly changing communications environments and in situations where message delivery must be completed in time frames much shorter than the minimum in 802.11-based infrastructure or ad-hoc networks [12]. The used frequency spectrum in an 802.11p network is divided into one control channel (CCH) and several service channels (SCHs), as shown in Figure 1. The CCH is dedicated for nodes to exchange network control messages while SCHs are used by nodes to exchange their data packets and Wave Short Messages (WSMs). The link bandwidth of these channels is further divided into transmission cycles, each comprising a control frame and a service frame. The draft standard suggests that the duration of a frame (either a control or a service frame) is set to 50 milliseconds.

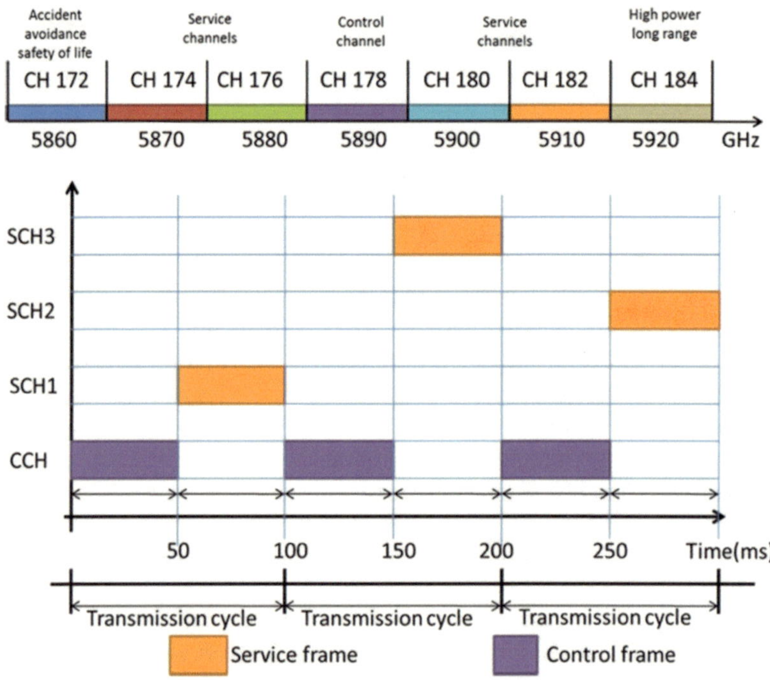

Fig. 1. Channels available for 802.11p

Authors in [13] evaluated the Packet Error Rate (PER) performance degradation of the WAVE PHY layer due to the time-varying channel and the Doppler Effect. They conclude that the estimation process is significantly affected by rapid changes of the channel, severely affecting the PER performance, while the Inter-Carrier Interference (ICI) has little or no impact on the performance at small data rates. WAVE also has a limited transmission range; simulations carried out by [14] show that only 1% of communication attempts at 750m are successful in a highway scenario presenting multipath shadowing.

The MAC layer in IEEE 802.11p has several significant drawbacks. For example, in vehicular scenarios, WAVE drops over 53% of packets sent according to simulation results [15]. Furthermore, results in [16] show that throughput decays as the number of vehicles increases. In fact, throughput decreases to almost zero with 20 concurrent transmissions.

The authors thus conclude that WAVE is not scalable. Additionally, IEEE 802.11p does not support QoS, which is essential in Vehicular Ad hoc Networks (VANETs).
Authors in [17] report that the probability of collisions grows significantly as the number of nodes sending Access Classes (AC3) increases. It is important to remember that a higher number of collisions can cause increased dead time, in which a channel is blocked and no useful data can be exchanged. Also, the continuous switching between the Control Chanel (CCH) and the Service Channel (SCH) use different packet cues, which amplifies the affects of collisions. As a result, packets destined for the CCH form longer cues which result in greater SCH intervals and higher end-to-end delay. WAVE technology can not ensure time critical message dissemination (e.g. collision warnings), especially in dense scenarios or in the case of filled MAC queues.
Authors in [18] developed a simulation framework for the Wireless Access in Vehicular Environments (WAVE) standard using the NS-2 simulator. Their framework included a handoff mechanism, but they did not employ realistic vehicular traffic models.
Recently, the IEEE 802.16-2004 taskforce [19, 20] actualized this standard to better permit it to handle QoS, mobility, and multi-hop relay communications. Networks using the Worldwide Interoperability for Microwave Access (WiMAX) -MAC layer now can potentially meet a wider range of demands, including VCN. WiMAX is a nonprofit consortium supported by over 400 companies dedicated to creating profiles based on the IEEE 802.16-2004 standard. The first IEEE 802.16-2004 standard considers fixed nodes in a straight line with line of sight between the base station and each fixed remote node [21]. Later, the IEEE 802.16e task force (TF) amended the original standard to provide mobility to end users (Mobile WiMAX [18]) in non-line-of- sight conditions. The most recent modification to IEEE 802.16e was in March, 2007. This modification, IEEE802.16j (approved in 2009), permits multi-hop relay communications [20].
IEEE 802.16j operates in both transparent and non-transparent modes. In transparent mode, mobile stations (MS) must decode the control messages relayed from the base station (BS). In other words, they must operate within the physical coverage radius of the BS. In non-transparent mode, one of the relay stations (RS) provides the control messages to the MS. The main difference between the transparent and the non-transparent mode architecture is that in transparent mode, RS increases network capacity while in non-transparent mode, the RS extends the BS range. Additionally, the RS can be classified according to mobility and can be fixed (FRS), nomads (NRS) or mobiles (MRS) [22].
In [23] the authors propose a cross-layer protocol called Coordinated-External Peer Communication (CEPEC) for Internet-Access services and peer communications for vehicular networks. Their simulation results show that the proposed CEPEC protocol provides higher throughput with guaranteed fairness in multi-hop data delivery in vehicular networks when compared with the purely IEEE 802.16-2004 based protocol.
Authors in [24] examine the IEEE 802.16j multi-hop relay (MR) technology that improves vehicle-to-infrastructure communications.
Finally, there are few studies that show comparative analysis between WiFi and WiMAX. In [25] authors report that while WiMAX can offer a longer communication range than WiFi, its latency can be significantly larger than that of WiFi at a short distance (e.g. less than 100m). Additionally, authors show the frame size´s value has a strong impact on the performance of WiMAX.

3. Scenario simulated and results obtained by simulations

In this section, we perform a simulation study to verify the efficiency and performance of both technologies as a communication media applied into a vehicle-to-infrastructure (V2I) communications environment. The performance of the two systems was evaluated for different vehicle speeds and traffic data rates.

In order to evaluate the performance of both technologies, we carried out several simulations in the NCTUNs network simulator and emulator [26], which is a free network simulation tool that runs on Linux. The reason behind the decision to use this tool is that NCTUns provides reproducible and traceable results. The simulator tightly integrates network and traffic simulations and provides a fast feedback loop between them. Simulation models for mobile WiMAX with the support of several features (such as PHY OFDMA, PMP and TDD modes, QoS scheduling services, among others) and for 802.11p that supports IEEE 802.11p On Board Units (OBUs) and Road Side Units (RSUs) are defined in NCTUns simulation tool.

As the metrics of performance evaluation, we mainly use the throughput, the packet loss rate and the packet end-to-end delay sent from the source node to the receptor node. The aim of these metrics is to quantitatively determine the packet loss average during the overall communication session, the average packet end-to-end delay sent from the source node to the receptor node and the average throughput obtained for the overall communication process.

3.1 Simulation environment and setting

As previously mentioned, for our simulations, we use the NCTUns network simulator. We ran the simulation to verify if 802.11p and WiMAX can ensure an acceptable performance for different speeds and traffic data rates.

To evaluate and compare the performance of both 802.11p and 802.16-2004 technologies in a Vehicular-to-Infrastructure (V2I) context, we consider the scenario shown in Figure 2. We define a 13km zone that is fully covered by base stations. For 802.11p technology, several Road-Side-Units (RSUs), which represent fixed devices with a Dedicated Short Range Communication (DSRC) radio, are mounted on road sides. For 802.16 technology, one base station (BS) is mounted in the scenario where twenty-five vehicles are randomly distributed on the road. The vehicles are equipped with a wireless communication device depending of the evaluated scenario either 802.11p or IEEE 802.16-2004.

As shown in Figure 2, for the case of 802.11p, each base station has a coverage area of 1000 m and a common coverage area of around 100m. The RSUs are connected to the router by means of links with a capacity of 100Mbits (to avoid any bottleneck outside the considered WiMAX/802.11p V2I networks). Each RSU is configured to provide the service in channel 174.

The movement of all vehicles on the road is generated randomly by the simulator. In NCTUns, each vehicle can be specified with different auto-driving behaviors. A driving behavior is defined by a car profile. After inserting vehicles, one can specify what kind of profile should be applied to an Intelligent Transportation System (ITS) car. We use the car profile tool included in NCTUns to define the behavior of the cars. An overview of the information of car profiles used for the simulation is shown in Table 2.

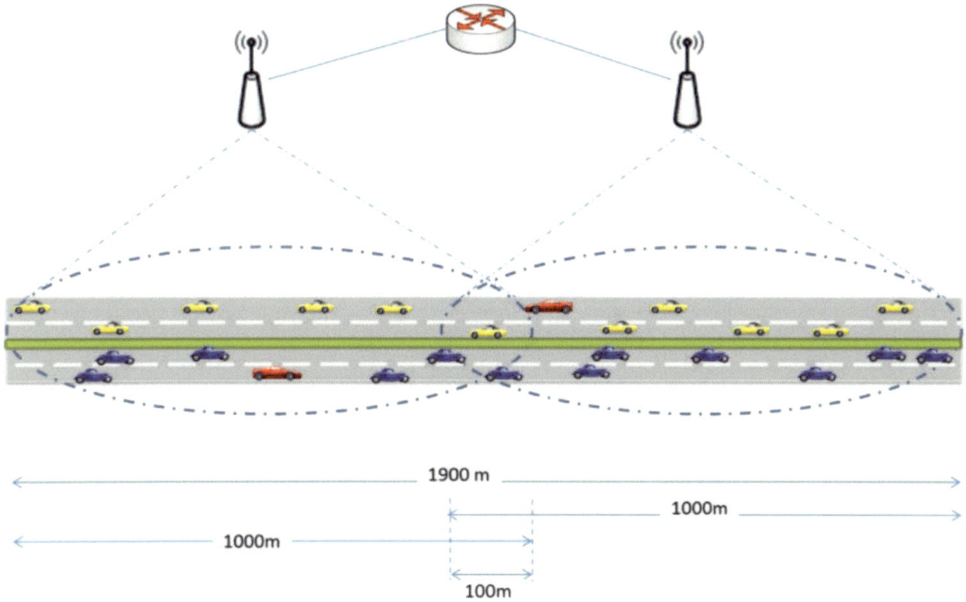

Fig. 2. Simulation scenario

	Profile1	Profile2	Profile3	Profile4	Profile5
Maximum speed	50km/h	60km/h	90km/h	100km/h	200km/h
Maximum acceleration	1.2 m/sec	1.34 m/sec	4.5m/sec	5m/sec	12m/sec
Maximum deceleration	2 m/sec	2.24 m/sec	8.9m/sec	10m/sec	20m/sec

Table 2. Car profiles

The simulation is divided into two scenarios. In both scenarios, we present a communication model between two vehicles moving in opposite directions.

Our first scenario studies the impact of vehicle speed on performance of both technologies into a V2I communication environment. This scenario consists of a constant bit-rate application running over User Datagram Protocol (UDP), and we have set the source data rate to 1Mbps. In this scenario we vary the average speed of the vehicles to 60km/h, 100km/h and 200km/h. The first speed reflects the typical mobility within an urban region, while the second and third speeds are meant to meet the IEEE 802.11p set requirements. In this scenario we examine the impact of varying the vehicle speed on the average throughput and the end-to-end delay.

The second scenario studies the impact of the source data rate on the performance of both technologies into a V2I communication environment. This second scenario consists of a variable bit-rate application with an increased traffic flow, consisting of exponential UDP traffic varying from 1Mbps to 6Mbps, in which we evaluate the impact of varying the source data rate on both the throughput and the end-to-end delay. In this scenario we set the average speed of the vehicles to 100 km/h, which is a realistic value of vehicles on the highway. An overview of the parameters used for the simulation configuration is shown in Table 3.

Parameter	802.11p	802.16-2004
Antenna settings		
RSU/BS Frequency	5.9GHz	5.4GHz
RSU/BS Transmission power	28.8 dBm	33dBm
RSU/BS Antenna height	2m.	30m
RSU/BS Antenna gain	3dBi	15dBi
Range	1000m	13Kms
Simulation environment		
Average building height	10m	10m
Street width	30m	30m
Road	13kms	13Kms
Nodes		
Channel bandwidth	10 MHz	10MHz
OBU transmission power	23dBm	23dBm
OBU antenna height	1m	1m
Type of antenna	Omnidirectional	Omnidirectional

Table 3. Simulation parameters

3.2 Simulation results

As previously mentioned, the first simulations study the impact of vehicle speed on performance of both technologies into a V2I communication environment. Figure 3 shows the impact of speed in the rate of packet loss. Results show there is an increase in the rate of packet loss when there is an increase in speed and the 802.11p technology is used. Results also show that when the vehicle speed increases, the connectivity time to the 802.11p RSU decreases, which, in turn, reduces the amount of data received by vehicles. As a consequence of the increased speed, the number of handovers increases, resulting in a greater number of packet losses. However, we observe that when the 802.16-2004 technology is used, as there are no handovers, the percentage of packet loss is almost null.

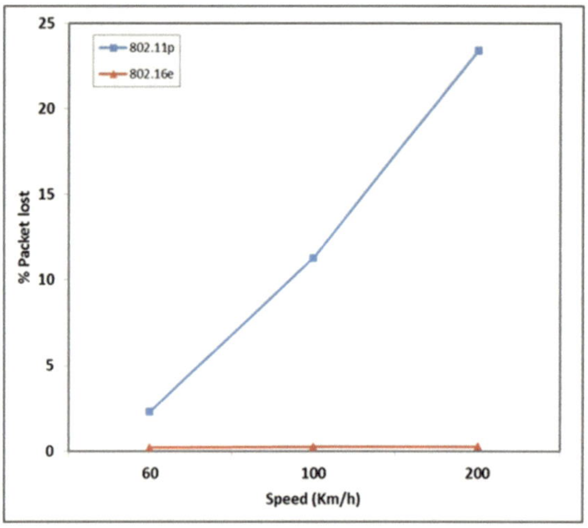

Fig. 3. Impact of vehicle speed on the rate of packet loss.

Figure 4 shows the impact of the vehicle speed on the average end-to-end delay. We can observe that for 802.11p, when vehicle speeds increase, difficulties related to handover increase, which then introduces a considerable delay in the communication process. However, in the case of 802.16-2004, there is no required time for the handover execution, which explains why it maintains a low average delay. The data presented in Figure 4 shows that the end-to-end delay for both technologies is less than 100 ms, which meets the minimum requirement of most ITS applications.

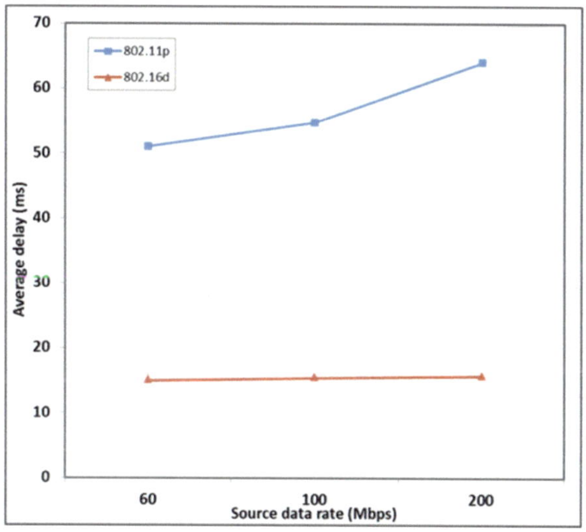

Fig. 4. Impact of vehicle speed on the average end-to-end delay.

Figure 5 shows the impact of different vehicle speeds on the average throughput. When 802.11p technology is used, the connectivity time to the RSUs decreases which, in turn, reduces the amount of data received by the vehicle. Additionally, a fraction of this period is

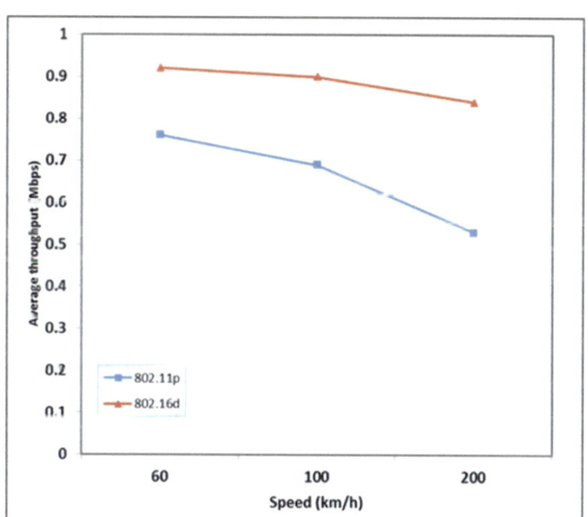

Fig. 5. Impact of vehicle speed on the average throughput.

required to switch from one RSU to another. As can be seen, the throughput is reduced up to 30% at higher speeds (200 km/h) in comparison to lower speeds (60km/h). On the other hand, in the case of 802.16-2004, the impact of speed on the average throughput is minimal (less than 9%).

The following section will analyze the impact of varying the source data rate on the performance of both technologies. Figure 6 presents the average throughput obtained for each communication technology. The results show that for 802.11p, the maximum throughput is approximately 1.2 Mbps while 802.16-2004 covers the maximum throughput demanded in each situation.

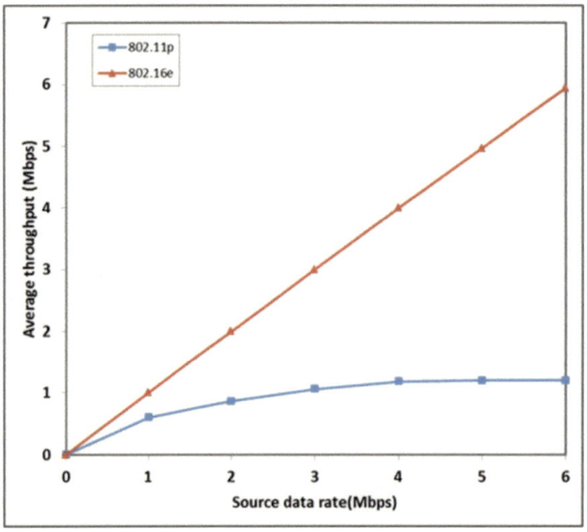

Fig. 6. Impact of the source data rate on the average throughput.

Finally, Figure 7 shows the impact of varying the source data rate on the average end-to-end delay. The results show that 802.11p technology experiences shorter delays when the source

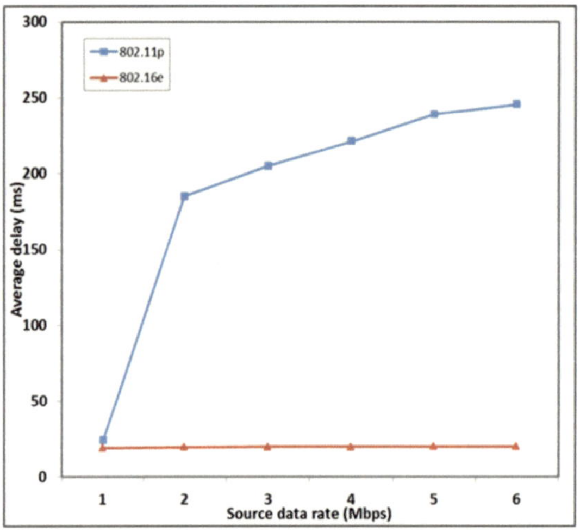

Fig. 7. Impact of the source data rate on the average end-to-end delay.

data rate is less than 1.5 (which we denominated as low data rate condition). However, when the source data rate increases to more than 1.5Mbps, delay increases by almost 600%, when compared with low data rate condition. As mentioned before, this result shows how the handover process greatly affects the performance of 802.11p.
On the other hand, in the case of 802.16-2004, the impact of various data rates on the average end-to-end delay is not very significant. The average delay for 802.16-2004 does not exceed 20 ms, which even meets the stringent demands of most emergency applications.

4. Conclusion

This work compared simulations of two important technologies for roadside communication in an urban scenario. This comparison is relevant because of the many obstacles that affect communications in cities or highly congested areas. Results show that IEEE 802.16-2004 technology outperforms IEEE 802.11p technology in terms of throughput (network capacity), packet loss (the percentage of information that is transmitted but not received) and end-to-end delay (the time necessary for the information to be delivered from the transmitter to the receiver). Therefore, we conclude that IEEE 802.16-2004 technology is more suitable for roadside applications including road safety, traffic efficiency and value-added services in urban settings, where there are many obstacles that can potentially cause attenuation or the entire loss of the transmission.
Our proposed future work will evaluate the Location-Based Routing Protocol with Cluster-Based Flooding (LORA-CBF) using IEEE 802.16-2004 technology.

5. Acknowledgment

This work was supported by the National Council for Science and Technology of the United Mexican States (CONACYT), under Grant 143582.

6. References

[1] M. L. Sichitiu and M. Kihl. Inter-vehicle communication systems: a survey. Communications Surveys & Tutorials, IEEE, 10(2):88-105, 2008.
[2] Michael J. Chu and Wayne E. Stark. Effect of Mobile Velocity on Communications in Fading Channels. IEEE Transactions on Vehicular Technology, vol. 49, no. 1, 2000.
[3] X. Yang, J. Liu, F. Zhao, and N. Vaidya. A Vehicle-to-Vehicle Communication Protocol for Cooperative Collision Warning. Proceedings of the 1st Annual International. Conference on Mobile and Ubiquitous Systems: Networking and Services, pp. 1–4, 2004
[4] J. Yin et al., "Performance Evaluation of Safety Applications over DSRC Vehicular Ad Hoc Networks," Proceedings of the 1st ACM Workshop on Vehicular Ad Hoc Networks, pp. 1 9, 2004
[5] R. Rajamani and S. Shladover. An Experimental Comparative Study of Autonomous and Co-Operative Vehicle-Follower Control Systems. Transportation Research Part C, vol. 9, pp. 15–31, 2001
[6] P. Varayia. Smart Cars on Smart Roads: Problems of Control. IEEE Trans. Automatic Control, vol. 38, no. 2, pp. 195–207, 1993

[7] A. Brown et al. Vehicle to Vehicle Communication Outage and Its Impact on Convoy Driving. Proceedings of the IEEE Intelligent Vehicle Symposium, 2000, pp. 528–33

[8] SAFESPOT, http://www.safespot-eu.org

[9] COMeSafety, http://www.comesafety.org

[10] Car2car Communication Consortium, http://www.car-tocar.org

[11] IEEE 802.11p,

[12] "IEEE 802.11p/D3.0," IEEE Standards Activities Department, July 2007.

[13] Iulia Ivan, Philippe Besnier, Matthieu Crussiere, M´hamed Drissi, Lois Le Danvic, Mickael Huard, and Eric Lardjane. Physical Layer Performance Analysis of V2V Communications in High Velocity Context. 9Th International Conference on Intelligent Transportation Systems Telecommunications, pp. 409-414, 2009.

[14] Yi Wang, Ahmed A, Krishnamachari B, and Psounis K. IEEE 802.11p performance evaluation and protocol enhancement. IEEE International Conference on Vehicular Electronics and Safety, ICVES 2008, pp. 317 - 322, 2008.

[15] Katrin Bilstrup, Elisabeth Uhlemann, Erik G. Strom, and Urban Bilstrup. Evaluation of the IEEE 802.11p MAC Method for Vehicle-to-Vehicle Communication, 2008. Vehicular Technology Conference, pp. 1-5, 2008. [Online].Available: http://dx.doi.org/10.1109/VETECF.2008.446

[16] L. Stibor, Y. Zang, and H.-J. Reumerman. Evaluation of communication distance of broadcast messages in a vehicular ad-hoc network using IEEE 802.11p. In Proceedings of the IEEE Wireless Communications and Networking Conference (WCNC '07), pp. 254-257, March 2007.

[17] Stephan Eichler. Performance Evaluation of the IEEE 802.11p WAVE Communication Standard. 66th IEEE Vehicular Technology Conference (VTC-2007 Fall), pp. 2199-2203, 2007.

[18] Balkrishna Sharma Gukhool and Soumaya Cherkaoui. Handoff in IEEE 802.11p-based vehicular networks. IFIP International Conference on Wireless and Optical Communications Networks, pp. 1-5, 2009.

[19] IEEE 802.16e-2005,

[20] IEEE 802.16j,

[21] IEEE 802.16-2004,

[22] The Future of WiMAX: Multihop Relaying with IEEE 802.16j

[23] Kung Yang, Shumao Ou, Hsiao-Hwa Chen, and Jianghua He. A Multihop Peer Communication Protocol with Fairness Guarantee for IEEE 802.16-based Vehicular Networks. IEEE Transactions on Vehicular Technology, vol. 56, no. 6, pp. 3358-3370, 2007.

[24] Yu Ge, Su Wen, Yew-Hock Ang, and Ying-Chang Liang. Optimal Relay Selection in IEEE 802.16j Multihop Relay vehicular Networks. IEEE Transactions on Vehicular Technology, vol. 59, no. 5, pp. 2198-2206, 2010.

[25] Chien-Ming Chou, Chen-Yuan Li, Wei-Min Chien, and Kun-Chan Lan. A Feasibility Study on Vehicle-to-Infraestruture Communication: WiFi Vs. WiMAX. Tenth International Conference on Mobile Data Management: Systems, Services and Middleware, pp. 397-398, 2009.

[26] SIMREAL technology Homepage, NCTUNS network simulator and emulator, http://nsl10.csie.nctu.edu.tw/.

10

The Role of Ad Hoc Networks in Mobile Telecommunication

Qurratul-Ain Minhas[1], Hasan Mahmood[1] and Hafiz Malik[2]
[1]*Department of Electronics, Quaid-i-Azam University, Islamabad*
[2]*Department of Electrical and Computer Engineering, University of Michigan – Dearborn, Dearborn, MI*
[1]*Pakistan*
[2]*USA*

1. Introduction

The installation, improvement, maintenance and operation of telecommunication networks are hampered by the ever changing trends in technology. The nature and design of existing telecommunication networks makes it difficult to integrate different segments, created by technological difference and fragments of various network operators and manufacturers. In addition to the enormous bandwidth and quality of service demands of telecommunications network users, they desire to ensure the availability of all the services and resources. Moreover, the users anticipate the availability of information services whenever or wherever they need it, while at the same time, satisfying temporal and spatial diversity considerations. In the midst of this ongoing progress in technology and continuing effort to integrate network resources from different providers, the ad hoc networks paradigm, with its unique benefits, is capable of easing the integration process and providing seamless access to information resources and connectivity in many practical scenarios (Remondo & Niemegeers, 2003). The ad hoc networks are self organizing networks and any kind of central management is not an essential requirement. In contrast to other prevailing telecommunication networks which support mobility, the case of ad hoc networks management is different and no stringent predefined infrastructure is critical to the establishment and operation of these networks. While the implementation of ad hoc networks seems to be simple at a first glance with apparent advantages in specific situations over existing mobile networks, there are significant challenges which inhibit the attachment and use in an integrated environment with existing mobile telecommunication networks (Freedman, 2009).

In this chapter, we discuss the role of ad hoc networks in mobile telecommunications. The trends in mobility are discussed with respect to the technological aspect, that is, from physical layer to application layer and social aspects which inherently influence the telecommunication network design and direct the future development trends in technology and its usage.

The existing network technologies and implementations are striving hard to accommodate the needs of users which indirectly govern the core design, life cycle, accessibility and integration, this also include person to person interaction, video on demand or streaming contents, music, pictures, gaming, information, and business applications (Wang et al., 2008). Many of these applications require providing content at a mobile location, which to some extent is covered by traditional telecommunication networks, nevertheless to take the accessibility to the limit, the ad hoc networks can play a vital role in providing connectivity and advance services, especially where last mile is very critical and difficult to manage. Another important aspect in providing the users the facility to access the information from various resources and multiple formats is the design and specifications of the end devices. These end devices are evolving rapidly and are capable of integrating ad hoc network protocols and technologies to make them compatible with existing networks and future generation of ad hoc network systems. For example, many smart phones and notebooks have access to ad hoc networks and to some extent, are capable of integrating various mobile ad hoc network versions. In some instances where the users are mobile in remote areas which lack access to telecommunication networks, the ad hoc networks or more precisely the mesh networks can provide connectivity and availability of vital network resources, which in situations like disaster areas, battle fields, conventions, and remote areas can be very useful. The inherent underlying advantages of mobile ad hoc networks can further be exploited in situations where we require mobile multimedia, fixed to mobile convergence, voice over Internet protocol mobile applications, and network technologies with short life cycle risk (Cricelli et al., 2011). The fragmentation in the already established networks can be overcome by using ad hoc networks as bridges. The ad hoc network in this case provides interfaces to different technologies while connecting different mobile telecommunication networks via virtual connections and in some cases virtual network backbones.

In this chapter, we also present the advantages, disadvantages, and consequences of ad hoc networks when integrated with prevailing and future mobile telecommunication networks. Traditionally, a telecommunication network is governed and operated by an entity. The services provided by these entities generate revenue and there is no concept of decentralized management as far as revenue is concerned or providing services which are free of cost. The risk of ad hoc networks being used by masses in situations where the cost to use a service is very competitive or even free, poses a great risk for existing telecommunication operators. We discuss the implications of these types of scenarios in the prospective of mobile telecommunications and its influence on users (Akyildiz et al., 2005).

The ad hoc networks are capable of adapting rapidly, therefore, in situations where behavior of different types of users such as entrepreneurs, students, pensioners, singles, families, is observed simultaneously, especially when they are mobile and use different type of technologies and request connectivity, the traditional mobile networks lack acceptability and in many occasions become too costly. We provide an insight into the research progress pertaining to ad hoc networks which show a promising future to amalgamate in existing technologies and systems (Zhao & Liang, 2010). In a struggle to cut the prices to levels which are not even imaginable by existing subscribers and users, the role of ad hoc networks is also presented, and methods and ways are discussed related to cost management (Hac, 2002).

In recent years, the Voice over Internet Protocol (VoIP) has played a vital role in the evolution, development, and regulation of mobile telecommunication networks (Chang et al., 2009). The VoIP platform eases the integration process and provides a standardized protocol stack for telecommunication equipment vendors and system designers. In this

chapter, we focus on the progress of ad hoc network and their compliance to VoIP applications, the benefits provided by the use of VoIP acceptability at all levels and the reduction in complexity in the integration process with future mobile telecommunication networks. We also discuss the technological aspects which exist at each layer in open systems interconnectivity layered mode (Smith, 2006). The medium access layer, network layer, transport layer, and session layer are emphasized because of their important role in the governance, management and integration of these networks.
With the diminishing number of available channels in radio spectrum, it is vital that the use of available recourses should be in an efficient manner. The use of cognitive radios in conjunction with ad hoc networks has shown substantial improvements in bandwidth utilization (Comeras et al., 2007). We also present the use of such technologies and their implications on the performance enhancement in light of mobile telecommunications networks and ad hoc networks. In addition, the new technologies and systems are vulnerable to hackers and sometimes are less secure (Kaosar & Sheltami, 2009). We also discuss the issues related to security in the ad hoc networks with emphasis on mobility and integration with existing networks (Bayer et al., 2010).
With the advent of modern technology, everyone desires to stay connected, everywhere, all the time. The ever increasing demand for keeping in touch with the outside world has not only triggered a never ending technological development, but is also affecting the social life of an ordinary person in various ways. Telephonic conversation is no longer the only requirement for telecommunication users, now they like to play interactive games, share videos and files, browse the Internet, use social networking sites, check emails, hold video conferences and use other multimedia applications. Moreover, the users own more than one communication device and there is a requirement for interconnectivity.
In a typical cellular system, each node (cell phone) completes its communication with the help of a base station which is the control authority. The base station manages the network in combination with the Mobile Switching Station (MSC) which is connected to a Public Switching Telephone Network (PSTN). The demand for maximum connectivity requires the cellular network to install greater number of base stations. The problem arises when expanding the infrastructure is not possible or is expensive. Moreover, in battle fields or during natural disasters, the infrastructure is disabled or destroyed. During these times, the need for connectivity also increases as communication becomes more crucial in emergency situations.

2. Ad hoc networks

Ad hoc networks were first employed in 1970s, when US Defense Advanced Research Projects Agency (DARPA) installed packet radio network. The past couple of decades have seen a huge progress in the field of ad hoc networking. With the introduction of Bluetooth in 1998, a proprietary open wireless technology standard for exchanging data over short distances, ad hoc networks were used in commercial applications with eight devices in a small network, called the piconet. Ad hoc networks are autonomous networks which are not managed or controlled by a centralized infrastructure. The nodes in ad hoc networks are responsible for establishing communication links as requested by source nodes. The participating nodes act as routers, gateways, and as transmitters or receivers. Moreover, these nodes are mobile which makes the topology of the network dynamic. The ad hoc networks are spontaneously formed with the minimal assistance of communication

hardware infrastructure; therefore, these types of networks are suitable for emergency situations.
Ad hoc networks can be deployed in a wide range of applications such as, disaster relief areas, networks at construction sites and other temporary installations, vehicular networks, connecting organizations and hospitals, home and office networks, conventions, networks for users at airports and shopping malls. The key advantage of ad hoc networks is that they can be employed in any technology and integrated with all kinds of environments due to their dynamic and flexible nature. Ad hoc networks can be connected to pre-existing technologies like cellular and access hosts which are not part of ad hoc system. The use of ad hoc networks in combination with the existing cellular network can not only improve the performance and services of existing cellular network but can also provide connectivity in areas where cellular network is not available or not feasible. Some of the advantages of ad hoc networks include independence, flexibility, low cost, self healing and zero set up cost.

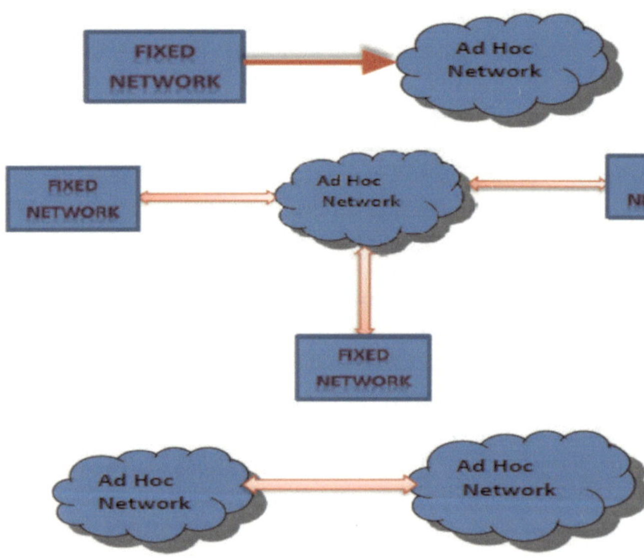

Fig. 1. Ad hoc networks employed in different configurations.

The cellular networks are organized in a master slave configuration and employ a centralized infrastructure, which in turn require base stations fixed in one location. They have a static network topology and stable connectivity. The installation requires careful planning and the setup cost is relatively high. On the other hand, ad hoc networks have less infrastructure requirements, and there is no need of a base station. These networks are highly dynamic with ever changing network topology which leads to unexpected interference and results in frequent disconnections. Ad hoc networks do not require any planning or setup of infrastructure and hence are very cost effective. Ad hoc networks when integrated with cellular networks can utilize and interface with cellular network hardware such as access points.
Ad hoc network topologies with minimal number of nodes operate more efficiently, whereas, networks with high number of nodes are relatively difficult to manage in terms of

achieving high spectral efficiency and securing the information propagating in the network. Although the issues are challenging as far as the scalability is concerned, the network still provides connectivity and enables the user to transmit urgent messages under various network and traffic conditions. Small networks are useful in applications when connecting devices in the home or office environment where a small network will be fast, efficient, and secure.

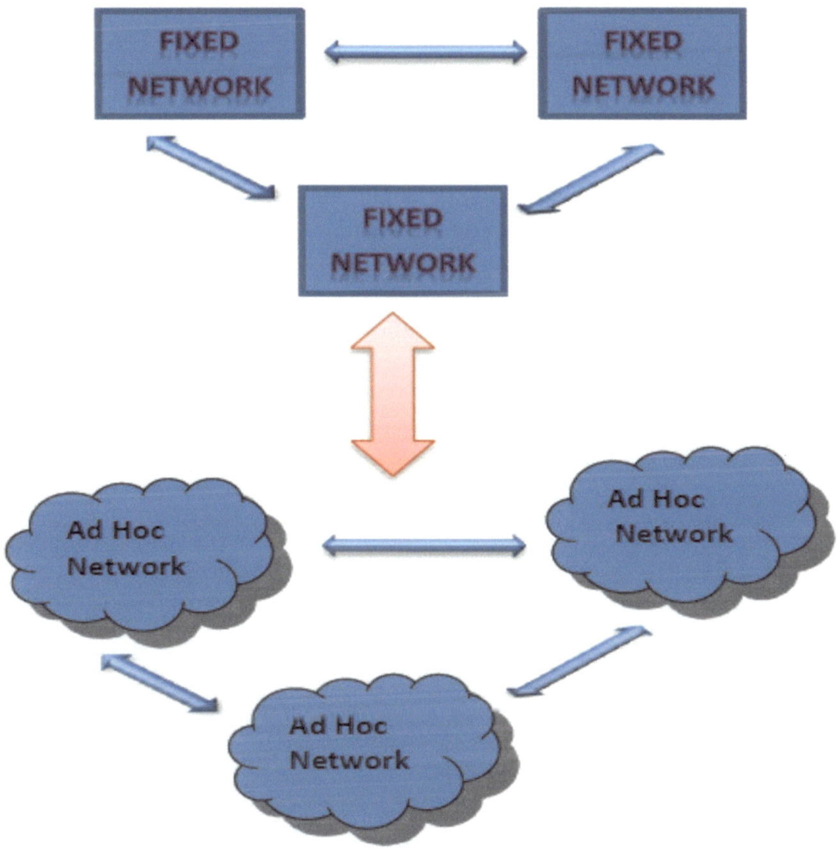

Fig. 2. Connecting fixed networks with ad hoc networks

2.1 Technologies employed in ad hoc networks

Ad hoc networks are realized using several prevailing technologies used in other wireless communication applications. The most popular physical layer standard is the IEEE 802.11. This is commonly used for WLAN services while Bluetooth is used for short range services. These technologies are successfully employed in ad hoc networks. The physical layer specifications in ad hoc networks are similar to the standards used in WLAN and Bluetooth, and 2G networks. For example, another existing and popular technology, GSM has mainly been used for voice services and offers a narrow range of services. With the development of 2.5G/3G generation of mobile communication, like GPRS, UMTS, CDMA etc., several advanced services like Internet and other multimedia applications are also provided. Several other schemes are also developed, such as IEEE 802.11/a/b/g (WLAN broadband)

and broadcast systems which include Digital Audio Broadcasting (DAB) and Digital Video Broadcasting (DVB) (Armeulles et al., 2004). All of these wireless communication technologies are desirable contenders for use in ad hoc networks. Although these technologies are designed for cellular and wireless local areas networks, with appropriate changes in system software can make these technologies useful for ad hoc networks applications. Bluetooth scheme is also implemented for use as short range ad hoc networks. The 4G networks include both the existing as well as future technologies and hope to provide good services with best connectivity everywhere at all times. The concept of applying 4G in military environment is called Fourth Generation on Warfare (4GM @ 4GW) which involves the use of ad hoc networks in combination with infrastructure based network. As new technologies are compatible with existing technologies, their application to ad hoc network is also anticipated (Szczodrak et al., 2007).

Recently, the main focus of research in the area of ad hoc network design is around optimizing energy consumption, the design of 3G and next generation systems also takes into consideration these requirements. The future generations of wireless equipment is also being developed on the same lines as required by ad hoc network devices. The objective of achieving higher data rate and lower transmission cost is also the main focus in the new specifications of mobile telecommunication equipments. Integration of ad hoc network with PSTN, 4G, 3G, GPRS, UMTS, 2G, and GSM is shown in Figure 3. Some of these configurations are in use while research is going on to integrate other technologies.

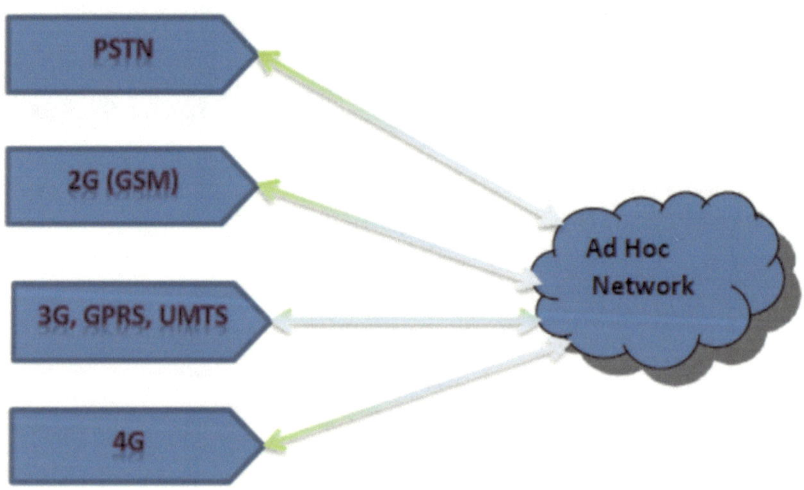

Fig. 3. Integrating ad hoc networks with the modern technologies.

Some of the features of existing and future technologies are presented below. The objective is to provide an insight into their features, capabilities, and suitability for use in ad hoc networks applications.

The 3G offers lower data rates than WLANs, where WLANs have a much smaller coverage area as compared to 3G. Wireless mesh networks (WMN) offer a solution with higher data rates and better coverage (Akildiz & Wang, 2005). The 3G system has limited support for mobility. The concept of MIMO and OFDM is used to increase the mobility, security and throughput of 4G. Space-time coding increases the reliability, high data capacity and spectral efficiency (Fakih et al., 2009). Hence, with an increase in the network capacity, the transmission rate also increases.

Standard Wi-Fi and WiMAX have point to point architecture. Wi-Fi devices are readily available in the market. The components used in these devices are available off the shelf, which makes them an ideal candidate for use in ad hoc networks. Certain applications like VoIP and video conferencing can take down the network. WiMAX networks are high speed and designed specifically for last mile distribution but these devices are expensive.

There has been research in the field of Cellular Aided Mobile Ad hoc (CAMA) networks, operated in places where MANETs overlap cellular networks. The servers or CAMA agents are part of cellular network and have a record of registered mobile ad hoc users. CAMA operates well in metropolitan areas. Besides the peer-to-peer communication, CAMA allows the ad hoc nodes to be connected to Internet access point, which provides a cheaper Internet access without using expensive cellular channel. Only control data is handled by base station, the rest is limited to ad hoc networks. For implementing CAMA in 2G(GSM) the number of channels can be reserved (Bhargava et al., 2004).

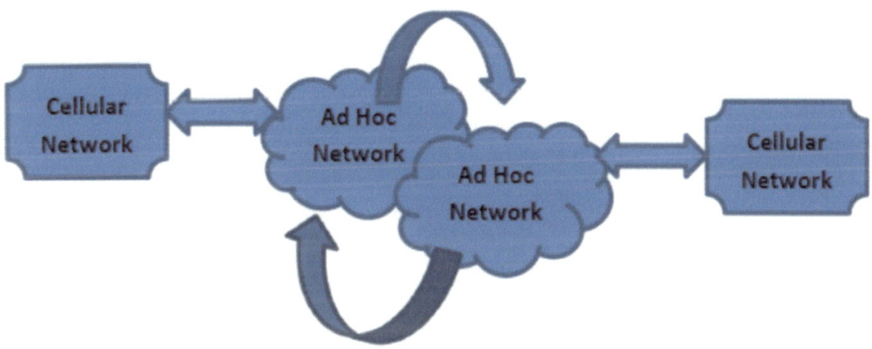

Fig. 4. Integrating ad hoc networks with cellular networks.

2.2 Applications of ad hoc networks

Ad hoc networking applications can be utilized in all kinds of networks due to their flexible nature. They are not limited to LAN or WAN but can be used in Body Area Networks (BAN), where connectivity among different types of wearable devices is required and Personal Area Networks (PAN), where connectivity between devices around a person is the objective. PDAs and notebooks can be automatically synchronized to transfer files, emails etc. A mobile phone from cellular network can connect to the PDA, allowing PDA to access cellular services without actually being part of the system. This can even lead to opening the door automatically upon your arrival at your home or office and adjust the air conditioning according to your preference. Airports can provide check in, boarding and seating services to customers as soon as they enter the airport and a lot of time can be saved by connecting to the customers' handheld device.

2.2.1 Audio applications

Certain applications like VoIP, video conferencing, file sharing etc., demand a close collaboration between network and application layers to ensure QoS. An important application is VoIP which is implemented by routing local telephone calls through the mesh.

There is no management in ad hoc networks that will ensure the QoS for services like VoIP, video conferencing and file sharing etc. Special protocols need to be designed to develop coordination between different Open System Interconnection (OSI) layers to implement VoIP services in ad hoc networks. Nodes in ad hoc networks need to recognize their Internet Protocol (IP) through Duplicate Address Detect (DAD) to ensure uniqueness. Since the IP is bound to change due to mobility, IP address cannot be used to initiate calls in ad hoc VoIP, instead a communication ID is used. To ensure security and authenticity, public and private key concept is used (Sun et al., 2005). VoIP can be used in many applications, including call center integration, directory services over telephones, IP video conferencing, fax over IP, and Radio/TV Broadcasting.

Push-to-talk (PTT) is like a walkie-talkie service defined by Open Mobile Alliance (OMA). Currently most operators are implementing PTT over cellular service. PTT performs an instant VoIP service in wireless ad hoc networks and is very useful in battle fields, earthquake and disaster relief. The growing popularity of services like Skype exploits the use of VoIP in a low cost manner. Now it is convenient for people to stay connected with other users across the world in an inexpensive way. The shortcoming is that despite the large bandwidth of IEEE 802.11, a limited number of calls can be completed successfully. In order to overcome this bottleneck, mesh networks technique can be exploited to improve routing so that the packet loss can be minimized. Here, the use of multihop ad hoc network can facilitate communication. In case of node failure, the re-routing process can result in delay or loss of packets which demand the routing process to be fast and efficient. The maintenance of end-to-end link during call is very important in all these applications (Chang et al., 2009).

2.2.2 Multimedia applications

Much work still needs to be done in the area of multimedia applications in ad hoc networks. Multimedia applications are relatively more complicated and must provide an acceptable video quality along with audio. These applications are not only used for entertainment purpose but are also useful for video conferencing. Video conferencing is beneficial for business meetings across the globes. Moreover, video conferencing can be really helpful for hearing impaired people who can communicate through sign language via video conference. Currently multimedia applications are expensive to use especially over portable devices. The use of ad hoc network in these areas can help reduce the cost and make the services more affordable for the ordinary user. Some cross layer protocols have been developed to optimize the video quality at the receiver. Since there is an increasing demand for multimedia applications, especially live streaming videos, there is a need for developing new and efficient protocols that are reliable.

2.3 Mesh networks

Wireless Mesh Networks (WMNs) are nodes organized in mesh topology consisting of mesh clients (cell phones, laptops, PDAs etc.), mesh routers and gateways, offering redundancy and reliability. Nodes themselves behave as router. WMN has a stable topology which may or may not be centralized. Non-centralized WMN can be considered a special type of ad hoc network with a more organized network topology, but unlike ad hoc networks, mesh routers are not limited in resources. The communication between other nodes is not affected by the sudden failure of a certain node. If a node fails, its neighboring nodes can find a new route among other nodes through the routing protocol. An interesting application of WMN is 66-satellite Iridium constellation with wireless links between adjacent satellites. Two satellites can communicate by routing through the mesh without the involvement of an earth station reducing the expenses as well as the time and distance the signal will otherwise have to travel back and forth from satellite to earth station, thus avoiding additional delays. That will also allow for lesser earth stations to be mounted saving precious resources (Akyldiz & Wang, 2005).

In WMN, nodes play their role as hosts and routers. WMNs are easy to deploy in ad hoc networks because of the inherent flexibility available in the latter. WMNs provide better connectivity and reliability, enhancing network performance and are easier to troubleshoot. Local networks can be developed and latency in communication can be reduced by communicating among local nodes without involving the main server all the time. For example, several devices in a home or office, like laptops, cell phones, PDAs, printers, TV, VCR, camcorders etc., can communicate with each other without access to the hub all the time. For integration of WMN with conventional networks, backward compatibility is also required.

Ad hoc networks are very useful in situations where a predefined infrastructure is unavailable, unreliable, or is destroyed due to a disaster. Since mobile devices have limited energy resources, they prefer establishing links to their nearest neighbors, which requires less setup time and energy, instead of long haul connections. Under this scenario, ad hoc mesh networks are a natural solution. Most modern technologies, for example, Rooftop communications (now part of Nokia), Mesh networks, and Radiant networks, use a multihop mesh-based architecture instead of 3G. Mesh networks are especially beneficial for last mile solutions (Wang & Lim, 2008).

Mesh networks can provide a natural solution for use of ad hoc networks in telecommunication applications. A mesh can be regarded similar to a cell which may or may not be centralized. Communication within a mesh network will be faster, efficient and cost effective. This method to connect communication devices also solves the basic connectivity issues faced by communication networks with mixed topology. Mesh networks are especially beneficial for multimedia applications among devices located in close proximity, (e.g., in an urban environment). Due to their scalable nature, the nodes can be added to or removed from the network.

2.4 Opportunistic networking

Opportunistic networking is an interesting and novel technique for multihop ad hoc networks. In this technique, a continuous, end to end path is not required between sender and receiver. Instead, the data is not lost upon disconnection due to topology change but is buffered in the node's local memory until a suitable forwarding opportunity is found. This is very useful, especially in data transmission applications like messaging, emails, data sharing etc., where immediate end to end connection is not required. However, in applications which require a continuous link between sender and receiver, like VoIP, gaming, video streaming, conference calling, this method is not desirable (Pelusi, 2006).

Opportunistic networking is a viable solution for data applications; however, it requires storage of data, which poses a memory capacity issue. This demands efficient data storage schemes leading to the encoding of data. This technique has also been shown to achieve minimum energy data transmission in which network coding is employed to fulfill this purpose. Network coding not only provides efficient data transmission, it also improves performance of noisy channels by adding redundancy. Network coding provides lower delays and less power consumption, hence it is suitable for ad hoc networks. In network coding, both source and relay nodes encode the data instead of repeating the source-coded data. As an alternative of receiving k packets, any set of k packets can be received and then decoded to retrieve original data. Data is organized into same length packets belonging to GF(2^s) over s bits. Assuming n packets ($M_0, M_1, \ldots, M_{(n-1)}$) of $l=L/s$ symbols, each node generates n coefficients $g_0, g_1, \ldots, g_{(n-1)}$ belonging to GF(2^s), outgoing packets x of size L are given as (Pelusi et al., 2009):

$$x = \sum_{i=0}^{n-1} g_i M_i$$

3. OSI model

Due to the independent and mobile nature of ad hoc networks, the OSI layers in ad hoc networks have a strong influence on each other. The physical layer should be able to handle the rapid changes in topology. Medium Access Control (MAC) layer should perform collision avoidance strategies. Network layer should be able to determine the optimum path for reliable communication. Transport layer handles the delay and loss of packets. Application layer should be able to handle frequent disconnections and reconnections as well as delay and packet loss. Any change in physical layer features like Signal to Interference plus Noise Ratio (SINR), modulation etc., affects the other layers of protocol stack, which need to be re-designed accordingly.

The common wireless protocol models include IEEE 802.11 PHY and MAC as well as ad hoc routing protocols like Ad hoc On-demand Distance Vector (AODV) and Distributed Source Routing (DSR). It is difficult to find an optimum solution for a network without taking into consideration all the network layers. In many circumstances, a cross layer solution is the most viable solution. At the link layer, the software design should consider scheduling, while at the network layer the design involves relaying of data or routing. Since the network resources are shared, the efficient management of resources is very important. Due to the lack of infrastructure scheduling and efficient resource, management of these networks is difficult and challenging. Interference mitigation techniques and distributed MAC protocols are used to handle the scheduling issues.

Self-Organizing Packet Radio Ad hoc Network with Overlay (SOPRANO) uses techniques to improve the capacity of cellular networks by involving ad hoc networks. It utilizes six steps of self-organization for physical, data and network layers to optimize bandwidth, routing and mobility management (Cavalcanti et al., 2005). The application layer supports network applications and manages network performance.

3.1 Physical layer

Physical layer effects include interference, noise, signal reception, fading, and path loss, which usually have an impact on routing protocols as well as on the performance of other higher layers. The physical layer handles data reception, modulation, data encryption, power management. Since there is no centralized mechanism to govern the ad hoc network, nodes are solely responsible to use their resources in the most efficient manner, therefore, power management is of pivotal importance. The protocol model is usually based on either SINR or Bit Error Rate (BER) (Rodriguez-Estrello et al., 2010).

Physical layer handles modulation/demodulation, error detection, decoding and is influenced by SINR present at the channel. Path loss models vary according to the environment in which the ad hoc networks are established (e.g., indoor environment or outdoor environment or free space, microcell or macrocell models). Physical layer preamble or signal header plays an important role in higher layer protocols as well. Common wire line medium defined by IEEE 802.3 standard requires only 8 bytes whereas Direct Sequence Spread Spectrum for IEEE 802.11 requires 192 microseconds for physical layer (Takai et al., 2001).

3.2 Data link layer

Data link layer is responsible for establishing reliable link, error control and security. It has two sub layers: Medium Access Control (MAC) and Logical Link Control (LLC). MAC is responsible for channel sharing, allotting time slots, frequency or code space among them. At the LLC, transmit power affects SINR and link capacity. The role of data link layer is important in multiplexing, frame detection, MAC and error control to ensure point-to-point and point-to-multipoint connections for reliable communication. Link layer also handles rate, power, and source/channel coding.

The main objective in MAC protocols is to optimize spectral reuse in order to maximize the channel utilization. Among many different MAC protocols, Carrier Sense Multiple Access/Collision Avoidance (CSMA/CA), IEEE CSMA/CA 802.11, also known as Wi-Fi is the most popular.

The data link layer allows nodes to use centralized MAC such as TDMA or CDMA when connecting to cellular networks, or random access scheme like CSMA/CA in IEEE 802.11. The cellular networks can achieve a data rate of 2.4 Mb/s while ad hoc 802.11b can provide up to 11 Mb/s.

3.2.1 MAC layer

MAC layer manages the framing, physical addressing, flow control, and error correction. This layer also assists in solving conflicts between nodes and ensures efficient and reliable transmission of data. The most commonly used scheme is CSMA/CA. In CSMA-based schemes, hidden and exposed node problems can exist. MAC protocol is designed to handle this issue. Several variations of MAC protocols exist. MAC protocols can be power aware to optimize power, or can use directional antennas which avoid interference and collisions in other direction and promote spatial reuse.

The contention based MAC protocols for ad hoc networks use random access protocols and collision resolution protocols. Random access schemes such as ALOHA allow channel use for nodes as soon as it is ready. Simultaneous transmissions can cause collisions. This works for less load and provides low throughput. Slotted ALOHA is a variation providing synchronized time slots similar to TDMA and in this scheme the throughput is doubled (Kumar et al., 2006).

Dynamic reservation collision resolution protocols, such as, Multiple Access Collision Avoidance (MACA) are used to solve hidden and exposed node problems in ad hoc networks by sending RTS/CTS packets to prevent collisions. MACA does not solve the hidden terminal and collision problem completely, especially for high network traffic. Transport layer will have to promote retransmission in case of transmission failure for any reason.

Another modification of MACA involves sending RTS-CTS-DS-DATA-ACK packets which provide better throughput, though it does not solve the hidden and exposed node problem completely. Floor Acquisition Multiple Access (FAMA) scheme solves the hidden node problem by repeatedly sending the CTS packet in response to a RTS packet, until all nodes can hear the CTS even if they cannot hear RTS (Drozda, 2005).

3.3 Network layer

Network layer is responsible for routing in ad hoc networks. To integrate ad hoc networks with the current telecommunication networks, the primary requirement is to develop a routing scheme between the two systems. With the appropriate implementation of network layer, the traffic flow can be ensured between the two types of networks. This enables the entire system to handle traffic between different types of systems in an efficient manner. A handoff between the two technologies can degrade the system performance which demands for application layer to handle the network requirements and ensure QoS requirements. The appropriate architecture of network layer is an important design consideration, which is not only necessary for providing connectivity, but it is also important for avoiding unnecessary bottlenecks or congestion points in order to restrict transmissions (Akyol et al., 2008).

3.3.1 Routing protocols

Routing in ad hoc network configurations is a challenging and important task in networks which lack central management. The ad hoc networks rely on efficient routing protocols which provide connectivity among various sources and destinations. Several routing protocols have been developed to serve different purposes and depend on network resources, changing network conditions and other parameters.

The routing protocols are broadly divided into table driven (or proactive) and on demand (or reactive) protocols. In addition, flow oriented protocols also exists which take into consideration the flow of data packets in the network. A combination of proactive and reactive protocols is classified as hybrid routing protocols. These protocols have functionality of both types of protocols incorporated in a single system. There are other types of protocols, which are basically variants of reactive or proactive protocols, which include hierarchical routing protocol, backpressure routing protocol, host specific routing protocols, power aware routing protocols, multicast routing protocol, geographical multicast routing (Geocasting) protocol, and optimized link state routing protocol.

The table driven (proactive) protocols maintain up-to-date information of route between all nodes. Routing information propagates through the network to keep the table recent. However, these protocols are not suitable for a highly dynamic network where node mobility is high. Table-driven protocols include Destination-Sequenced-Distance-Vector (DSDV) and Wireless Routing Protocol (WRP). DSDV is a routing protocol where every node maintains a routing table with a destination sequence number (Abolhassan et al., 2004). Cluster head Gateway Switch Routing (CGSR) is a special case of DSDV and uses clustered multihop mobile network for cluster-head-to-gateway routing. In WRP, each node maintains four tables: distance table, routing table, link cost and Message Retransmission List (MRL) tables. These tables are updated by sharing of information among neighboring nodes (Royer & Toh, 1999).

Source-initiated or on-demand routing protocols include AODV, DSR, Temporally Ordered Routing Algorithm (TORA) and Signal Stability Routing (SSR) protocols. Dedicated Source

Routing (DSR) involves route discovery and route maintenance. In route discovery, node broadcasts route request to its neighbors. The neighboring nodes in range add their ID to route request and initiate a rebroadcast which eventually reaches the destination or to the node that has a recent route to it. Nodes maintain a route cache, if a route is found in route cache, the node will return a route reply to source and forward the packet through cached route to the destination node.

AODV discovers routes on demand instead of keep updating the routing information, thereby reducing the number of broadcast messages. When a source node desires to send a packet, it checks its routing table for a route to the destination and then transmits. If route is not found, it broadcasts a route request to its neighbors which rebroadcasts it until a route is eventually discovered. Route maintenance is also important. If a node moves away, it sends a link failure notification to its neighbors to ensure deletion of that route and the source can reinitiate the route discovery process (Abolhasan et al., 2004).

TORA finds multiple routes in a highly dynamic network by making the nodes keep the routing information for one hop neighbor. It has three basic functions: route creation, route maintenance and route erasure and uses a height metric to create a Directed Acyclic Graph (DAG) (Royer & Toh, 1999).

3.4 Transport layer

The existing telecommunication applications use Transport Layer Protocol (TCP). In TCP, network congestion is detected by the packet loss. This scheme works well for wired and fixed topology networks, but is not suitable for ad hoc networks due to the continuous change in network topology, higher error rate and frequent disconnections. Moreover, in ad hoc networks, transport layer cannot be treated isolated for better network performance, since physical and MAC layers and routing protocols also have a significant effect on the transport layer. Packet loss may not be due to congestion in transport layer but due to disconnection, node movement, lower SINR or route change. TCP misinterprets route failures and wireless errors as congestion while delay causes it to initiate unnecessary retransmissions. Packet loss due to mobility is also detected as congestion.

Performance degradation of TCP, the transport layer protocol, is responsible for packet loss and congestion, delay and overlooked handoffs in cellular networks and is sometimes used with some modifications for ad hoc networks as well. Two kinds of handoffs exist; horizontal handoffs take place between the access points of the same network, while vertical handoffs are between different networks. Cellular networks handle horizontal handoffs at link layer and network layer handles mobility management. Vertical handoffs have to take into account when to start and how to regulate traffic for the handoff. When TCP is implemented in ad hoc networks, channel errors can be mistakenly termed as congestion and result in retransmissions causing degraded system performance especially in multihop networks. In single hop ad hoc networks, collisions are reduced but channel utilization is inefficient.

TCP performance also depends on the routing protocol used. Liu (Liu et al., 2011) demonstrated the effect of different routing protocols like AODV and OLSR. According to their study, OLSR provides better TCP performance for relatively slower node mobility as compared to AODV.

Ad hoc Transport Protocol (ATP) is specially designed for MANETs. It uses lower layer information, e.g., initial rate feedback, regular rate-based feedback from neighbors to control sending rate and path failure notification in case of route failure. Unlike conventional TCP which uses window-based transmission, ATP uses a rate-based transmission. ATP segregates congestion and reliability and an ACK signal is not required to clock the packet transmissions. ATP uses Selective Acknowledgment (SACK) reported periodically to detect packet loss and ensure reliability besides reducing traffic load on reverse path. ATP avoids bottleneck by accurate knowledge of maximum transmission rate obtained by feedback from neighboring nodes. ATP is incompatible with the TCP protocols and requires major modifications and redevelopment in the applications that use the older TCP version.

3.5 Power management

Since nodes in ad hoc networks are mobile, they depend on battery power and thus have limited energy. While the idea of running several multimedia applications is quite alluring, it also requires huge battery power, therefore, power management is an important issue in ad hoc networks.

Power management issue involves all the layers of protocol stack from physical to network layer. The energy efficient routing protocols are designed to reduce power consumption and take into account the remaining power of nodes in making routing decisions. To conserve power, links that are not used are shut off; network is properly scheduled to avoid wastage of energy due to congestion and collisions and by avoiding redundant transmissions (Goldsmith, 2002).

Most transmissions are performed in ad hoc networks through broadcasting. There are several broadcasting techniques that can be used, for example, blind flooding (each node rebroadcasts), probabilistic broadcasting and broadcasting based on area or neighbor knowledge. While blind flooding is the simplest approach, it has lot of redundancy and wastes too much power. Cluster based and tree based (source dependent or source independent) broadcast methods are also used where nodes transmit broadcasts messages to other members of clusters or trees. These methods belong to proactive routing.

Power management is a cross layer issue in ad hoc networks. At data link layer, power can be optimized by putting the unused nodes in standby mode, choosing consecutive slots for transmission and reception, avoiding collisions and retransmissions. At network layer, power consumption is minimized by using multihop routing, optimizing and lowering the number of control messages and using efficient routing techniques. The transport layer helps minimize power by controlling packet loss and retransmissions. Application layer adopts a dynamic QoS framework, helps power control by caching frequently used data/information and suppressing unimportant data to allow transmission of important information (Kawadia & Kumar, 2005).

4. Cognitive radio

Cognitive radio is a dynamic spectrum access technique and provides high bandwidth by opportunistically sharing the wireless channel with licensed users. The use of cognitive radio approach in ad hoc networks and the utilization of Dynamic Spectrum Access (DSA) technique is a promising method in further increasing the efficiency of radio frequency usage (Akyildiz et al., 2009). Cognitive Radio (CR) must be able to determine when and which portion of spectrum is available and select the best available channel, coordinate access to this channel with other ad hoc network users and vacate the channel when licensed user is detected. The cognitive radio detects an unused spectrum and allocates a channel to users. If more than one users need to share the spectrum, it coordinates the spectrum access to avoid collisions. If a primary user needs to access the spectrum, CR users have to vacate that spectrum and communication is continued in some other vacant portion of spectrum. CR network is not aware of the exact location of other nodes or the amount of interference offered by other nodes. There is a maximum interference limit (usually termed interference temperature) on each receiver which they can tolerate. If the CR users do not exceed this limit, they can use the channel. When a Primary User (PU) wants the channel, the CR users change their frequency of operation through a spectrum handoff. Each time a handoff occurs, the network needs to modify its protocols to avoid performance degradation during handoff. Handoff may be due to PU, node mobility or degradation in QoS. A temporary communication break-off cannot be avoided during a handoff since the CR user has to search for the next available band. The spectrum discovery process can take time and create longer switching time. The transmitter has to determine a new route according to the new spectral band to allow the commencement of interrupted communication.

Independent spectrum management involves opportunistic spectrum use (use of primary user's spectrum to avoid additional interference) or dynamic spectrum access. Regulated spectrum management involves pooling (two or more parties decide to pool resources), leasing, sharing or negotiating. *Spectrum sensing, spectrum decision* and *spectrum sharing* are the three important steps in designing a cognitive radio (Akyildiz et al., 2009).

Spectrum sensing is responsible for determining the available spectrum to be used by CR users. At the same time it should be capable of providing alternate vacant spectrum in case a primary user appears in the scenario. The CR users have to search for the available spectrum individually. Since the observation range of each user is limited, they have limited opportunity to access the spectrum. Cooperation among CR users can help to increase the efficiency of spectrum sensing and provides more accurate results.

Spectrum decision is also a crucial step in CR, since it enables the CR users to choose the best available spectrum band. It depends on the behavior of primary users as well as channel characteristics. It involves *spectrum characterization* (characteristics of spectrum as well as PU activity model), *spectrum selection* (choosing the best spectrum to ensure QoS) and *reconfiguration* (modifying protocols and hardware according to CR users' requirements) Unlike traditional ad hoc networks, CR users maintain a heterogeneous spectrum over time and space. Hence the resource allocation is dynamic and QoS parameters change with the change in spectral band. Moreover, the spectrum selection and decision is highly correlated with the routing protocols. Since the topology of CR can change due to their ad hoc nature, spectrum switching demands a whole new routing protocol and existing protocols for ad hoc networks may not be efficient or even working.

Spectrum sharing is another important development in CR networks, where CR users share the opportunistically obtained spectrum with each other. The users have to cooperate and avoid causing interference to other CR users. Spectrum sharing is performed within a band during a communication session. This involves proper resource allocation and designing a suitable MAC protocol to allow the coexistence of CR users among themselves (intra-network) and/or with other CR networks (inter networks). MAC protocols in case of CR have to take into account the spectrum sensing and sharing. Random access, time slotted and hybrid approaches are used to design MAC protocols in case of CR. Random access scheme is based on CSMA/CA. Dynamic Open Spectrum Sharing (DOSS) MAC protocol solves the hidden and exposed node problem. It is complex and uses spectrum inefficiently.

Time slotted MAC protocol such as Cognitive MAC (C-MAC), define slotted beaconing periods. Protocol determines the best available channel based on the beacon. C-MAC defines super frames for *data transfer period* (DTP), *beacon period* (BP) and *quiet period* (QP) (Akyildiz et al., 2009).

Primary user detection is very important in cognitive radio, as every cognitive user must vacate the spectrum upon detection of primary user. Primary user detection involves detection of transmitter, receiver (PU transmitting or receiving data in the communication range of CR users) and interference temperature management. Methods like matched filter, energy detection and feature detection are used to detect transmitter. Matched filter technique requires the CR users to have complete knowledge of primary user characteristics. Energy detection requires noise power. Energy detector can determine the presence of signal but signal type cannot be differentiated. Primary and CR users both cannot be distinguished, resulting in false alarm. Feature detection is based on determining the features of primary users like symbol rate, spreading codes, modulation type and cyclic prefix etc. This scheme is robust and has optimal performance, therefore, makes it most suitable for the case of ad hoc networks.

Hybrid protocols use partial slotting with synchronized control signaling. The access to channel may be completely random with the control or data durations. Resource allocation is another important factor for CR users, which can benefit by intelligently managing and utilizing their limited resources. Game theoretical concepts have been invoked to ensure the best possible strategies for CR users as well as to develop cooperation among them.

The Common Control Channel (CCC) coordinates transmission and provides spectrum information exchange between users. Since the parameters like channel quality, network load, access time are not initially known to CR users, CCC has to be designed with almost no information. Moreover, a primary user can make the spectrum unavailable for CR user at any point, hence an always on CCC is required to reliably send the control information to new spectrum. CCC can be in-band or out-of-band. In-band CCC is temporary and for a specific purpose only and uses the same data channel, avoiding additional spectrum switching cost. Out-of-band CCC has separate data and control signaling and CCC spectrum reservation may be permanent or for short duration.

CR users can simultaneously transmit to different users by tuning each transceiver to different spectral bands. This allows less power to be used in each band minimizing the interference. Knowledge of primary user activity is again very important as it helps to minimize the conflict and allows CR users to devise a strategy according to the PU activity.

Transport layer protocols must be designed to make them aware of PU activity as well as channel. Akyildiz (Akyildiz et al., 2009) developed a transport layer protocol for CR ad hoc networks named TP-CRAHN, which involves six states:

- Connection establishment (involves a three-way handshake for connection setup)
- Normal state (information is piggybacked over incoming ACK to aware the source)
- Spectrum sensing (minimize spectrum sensing time by keeping track of PU history)
- Spectrum change (source TCP state is frozen and after spectrum selection bandwidth is estimated and communicated to source)
- Mobility prediction (next hop is predicted using Kalman filtering)
- Route failure

5. Security

Security is an important concern in communication. In ad hoc networks, the problem is further complicated due to the absence of infrastructure and dynamic topology. The lack of centralized control makes the system more vulnerable to attacks and threats. Nodes can enter or leave the system at any time, making it difficult to identify a malicious node which might be re-joining the system and launching attacks from different locations repeatedly. The wireless channel, mobility of nodes, dynamic network topology and dependence on routing protocols all make the ad hoc networks more prone to attacks.

Security concerns include attacks from malicious users, availability, authentication, confidentiality, integrity, non-repudiation, routing protocol protection. Attacks may be *external* (outside network or denial of service from nodes) or *internal* (nodes inside network are turned malicious), *Passive* (eavesdropping) or *Active* (replication, modification and deletion of data). Passive attacks are more concerned with confidentiality and are relatively easier to combat. Active attacks are much more threatening to all security areas from authenticity to non-repudiation and require sophisticated protection schemes on all fronts (Yang et al., 2004). Attacks can be made directly on the routing to disrupt the source-destination path or attacks can be made on forwarding to cause packet loss.

Hackers can affect the security and reliability of network by deleting packets, manipulating the routing table, sending erroneous messages to nodes, jeopardizing the availability, integrity and authenticity of network. Attackers can also turn nodes inside the network into malicious nodes which can launch attacks from within. These nodes may appear to work correctly, but they may modify or misuse the routing protocols to undetectably disrupt routing, causing disconnections. These nodes can also inject more malicious nodes in the routing strategy and use the whole network for their malicious objectives. They can launch denial of service attacks by additional transmissions or unnecessary computations to exhaust battery power of nodes (Zhou & Haas, 1999).

Authentication is difficult to maintain due to the lack of centralized authority issuing certificate. Integrity of data involves key which may be symmetric or asymmetric (all nodes need to have others' public keys). Confidentiality can be ensured by changing the key every time a node enters or leaves the system. Availability involves service continuity despite Denial of Service (DoS) attacks. These attacks can be launched from any layer, physical, MAC or Network layer. Confidentiality means ensuring the protection of data from unintended users. This is especially important in military applications where information leakage to enemy can be detrimental. Integrity involves making sure that transmitted data is never corrupted, either due to channel conditions or malicious node involvement. Authentication protects the identity of a node and helps avoid attackers. Non-repudiation ensures that source node (or relay) cannot deny the initiation of message.

Security issue can be solved by protecting the MAC layer; encrypting everything, verification of authenticity and making sure the private keys are secure. Keeping too many keys and different security levels at every layer assists in solving security issues. Public Key Infrastructure (PKI) can be used but asymmetric cryptography is expensive. In PKI, every node has a public/private key pair. The public part is made known to all other nodes, keeping the private key confidential. A Certificate Authority (CA) manages the key by distributing its own public key and signing certificate binding keys to nodes. Usually, more than one CA is used to ensure protection in case of unavailability of CA or in cases when CA itself might be compromised. Malfunctioning of CA can threaten the security of the whole network. CA technique can only be implemented in ad hoc networks if these are integrated with cellular systems where they can benefit from the centralized management of cellular networks. When working in isolation, MANETs cannot use central key management systems, rather a distributed key management scheme is used that relies on secret sharing where key is distributed using cryptography. Another method is that each node builds its trust chart based on the nodes it trusts and matches it with the node it wants to communicate with.

Security can be maintained by verifying the authenticity of message between communicating nodes. Another approach to improve security is through Intrusion Detection (ID). To ensure route security, routing protocols use multi-hopping and authenticity is verified on hop-to-hop basis. All intermediate nodes validate digital signatures cryptographically.

Byzantine robustness, which is a strict self-stabilization policy, ensures the proper functioning of routing protocol even in the presence of some malicious nodes. Routing protocols should be designed so that they not only recover from attacks but also ensure the normal functioning of system even during attacks. Trust maintenance is also very important in ad hoc networks.

Secure routing protocols used in MANETs include Secure Efficient Ad hoc Distance vector (SEAD) where source selects a random seed and sets the maximum hop count. Using a hash function, source computes hash value *h*(seed). Fuzzy logic is employed in protocols to improve security. In FL-SAODV protocol, each mobile host uses a secure key with its neighbor nodes and relies on the secret key and node's environment. FL-SAODV uses security level as one of the parameters while choosing a route. Moreover, it aims to minimize the transmitting time by choosing shortest path so that attack time is reduced. Dynamic Destination Multicast (DDM) protocol is another security protocol where only sender has the authority to control information transmission. This protocol is based on grouping and does not allow arbitrary nodes from outside to enter the group without prior authorization.

Another protocol to improve security is *i-key* protocol where secret key is generated using previous data as seed which ensures that the decryption can only be performed by sender or authorized client. Alliance of Remote Instructional Authoring and Distributed Networks for Europe (ARIADNE) is another secure routing protocol that uses shared secrets or digital signatures and utilizes DSR and symmetric cryptography. This works for one attacker only. SEcure Neighbour Discovery (SEND) Vector has lower processing time and uses one way public key signed hash function. Authenticated Routing for Ad hoc Network (ARAN) protects against malicious nodes using pre-determined public key.

SRP is another scheme that provides better security by splitting a message into fragments and transmitting them along different paths. The original message can be reconstructed if a certain number of fragments are correctly received.

Security also depends on number of neighbors and key length. Frequently changing the key also helps in improving security by making it less vulnerable. It has been found that security level is directly proportional to the key changing frequency and key length and inversely proportional to the number of neighboring nodes. Security level is also dependent on the application. Some applications are more sensitive and require higher level of security, for example mobile banking and E-commerce applications. Other applications like VoIP and video conferencing require privacy as well as maintain QoS.

5.1 Intrusion detection

Intrusion detection involves in determining whether the system is under attack or not. Two types of Intrusion Detection Schemes (IDS) are used: host-based or network based. Network based IDS works at gateways and examines network packets. Host based IDS uses operating system data to analyze program or user activities. IDS are categorized into *misuse detection* and *anomaly detection*. Misuse detection identifies intrusion using patterns of known attacks or weak spots. Anomaly detection detects deviations from normal activities.

6. Social impacts

The integration of ad hoc networks in the current telecommunication system will greatly affect the lifestyle of an ordinary person. While the people already involved in technology will feel facilitated by this new system, an ordinary person will also find these services useful as well as entertaining.

At home ad hoc networks can provide connectivity to different personal devices such as notebooks, PDAs, camcorders, TV, cell phones and even appliances which include air conditioners, microwave oven, intercom etc. It also helps people in getting organized. Since no cost is required to establish the infrastructure, the services offered are quite cheap and are attractive for masses.

Staying connected is also very important for business professionals as it helps broaden their horizon by exploring new markets. Important business transactions can be completed and meetings with worldwide companies can be held through video conferencing. Improved telecommunication has been linked to economic growth.

With the growing popularity of social networking sites like Facebook and Twitter, especially among the younger generation, the demand to have an all time Internet access everywhere is increasing. Ad hoc networks can provide connectivity even in remote areas at a very affordable price.

Due to the access to multimedia applications and availability of live streaming, people can watch TV on their potable hand held devices. This is especially very attractive for people who are interested in sports and don't want to miss their favorite games. Now people don't have to walk in a music store or theater to get a CD of their favorite songs or watch their favorite movie.

6.1 Impact on business players

Several stake holders are involved from device manufacturers, infrastructure equipment vendor, operating system providers, application developers, service providers, and customers. It is difficult to replace the existing infrastructure so the current network operators will still be in business. Skills of technical people will still be required for the new system development. Operators can support ad hoc networking in public venues like airports, hospitals, hotels, stadium, and theatres.

Device manufacturers face more responsibility and challenge, because ad hoc devices differ from current devices. Ad hoc network devices are capable of routing. Some of the functions available in mobile devices can be replaced or moved to the device itself. For example, vehicular applications can be mounted on the vehicle itself instead of integrating with the hand held device.

With the increase in devices and also the available applications, the operating system needs to be modified and improved. This opens up more business opportunities for providers of operating systems. The operating systems have to deal with routing and security issues of ad hoc networks and handle the bottle necks in the network.

Application developers can take advantage of technological developments, posing no threat to existing markets and creating new frontiers. Applications can be provided to customers without the involvement of mobile operators. Application service providers have more benefit in large centralized networks than smaller ad hoc networks. But they can offer variety to customers and can gain from services which require the users to be in close proximity of each other.

Network operators can be affected in a variety of ways. Although the demands for multiple services are on the increase, most facilities can be provided free of cost in ad hoc networks. The monopoly of network operators will be affected and new players will come forward in a bargaining position. Network operators will cease to be in absolute authority or they may chose to provide ad hoc networking services themselves.

Mobile portal providers also have market in large networks but new demands may emerge for applications. Supporting services providers will have to broaden their services to include security and cater to the needs of operators, portal providers and application service providers. Operators of venues can provide services to their customers without the involvement of traditional network operators. For example, airport or train stations can track lost or stolen baggage using ad hoc networks (Stanoevska-Slabeva & Heitmann, 2003).

7. Conclusion

In this chapter we give the readers in depth information about the role of ad hoc network in mobile telecommunications. Various aspects of these networks are discussed, ranging from the integration issues, adaptation to radical changes, and layered implementation of ad hoc networks. Special emphasis has been placed on the routing protocols as they play a very crucial role in the otherwise autonomous ad hoc networks. The use of mesh networking has been discussed to improve the connectivity and resource management in small area networks. We argue that the ad hoc networks have a definite advantage in specific situations over other existing network technologies and systems, and therefore, can be used in wide range of applications. It has been discussed that ad hoc networks can also improve the performance of existing networks and provide a wider range of services if integrated in the existing networks. The use of other existing and upcoming technologies in ad hoc networks is also discussed. The implementation of cognitive radio technology has been specifically presented to show how the available spectrum can be used efficiently. The security aspect of ad hoc networks is shown and methods are proposed to improve the security of the vulnerable ad hoc networks. We also present the future trends and social issues of the use of ad hoc networks from the perspective of users as well as from business point of view.

8. References

Abolhasan, M., Wysocki, T. & Dutkiewicz, E. (2004). A Review of Routing Protocols for Mobile Ad hoc Networks, *Ad Hoc Networks,* Vol. 2, No. 1, pp. 1-22

Akyol, U., Andrews, M., Gupta, P., Hobby, J., Saniee, I. & Stolyar, A. (2008). Joint Scheduling and Congestion Control in Mobile Ad-Hoc Networks, *INFOCOM 2008, The 27th IEEE Conference on Computer Communications,* April 2008, pp. 619-627

Akyildiz, I., Lee, W. & Chowdhury, K. (2009). CRAHNs: Cognitive Radio Ad Hoc Networks, *Ad hoc Networks,* Vol. 7, No. 5, 2009, pp. 810-836

Akyildiz, I., Wang, X. & Wang, W. (2005). Wireless Mesh Networks: a Survey, *Computer Networks,* Vol. 47, No. 4, March 2005, pp. 445-487

Akyildiz, I., Lee, W., Vuran, M. & Mohanty, S. (2008). A Survey on Spectrum Management in Cognitive Radio Networks, *IEEE Communications Magazine,* Vol. 46, No. 4, pp. 40-48

Armuelles, A., Robles, T., Chaouchi, H., Ganchev, I, O'droma, M. & Sierbet, M. (2004). On Ad Hoc Networks in the 4G Integration Process, *The Third Annual Mediterranean Ad Hoc Networking Workshop (Med-Hoc-Net 2004),* Bodrum, Turkey, 2004, pp. 45-56

Bayer, N., Xu, B., Rakocevic, V. & Habermann, J. (2010). Application-Aware Scheduling for VoIP in Wireless Mesh Networks, *Computer Networks,* Vol. 54, No. 2, February 2010, pp. 257-277

Bhargava, B., Wu, X., Lu, Y. & Wang, W. (2004). Integrating Heterogeneous Wireless Technologies: A Cellular Aided Mobile Ad hoc Network (CAMA), *Mobile Networks and Applications,* Vol. 9, No. 4, August 2004, pp. 393-408

Pelusi, L., Passarella, A. & Conti, M. (2009). Encoding for Efficient Data Distribution in Multihop Ad Hoc Networks, In: *Algorithms and Protocols for Wireless and Mobile Ad hoc Networks,* Boukerche, A. (Ed.) (2009), John Wiley & Sons, Inc., ISBN 978-0-470-38358-2, New Jersey.

Cavalcanti, D., Agrawal, D., Cordeiro, C., Xie, B. & Kumar, A. (2005). Issues in Integrating Cellular Networks, WLANs and MANETs: A Futuristic Heterogeneous Wireless Network, *IEEE Wireless Communications,* Vol. 12, No. 3, June 2005, pp. 30-41

Cricellia, L., Grimaldia, M. & Ghiron, N. (2011). The Competition Among Mobile Network Operators in the Telecommunication Supply Chain, *Int. J. Production Economics,* 2011, pp. 22–29

Chang, L., Sung, C., Chu, H. & Liaw, J. (2009). Design and Implementation of the Push-to-Talk Service in Ad hoc VoIP Network Communications, *IET,* Vol. 3, No. 5, 2009, pp. 740-751

Comeras, M., Bafalluy, J. & Suriol, M. (2007). Performance Issues for VoIP Call Routing in a Hybrid Ad hoc Office Environment, *The 16th IST Mobile and Wireless Communications Summit,* 2007, pp. 1-5

Drozda, M. (2005). *Performance Analysis of Wireless Ad hoc Networks.* Ph.D. Dissertation.

Fakih, K., Diouris, J. & Andrieux, G. (2009). Transmission Strategies in MIMO Ad hoc Networks, *EURASIP Journal on Wireless Communications and Networking,* Vol. 2009, 2009

Freedman, A. (2009). International Mobile Telecommunications - Advanced Tutorial Highlights, *Microwaves, Communications, Antennas and Electronics Systems, COMCAS, International Conference IEEE,* 2009, pp. 1-2

Goldsmith, A. (2002). Design Challenges for Energy Constrained Ad Hoc Wireless Networks, *IEEE Wireless Communications,* In: *The Handbook of Ad Hoc Networks,* Ilyas, M. (Ed.), 2003, CRC Press LLC, ISBN 0-8493- 1332-5, Danvers, USA

Hac, A. (2002). *Mobile Telecommunications Protocols for Data Networks,* John Wiley, 2002, ISBN: 047-085-056-6

Kawadia, V. & Kumar, P. (2005). Principles and Protocols for Power Control in Wireless Ad Hoc Networks, *IEEE Journal on Selected Areas in Communications,* Vol. 23, No. 1, 2005

Kumar, S., Raghavan, V. & Deng, J. (2006). Medium Access Control Protocols for Ad hoc Wireless Networks: A Survey, *Ad Hoc Networks,* Vol. 4, No. 3, May 2006, pp. 326-358

Kaosar, M. & Sheltami, T. (2009). Voice Transmission Over Ad hoc Network Adapting Optimum Approaches to Maximize the Performance, *Computer Communications,* Vol. 32, No. 4, March 2009, pp. 634-639

Liu, P., Chen, D., Hu, C., Sun, W., Lee, J., Chou, C. & Shih, W. (2011). Analyzing the TCP Performance on Mobile Ad-hoc Networks, *ICACT,* Feb. 13~16, 2011, ISBN 978-89-5519-155-4143

Pelusi, L., Passarella, A. & Conti, M. (2006). Opportunistic Networking: Data Forwarding in Disconnected Mobile Ad hoc Networks, *IEEE Communication Magazine,* 2006, pp. 134-141

Remondo, D. & Niemegeers, I. (2003). Ad hoc Networking in Future Wireless Communications, *Computer Communications,* Vol. 26, No. 1, January 2003, pp. 36-40

Royer, E. & Toh, C. (1999). A Review of Current Routing Protocols Ad hoc Mobile Wireless Networks, *Personal Communications, IEEE,* Vol. 6, No. 2, April 1999, pp. 46-55

Rodriguez-Estrello, C., Valdez, G. & Perez, F. (2010). Performance Modeling and Analysis of Mobile Wireless Networks, In: *Mobile and Wireless Communications Physical Layer Development and Implementation,* Fares, S. & Adachi, F. (Eds.), ISBN: 978-953-307-043-8, InTech, Available from: http://www.intechopen.com/articles/show/title/performance-modeling-and-analysis-of-mobile-wireless-networks

Sarkar, S., Basavaraju, T. & Puttamadappa, C. (2007). *Ad Hoc Mobile Wireless Networks,* Taylor & Francis Ltd (United Kingdom), 2007, ISBN 13 9781420062212

Smith, N. J. (2006). Suitability of UMTS to Act as an Ad Hoc Network Gateway for VoIP Services, *Wireless Telecommunications Symposium, WTS,* 2006, pp. 1-6

Stanoevska-Slabeva, K. & Heitmann, M. (2003). Impact of Mobile Ad Hoc Networks on the Mobile Value System, *2nd Conference on m-Business,* Vienna, June 2003

Sun, Y., Fang, G., & Shi, J. (2005). *Research on the Implementation of VoIP Service in Mobile Ad-hoc Network.* Auswireless Conference, 2006

Szczodrak, M., Kim, J., & Baek, Y. (2007). *4GM@4GW: Implementing 4G in the Military Mobile Ad-Hoc Network Environment.* 2007. IJCSNS International Journal of Computer Science and Network Security, Vol. 7, No. 4, April 2007

Takai, M., Martin, J. & Bagrodia, R. (2001). Effects of Wireless Physical Layer Modeling in Mobile Ad hoc Networks, *Proceedings of the 2001 ACM Symposium on Mobile Ad Hoc Networking & Computing, New York, ACM Press,* pp. 87-94

Wang, X. (2008). Wireless Mesh Networks, *Journal of Telemedicine and Telecare,* Vol. 14, No. 8, pp. 401–403

Wang, X. & Lim, A. (2008). IEEE 802.11s Wireless Mesh Networks: Framework and Challenges, *Ad Hoc Networks,* Vol. 6, pp. 970-984

Wang, Y., Sanguansintukul, S. & Lursinsap, C. (2008). The Customer Lifetime Value Prediction in Mobile Telecommunications, Management of Innovation and Technology, *ICMIT, IEEE 4th International Conference,* 2008, pp. 565-569

Yang, H., Lou, H., Ye, F., Lu, S. & Zhang, L. (2004). Security in Mobile Ad Hoc Networks: Challenges and Solutions, *IEEE Wireless Communications,* 2004

Zhou, L. & Haas, Z. (1999). Securing Ad Hoc Networks, *IEEE Network,* Vol. 13, No. 6, Nov/Dec. 1999, pp. 24-30

Zhao, X., and Liang, H. (2010). Capacity Dimensioning for Wireless Communications System, In: *Mobile and Wireless Communications Physical Layer Development and Implementation,* Fares, S. & Adachi, F. (Eds.), ISBN: 978-953-307-043-8, InTech, Available from: http://www.intechopen.com/articles/show/title/capacity-dimensioning-for-wireless-communications-system

11

Beamforming in 3G and 4G Mobile Communications: The Switched-Beam Approach

Konstantinos A. Gotsis[1] and John N. Sahalos[2]
[1]*Aristotle University of Thessaloniki*
[2]*University of Nicosia*
[1]*Greece*
[2]*Cyprus*

1. Introduction

The technology and deployment of modern mobile communications systems, should adapt to the continuous and rapid growth of wireless data traffic. Besides the increase in bandwidth, the addition of cell sites and sectors, and the enhancement of air interface capabilities, smart antennas play also a substantial role in the improvement of wireless systems' performance. When we refer to smart antennas, we mean structures of multiple antenna elements at the transmitting and/or the receiving side of the radio link, whose signals are properly processed, in order to better exploit the mobile radio channel and enhance the communications performance. During the last decade there has been intensive research on Multiple Input-Multiple Output (MIMO) systems, which are antenna formations that involve processing at both sides of the link (Jensen & Wallace, 2004).
Depending on the signal processing methods and the adaptive schemes used, smart antenna techniques can be separated into three broad categories: a) Diversity, b) Spatial Multiplexing (SM), and c) Beamforming. Roughly speaking, beamforming aims at improving Signal to Interference plus Noise Ratio (SINR), diversity aims at reducing the variations in the SINR experienced by the receiver, while SM aims at sharing SINR in high SINR scenarios (3G-Americas, 2009).
Transmit and receive diversity are used in order to mitigate the problem of multipath fading, enhancing the reliability of a wireless link. In SM, which is the most popular transmission scheme of MIMO systems, multiple data streams are transmitted in parallel, increasing the data transmission rate. Beamforming uses an antenna array[1] to transmit/receive energy in a specific direction, increasing the cellular capacity and coverage. Although it is an early smart antenna technology, beamforming is still supported by the latest 3GPP releases (3rd Generation Partnership Project - www.3gpp.org), namely the LTE (Long Term Evolution) and LTE advanced. Operators though seem reluctant to incorporate smart beamforming techniques into base stations, mainly due to the cost and complexity of such implementations. The current research concerning 4G mobile systems is mostly concentrated on MIMO processing and space-time coding algorithms. However, antenna arrays and beamforming have still the potential to contribute in the enhancement of modern cellular systems.

[1] Antenna array is called the aggregation of radiating elements in a certain electrical and geometrical arrangement (Balanis, 2005).

Beamforming systems are generally classified as either Switched-Beam Systems (SBS) or Adaptive Array Systems (AAS). A SBS relies on a fixed BeamForming Network (BFN) that produces a set of predefined beams. Probably the most popular solution for fixed BFN is a Butler Matrix (BM) (Butler & Lowe, 1961). A BM in its standard form is a $M \times M$ network, which consists of hybrid couplers, phase shifters and crossovers. M is the number of input/output ports that give a set of M different beams. In (Kaifas & Sahalos, 2006), the reader can find a comprehensive review of the BM functionality and its implementation issues.
A SBS needs a Switching Network (SN) in order to select the appropriate beam to receive the signal from a particular Mobile Station (MS). As it is shown in Fig. 1a, the maximum of the selected beam might not point at the desired direction. Moreover, typically a beam serves more than one MS. On the contrary, an AAS has the possibility to form a special beam for each user Fig. 1b. This is accomplished by a series of adaptive array processors that apply weight vectors to the received and transmitted signals, in order to control the relative phase between the antenna elements and their amplitude distribution. In this way specific beam patterns can be produced, directing the main lobe towards the desired MS and nulls towards the interfering signals. Thus, Direction of Arrival (DoA) estimation of signals impinging on an antenna array is a very important issue for cellular communications.

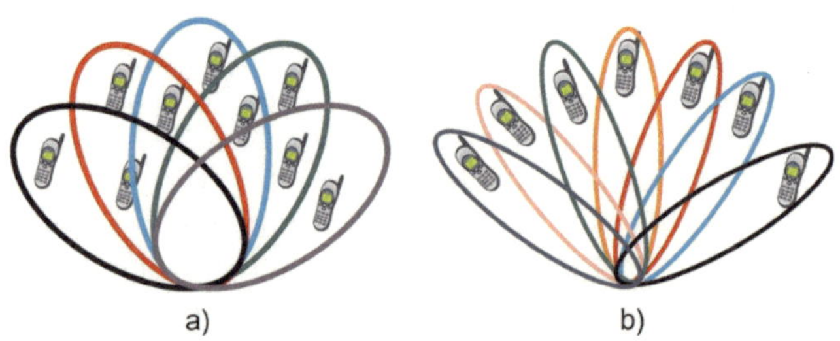

Fig. 1. Beam coverage of a) switched-beam system and b) adaptive array system.

Adaptive beamforming presupposes that the Base Station (BS) updates the localization of the MS. However, accurate localization is not an easy task, since a big number of simultaneous mobile users can overload the process. Therefore, although many popular DoA estimation methods and adaptive beamforming algorithms have been developed (Godara, 2004), no particular standardization has been established. The implementation of an adaptive system is much more complex than a switched-beam one. On the other hand, ideal adaptive beam pointing minimizes the interference between users and exploits much better the available power resources. In any case there are advantages and disadvantages which have to be considered (Baumgartner, 2003; Baumgartner & Bonek, 2006; Osseiran et al., 2001; Pedersen et al., 2003).
In order to exploit the simplicity of SBS, and promote effective smart antenna beamforming, the authors in (Gotsis et al., 2009) developed a Neural Network (NN) DoA estimation methodology, which has been especially designed for an SBS and applied to a Direct Sequence Code Division Multiple Access (DS-CDMA) scheme. As a succession of the above paper, an improved BM based beamforming network has been presented in (Gotsis et al., 2010). The proposed structure provides enhanced beamforming flexibility compared to a typical BM and also has the possibility to efficiently work in conjunction with the NN-DoA estimation technique.
The main scope of this chapter is the completion and extension of the work presented in the above references. A basic subject is the introduction of 'DoA-based Switching (DoAS)'

and its incorporation into a 3G mobile communications framework, like the Universal Mobile Telecommunications System (UMTS). DoAS is evaluated and compared to the so called in this work 'Typical Switching (TS)'. The term 'TS' refers to the typical operation of a SBS, which is determined by the highest uplink SINR or by the highest mean received power.
In (3G-Americas, 2009; 2010) one can find various propositions for 4G systems that combine MIMO techniques and fixed beamforming (using BM) into a single structure. Generally, BM still attracts the interest of researchers (Chia-Chan et al., 2010; Peng et al., 2009). Therefore, the 3G case is followed by a discussion of how the switched-beam approach using BM could be useful in the context of next generation systems.

2. The improved switched-beam system

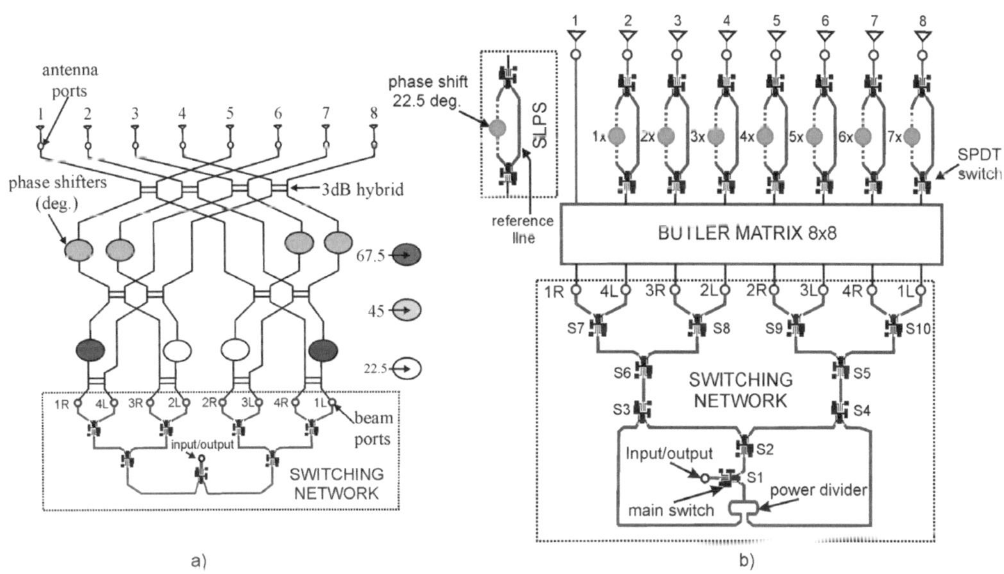

Fig. 2. a) Typical schematic of a Butler Matrix (BM) 8 × 8 and its SN. b) Schematic of a beamforming network with a typical BM 8 × 8 enhanced by the appropriate SN and a group of SLPS.

2.1 The beamforming network

A concentrated illustration of the beamforming network proposed in (Gotsis et al., 2010), is depicted in Fig. 2b. In Fig. 2a, a typical schematic of a BM 8 × 8 and its SN is shown. The BFN is constituted of three main blocks: a) a BM 8 × 8, b) a switching network and c) a block of Switched Line Phase Shifters (SLPS). Regarding the BM, taking into account the radiation pattern characteristics, the degrees of freedom offered and the complexity of the implementation, the 8 × 8 dimensions are the most appropriate to cover an angular sector of 120°.

2.1.1 Single and combined BM port excitation

The SN consists of a circuit of Single Pole Double Throw (SPDT) switches (Pozar, 1990) in dendroid structure and a 3dB power divider. A digital word determines the switches' state ('0'

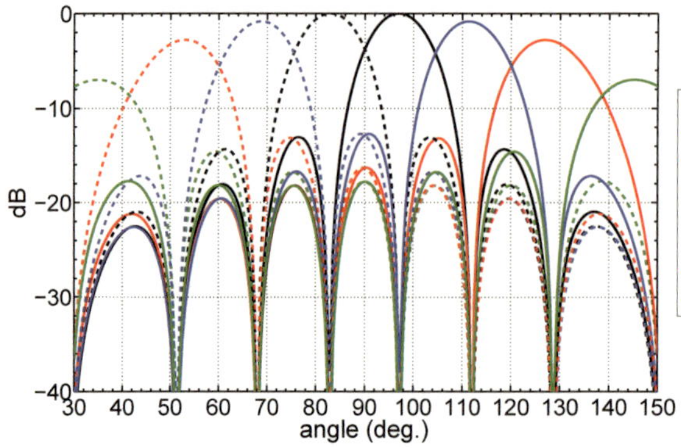

(a)

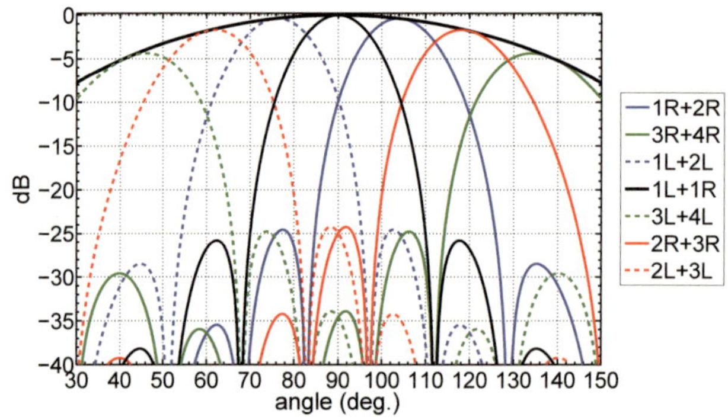

(b)

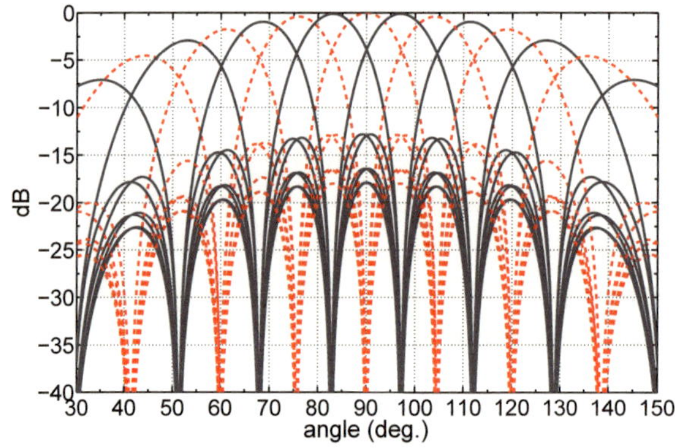

(c)

Fig. 3. Radiation patterns produced by the proposed SBS: a) eight orthogonal beams from the single (1:0) beam port excitation, b) seven beams from the combined (1:1) port excitation plus a sector beam from a single antenna element, and c) enhanced grid of 15 beams resulting from the addition of the SLPS at the output of the BM 8 $\times$ 8. The dotted lines correspond to the added beams.

or '1') and thus, the signal's route through the network. The first bit of the word corresponds to the state of the main switch (S1), which gives two options. The first option is the typical single port excitation (1:0) of the BM, that leads to the uniform illumination of the antenna elements. The result is a grid of eight orthogonal beams, as shown in Fig. 3a. The beams are called orthogonal in the sense that at one beam's maximum, all the other beam patterns have a minimum. There are four beams at the right (R) and four at the left (L) of the broadside. The beam ports take their name after the position of the corresponding beam in relation to the broadside. The second option is the combined equal excitation (1:1) of the BM ports that correspond to pairs of adjacent in space orthogonal beams. This results to the cosine illumination of the antenna elements and a grid of seven beams, as shown in Fig. 3b. In the same diagram the sector beam of a single antenna element (a rectangular microstrip patch) is also illustrated. The reader observes that 1:1 provides beams with lower Side Lobe Level (SLL), higher Crossover Level (CL) and wider Half-Power BeamWidth (HPBW) than the 1:0 case. This is due to the cosine versus the uniform illumination of the antenna elements, that confirms the following rule of thumb: "in linear antenna array synthesis the array with the smoothest amplitude distribution has the smallest side lobes and the larger HPBW (and vice versa)" (Balanis, 2005). Details about the characteristics of the discussed radiation patterns in (Gotsis et al., 2009).

2.1.2 Extra beams from single port excitation

Extra beams in a BM can be produced if a block of SLPS is added at its output (see Fig. 2). The SLPS do not modify the layout of the typical BM. They just interconnect the last row of the hybrids and the antenna ports of the BM. A SLPS is a simple structure, which uses two SPDT switches to drive the signal between the one of two microstrip lines of different length (Pozar, 1990). The length difference determines the phase difference between the two paths. The one path (solid reference line) of the proposed structure, is the path of the typical BM that gives the orthogonal beams. The length of the other line (dotted in Fig. 2b), has been chosen so that an extra phase $\Delta\phi = (n-1)\cdot 22.5^\circ$ is added at the nth antenna element[2]. Thus, the progressive phase difference produced by the excitation of each beam port is increased by 22.5°. For example the beam port 1R produces either the predefined phase difference $\beta = 22.5^\circ$ or $\beta = 45^\circ$. The new phase difference corresponds to a new beam that lies between 1R and 2R. All possible single port excitations create an enhanced grid of 15 beams. This grid is constituted from the standard eight orthogonal beams, plus the seven beams (also orthogonal between them) that come from the use of the added SLPS (Fig. 3c). Table 1 summarizes the features of the resulting pattern.

Compared to Fig. 3a the CL is much higher. Even between the edge beams, the CL is higher than -3dB. Between the rest of the beams the CL ranges from -1.9dB to -0.9dB, whereas in the typical eight beams pattern ranges from -4.7dB to -3.8dB (Gotsis et al., 2009). This means that anywhere in the 120° sector a MS can be served by nearly the maximum of one of the main lobes, which has the narrowest possible HPBW. For example the 1:0 beam with maximum at $\theta_0 = 61^\circ$ has $HPBW = 14.2^\circ$, while the 1:1 beam with maximum at $\theta_0 = 62^\circ$ has $HPBW = 18.5^\circ$. The 1:0 excitations give narrower beamwidths than all the possible corresponding combined excitations $1 : x$, where $0 < x < 1$. On the other hand, the 1:0 beams have the highest SLL.

Besides producing the desired radiation patterns, the SLPS addition is practical and easy to implement. Their integration in a single layer microstrip structure is simple and the SPDT

[2] It should be noted that for the 1st element there is no phase shifter, since its phase is taken as a reference.

Beam	$\beta(^\circ)$	HPBW($^\circ$)	SLL (dB)	θ_0 ($^\circ$)	CL (dB)
7L	-157.5	19.2	-7.6	35	6L-7L: -2.6
6L	-135.0	17.2	-9.3	44.5	5L-6L: -1.9
5L	-112.5	15.5	-10.4	53	4L-5L: -1.6
4L	-90.0	14.2	-11.2	61	3L-4L: -1.3
3L	-67.5	13.4	-11.9	68.5	2L-3L: -1.1
2L	-45.0	13.0	-12.5	76	1L-2L: -1.0
1L	-22.5	12.8	-13.1	83	1L-0: -0.9
0	0	12.6	-13.7	90	1R-0: -0.9
1R	22.5	12.8	-13.1	97	1R-2R: -1.0
2R	45.0	13.0	-12.5	104	2R-3R: -1.1
3R	67.5	13.4	-11.9	111.5	3R-4R: -1.3
4R	90.0	14.2	-11.2	119	4R-5R: -1.6
5R	112.5	15.5	-10.4	127	5R-6R: -1.9
6R	135.0	17.2	-9.3	135.5	6R-7R: -2.6
7R	157.5	19.2	-7.6	145	

Table 1. Radiation pattern characteristics of the enhanced grid of 15 beams resulting from the addition of the SLPS.

switches offer very fast (nanosecond) switching between their two states. An important feature of the SLPS is their equivalent operation for both transmission and reception. The proposed way of their integration reserves the typical BM functionality and does not modify the SN, which has been designed for simplicity and minimization of losses.

2.2 The neural network direction of arrival estimation method

Several algorithms have been proposed concerning DoA estimation (Godara, 2004). In principle, most of these have been designed for adaptive systems. Very popular are the super resolution algorithms, MUSIC (Schmidt, 1986), ESPRIT (Ray & Kailath, 1989), and their variants (e.g. ROOT MUSIC). Over two decades later, many research proposals are still based on the concept of these classic methods, e.g. (Ying & Boon, 2010). However, the drawback of these approaches is the need for intensive signal processing, like eigenvalue decomposition and signal autocorrelation matrix calculations. In order to avoid eigenvalue decomposition, NN-DoA finding procedures have been developed, which basically apply the mapping of the signal autocorrelation matrix with the signals' angles of arrival (AoA) (Christodoulou & Georgiopoulos, 2001; El Zooghby et al., 2000). A DoA estimation methodology has been firstly presented in (Gotsis et al., 2007), based on the mapping between the signals' AoA and the power measured at the input/output of the BFN. The mapping is exploited through the supervised learning of NNs. The main novelty of this approach has been the special design for SBS in conjunction with NNs, without any other complex signal processing techniques. The advantage of NNs is that although their training may be time consuming, the response of their application is instant. This makes them appropriate for real time applications like DoA estimation and beamforming.

In (Gotsis et al., 2008) the signal model has been given and the generic concept of the method has been extended to a DS-CDMA mobile communications scheme, focusing on the DoA estimation of the desired mobile user at the presence of many other interfering signals that constitute the Multiple Access Interference (MAI). In (Gotsis et al., 2009) the method has been described in detail and studied in depth through extensive simulations. It has been shown

that the desired signal's DoA can be extracted with a less than one degree Root Mean Square Error (RMSE), even if its power level is 6dB less than each one of 39 interfering signals. A basic conclusion was that the 1:1 excitation of the BM should be used for the uplink communication between the MS and the base station. Due to the non-orthogonality and the high CL of the seven beams pattern, the 1:1 mode gives much better DoA estimation results than the typical 1:0. The orthogonality and higher directivity of the eight beams pattern are better for the downlink transmission towards the desired MS. That is why the SN of the SBS has been designed to support both excitation modes.
In the next section various beamforming possibilities involving the proposed SBS and the NN-DoA estimation method are presented and evaluated in terms of their performance in the framework of UMTS.

3. Smart switched beamforming in UMTS

UMTS has been standardized by the 3rd Generation Partnership Project (3GPP), which has been a collaboration of various telecommunications associations from Europe, Japan, Korea, USA and China. The original scope of 3GPP was to produce technical specifications and technical reports for a 3G and beyond mobile system. Release 99 was the first release of 3G specifications and it was essentially a consolidation of the underlying Global System for Mobile Communications (GSM) and the development of the new Universal Terrestrial Radio Access Network (UTRAN). There have been many steps till 3GPP reached the Long Term Evolution (LTE), which is a pre-4G standard, one step behind the LTE Advanced that fully complies with the IMT-Advanced (International Mobile Telecommunications Advanced) requirements for 4G standards (www.3gpp.org).

3.1 Beam switching modes

Base station beamforming for UMTS has followed the general classification described in the introduction. Particularly, UMTS Release 6 (3GPP, 2004) specified three possible modes: "none", "flexible beamforming", and "grid of fixed beams". The "none" mode signifies that beamforming is optional. "Flexible beamforming" corresponds to the adaptive beamforming used by an AAS, whereas the "grid of fixed beams" corresponds to a set of predefined beams used by a SBS. As it was mentioned in the introduction, the SBS may operate either in the TS or the DoAS mode.

3.1.1 Typical Switching (TS)

In (3GPP, 2004) it has been specified that the operation of a SBS depends on the so called 'best cell portion measurement', which corresponds to the TS operation mode. A cell portion has been "the part of a cell that is covered by a specific beam antenna radiation pattern, which can be created using a grid of fixed beam directions". In this context a cell is a "logical cell", which is the area covered by the Primary Common Pilot Channel (P-CPICH) (Pedersen et al., 2003). When a call is initiated, the User Equipment (UE)[3] is being accepted by the beam that measures the highest SINR for the Physical Random Access Channel (PRACH). During the communication the switching is activated depending on the SINR measurement for the Dedicated Physical Control Channel (DPCCH) and it is reported to the Radio Network Controller (RNC). The difference between common and dedicated channels is that in the first case the channels' resources are shared to all or a group of cell users, whereas in the second

[3] In UMTS terminology the mobile station is called User Equipment and the base station Node-B.

case the channels are dedicated, through a particular code and frequency, to only one user. The RNC which is the governing element in UTRAN and controls the Node-Bs that are connected to it, decides for the best downlink beam. When a user leaves the coverage region of a fixed beam and enters another beam region of the same cell, the UE should get informed about this change. The information is done by the Radio Resource Control (RRC) messages, like the RRC physical channel reconfiguration message (3GPP, 2008). The detailed description of UMTS communication channels and their role in beamforming is out of the scope of this chapter; these can be found in the literature (Baumgartner, 2003; Holma & Toskala, 2007; Pedersen et al., 2003). Basic procedures are roughly given in order to frame our study.

3.1.2 DoA-based Switching (DoAS)

DoA-based switching can be applied using the NN-DoA estimation technique described earlier. Figure 4 illustrates an example of DoAS operation. Three SBS are established in a trihedral form at the Node-B of an UTRA network. Each system feeds a linear array of eight microstrip patches and covers a sector-cell of 120°. The grid of seven cosine illumination beams is used for the uplink. One of the array's elements takes on the transmission of a sector beam for the service of the common communication channels (e.g. the P-CPICH). The P-CPICH reception quality, determines the service sectors of the UE. The data transmission/reception between the Node-B and the UE takes place through one of the fixed beams.

Since the NN-DoA estimation is based only on power measurements, no modifications of the UMTS specifications are needed. According to the concept of TS, the method is applied at the PRACH control bits when the communication is initialized, whereas during the communication DoA estimation is applied at the DPCCH bits. The only difference is that the Node-B's power measurements are not instantly transmitted to the RNC, but they are first processed and fed to the appropriate NN, which gives as output the DoA of the desired signal. The NN training should follow the guidelines of (Gotsis et al., 2009), taking into account the restrictions and requirements of the particular communication scenario (e.g. processing gain, maximum number of simultaneous users, SIR variation range etc). Then, in order to activate beam switching, the standard UMTS signalling procedure between Node-B and the RNC is

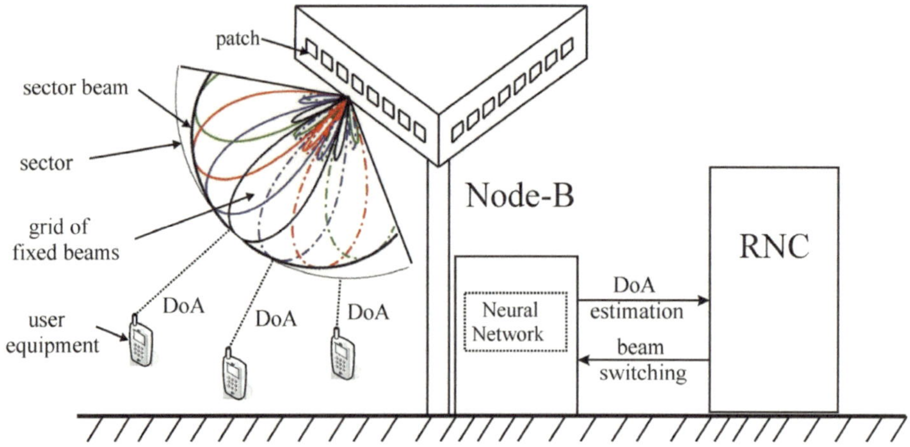

Fig. 4. UMTS DoA-based Switching operation of the SBS .

realized. However, the "best cell portion measurement" is not determined by the highest SINR, but by the estimated actual DoA of the signal. The choice of the downlink beam is based on the uplink DoA estimation, which is a realistic approach that is used in wireless communications. This has been shown experimentally for the communication between a BS with an eight element linear antenna array and a single element MS (Pedersen et al., 1999).

3.2 Performance of TS and DoAS

Depending on the switches' state, the proposed SBS (Fig. 2b) supports various switched-beam configurations. The simplest one is the typical use of the eight orthogonal beams (1:0) for both the Up and Down (UD) link. This configuration should work in the TS mode and is called '8UD'. An advanced choice for the TS mode, which is a proposition of this work, is the use of fifteen beams (1:0) for both the uplink and downlink (i.e. '15UD' configuration). Finally, a novel structure called '7U15D', works in the DoAS mode using the seven beams (1:1) for the uplink and the fifteen beams (1:0) for the downlink. A simulation model has been developed in order to perform a comparison between 15UD and 7U15D with the conventional 8UD. The model makes use of the Monte Carlo approach. N mobile stations are randomly located in an angular sector of 120° with radius R_0 (the logical cell of the communication scenario) (Fig. 5). The stations are served by a single SBS that covers the sector and it is assumed that there is no interference from other adjacent cells. In the case of DoAS mode the downlink beam is chosen from a predefined lookup table, depending on the estimated DoA of the desired signal on the uplink. In the TS mode the NN-DoA algorithm is not involved and the downlink beam is the one that has measured the highest uplink received power.

Since the downlink beam has been chosen, power control is assumed and a particular $SINR_t$ target is set, in order to calculate the transmit power required for the desired mobile user. The SINR for the nth UE is calculated by

$$\lambda_n = \frac{\frac{P_{n,b}}{L_{n,b}} Q}{N_0 + (1-\alpha) I_{n,b}} = \frac{\frac{P_{n,b}}{L_{n,b}} Q}{I_n^{tot}} \tag{1}$$

where $P_{n,b}$ is the transmit power of the beam b that serves the mobile n, Q is the processing gain, $L_{n,b}$ is the path loss between the nth station and beam b, $I_{n,b}$ is the received power at the nth station due to all other stations served by the same beam, N_0 is the thermal noise, and I_n^{tot} is the total interference at station n. The non-orthogonality factor $0 \leq \alpha \leq 1$ takes into

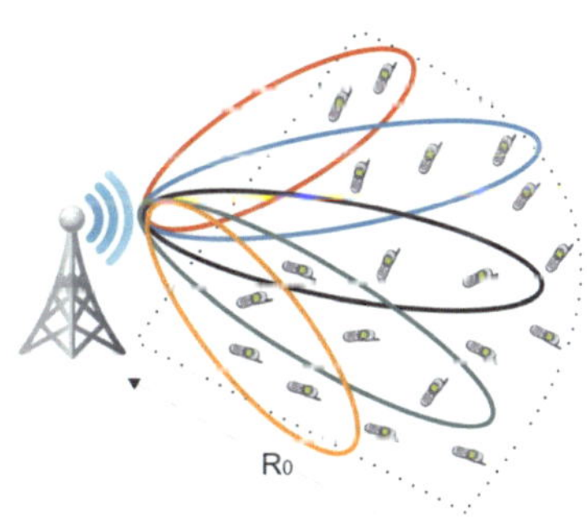

Fig. 5. The SBS covers an angular sector of 120° and serves N randomly located mobile stations.

account that signals spread with different code-words are not fully orthogonal after travelling through a radio channel with more than one delay tap (Baumgartner, 2003). The path loss in dB is calculated using the COST Walfisch-Ikegami-Model (COST-WI), for the line-of-sight (LOS) case between the base station and the mobile antennas

$$L_{n,b}[\mathrm{dB}] = 42.6 + 26\log(R_n/\mathrm{km}) + 20\log(f/\mathrm{MHz}) - D_b(\theta_n), R_n \geq 20\mathrm{m} \tag{2}$$

where R_n is the distance between the nth UE and the base station, f is the downlink operation frequency and $D_b(\theta_n)$ is the directivity pattern of beam b towards the angle θ_n of the station. The directivity patterns come from the radiation patterns, taking also into account the directivity of each beam calculated by the ORAMA computer tool (Sahalos, 2006). Given that moving towards the edges of the sector the beams have wider HPBW, the directivity patterns give even weaker coverage at these angular regions, compared to the corresponding radiation patterns. The received interference at the nth UE, served by beam b, due to the M stations served by the same beam is

$$I_{n,b} = \sum_{m=1\neq n,b}^{M} \frac{P_{m,b}}{L_{n,b}} = \sum_{m=1\neq n,b}^{M} 10^{0.1(P_{m,b}[\mathrm{dBm}]-L_{n,b}[\mathrm{dB}])}\ [\mathrm{mW}] \tag{3}$$

where $P_{m,b}$ is the transmit power of the beam b that serves the mobile m. $P_{m,b}$ is calculated so as to achieve at the corresponding UE a particular SNR_t. Therefore

$$P_{m,b}[\mathrm{dBm}] = N_0[\mathrm{dBm}] + L_{m,b}[\mathrm{dB}] + SNR_t[\mathrm{dB}] \tag{4}$$

The required transmit power $P^r_{n,b}$ is given by

$$P^r_{n,b}[\mathrm{dBm}] = I^{tot}_n[\mathrm{dBm}] + L_{m(n),b}[\mathrm{dB}] + SINR_t[\mathrm{dB}] \ - 10\log Q \tag{5}$$

where

$$I^{tot}_n[\mathrm{dBm}] = 10\log\{10^{0.1N_0[\mathrm{dBm}]}[\mathrm{mW}] + (1-\alpha)I_{n,b}[\mathrm{mW}]\} \tag{6}$$

The simulation equations are followed by a specific numerical application in order to quantify the performance of each beam configuration. Thermal noise is calculated for $N_0 = -99\mathrm{dBm}$ (3GPP, 2007). The non-orthogonality factor is taken for $\alpha = 0.5$, the cell radius $R_0 = 1\mathrm{km}$, the processing gain for both uplink and downlink $Q = 128$, the downlink operation frequency $f = 2000\mathrm{MHz}$, the SINR target $SINR_t = 20\mathrm{dB}$, the SNR target $SNR_t = 10\mathrm{dB}$ and the total number of simultaneous users ranges from 10 to 80.
Depending on the type of switching the downlink beam is chosen for each one of the three beam configurations and a mean $P^r_{n,b}$ is calculated for the desired user, for each beamforming case and number of users. The simulations results (after 10000 independent runs) show that when 15 downlink beams are used (either TS or DoAS), the performance is almost identical and there is a power gain of about 5 dB per user compared to the 8 beams case. This is depicted in Fig. 6a, where the mean required transmit power towards the desired user, is plotted for each switched-beam structure. The improvement comes from a) the fact of fewer interfering signals per beam and b) the more directive transmission due to the bigger number of available beams. The coincidence between 15UD and 7U15D happens since normally the DoAS mode activates the same downlink beam as the typical one. This stands even when the number of

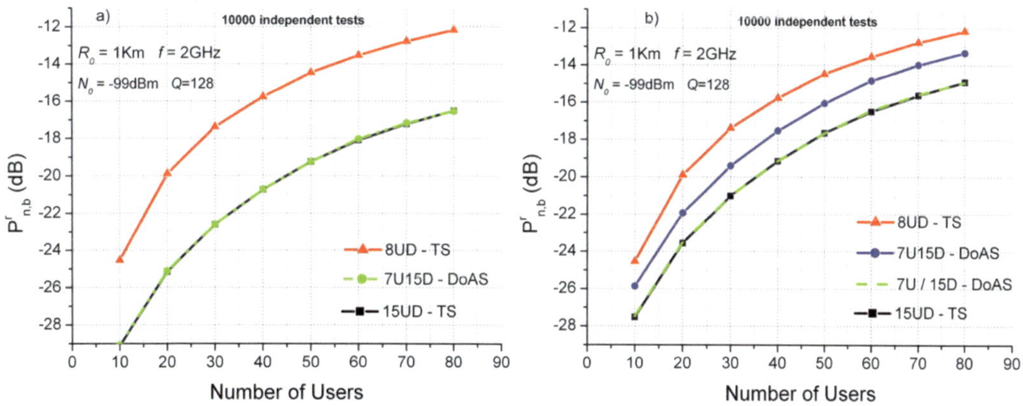

Fig. 6. a) Mean transmit power required for each switched-beam configuration in order to achieve at the desired user an $SINR_t = 20$dB. In b) the results take also into account the BFN power losses due to the SPDT switches.

simultaneous users increases and the DoA estimation performance deteriorates, mostly at the edges of the sector where the coverage is weaker.

The 5 dB power gain does not take into account the BFN losses due to the SPDT switches. In order to have a more 'complete' evaluation of the three switched-beam configurations, the performance study considers three separate SBS, one for each configuration. Figure 2b shows the 7U15D SBS, whose SN has both the 1:0 and 1:1 excitation possibility and at the output of the BM there is a group of SLPS for the extra seven 1:0 orthogonal beams. In Fig. 2a, it is shown that a typical 8UD SBS needs only a simple SN (without a 3 dB divider) for the eight 1:0 orthogonal beams. Finally, a 15UD SBS needs the same SN as the 8UD SBS, plus the group of SLPS. It is obvious that the 8UD system uses the fewer SPDT switches, whereas the 7U15D the more. Figure 6b gives the mean required transmit power taking also into account that each switch has an insertion loss of about 0.8 dB. The power level reference for the curves of Fig. 6 is the required transmit power of a sector beam that comes from one microstrip patch (at the presence of 40 users).

Both modes that use fifteen downlink beams perform better than the typical eight beams case. The 7U15D (DoAS) has less gain than the 15UD (TS), due to the extra switches of the SN. However, DoAS uses only seven instead of fifteen uplink beams, and thus there is less administrative cost. The above lead us to another possible configuration that consists of two separate SBS, one for the uplink and one for the downlink. The uplink SBS works with an SN that supports only the 1:1 mode and it is responsible for the DoA estimation using the grid of 7 beams. The downlink SBS works only in the 1:0 mode and uses the enhanced grid of 15 beams to transmit towards the mobile users. This structure, symbolised as 7U / 15D, uses less uplink beams, having the same downlink performance with the 15UD case (dashed line in Fig. 6).

4. Beamforming in beyond 3G and 4G mobile communications

An extension of the DoAS concept towards next generation beamforming, would require the application of the proposed DoA estimation method to an Orthogonal Frequency-Division Multiple Access (OFDMA) scheme (instead of DS-CDMA), since OFDMA is the dominant multiple access technique for 4G mobile communications systems. However, according to the latest 3GPP releases, base station multi-antenna configurations should be able to choose the appropriate smart antenna technique, depending on the radiocommunications requirements and the channel characteristics. Thus, it is important to consider beamforming in an overall smart antenna context, together with spatial multiplexing and diversity. Following this context, a practical example of combining SM and beamforming will be described in this section.

4.1 Transmission rank

Spatial multiplexing is a MIMO scheme that employs multiple antennas at both sides of the radio link, in order to create multiple parallel channels that share the overall SINR and increase the data transmission rate. The number of simultaneously transmitted parallel streams is termed as 'transmission rank'. Generally speaking, the transmission rank can be defined as the number of independent symbols transmitted per time-frequency resource (3G-Americas, 2009). Although the need for this definition comes from spatial multiplexing (transmission ranks higher than one), it can be also used for beamforming and transmit diversity, which obviously are considered as single rank schemes. As it has been mentioned in the introductory section, beamforming and diversity aim at improving SINR, whereas SM aims at sharing SINR. Thus, generally speaking, rank-one transmissions increase coverage and keep a steady quality, whereas higher ranks improve data rates.

4.2 Codebook & non-codebook based spatial precoding

4.2.1 Spatial multiplexing

The various ways an antenna array transmits the modulated symbols with spatial multiplexing precoding, can be given by a general expression for the transmit vector **T**

$$\mathbf{T} = \begin{bmatrix} w_{11} & \cdots & w_{1K} \\ \vdots & \cdots & \vdots \\ w_{N1} & \cdots & w_{NK} \end{bmatrix} \begin{bmatrix} s_1 \\ s_2 \\ \vdots \\ s_K \end{bmatrix} = [\mathbf{w}_1\ \mathbf{w}_2\ \ldots\ \mathbf{w}_K] \begin{bmatrix} s_1 \\ s_2 \\ \vdots \\ s_K \end{bmatrix} = \mathbf{Ws} \quad (7)$$

where w_{nk} is the nth element of the kth weight vector $\mathbf{w}_k$; $\mathbf{w}_k$ weights the symbol s_k, which corresponds to the kth symbol stream. The weight vectors may be dependent or independent of the channel, depending on the availability of channel information on the transmit side. Moreover, when the weight matrix **W** is chosen from a pre-fixed set of matrices we have the so-called 'codebook-based precoding', whereas when the weight choice is free we have the 'non-codebook-based precoding'.

Spatial multiplexing needs low spatial correlation on both the transmitter and receiver side. This can be accomplished by either co-polarized antenna elements with large distance between them, or cross-polarized elements with small inter-element distance. The definitions of 'large' and 'small' depend on the angular spread and the wavelength, however for a base station a typical small inter-antenna distance may be taken as half a wavelength, whereas a large as four to ten wavelengths.

4.2.2 Beamforming

Unlike SM, beamforming needs strong spatial correlation, which typically means a half-wavelength spaced antenna array of co-polarized antenna elements. Beamforming can be expressed by a vector $\mathbf{T}_{bf}$, which comes from a reduced form of (7)

$$\mathbf{T}_{bf} = \mathbf{w}_1 s_1 = \begin{bmatrix} w_1 s_1 \\ w_2 s_1 \\ \vdots \\ w_N s_1 \end{bmatrix} \tag{8}$$

where only a single symbol s_1 is multiplied by a weight vector $\mathbf{w}_1$. The nth element of the weight vector (w_n) controls the phase and the amplitude of the nth antenna element. If the weight vector is chosen from a set of predefined vectors we have 'codebook-based beamforming', whereas when the weight choice is not restricted and dynamically adapts to channel's variations we refer to 'non-codebook-based beamforming'. Beam switching using a set of fixed beams may be considered as a simple form of codebook-based beamforming, since each beam corresponds to a predefined weight vector.

4.3 Beam switching combined with spatial multiplexing

Consider two antenna arrays separated by a distance of several wavelengths (e.g. four). Each array consists of two half-wavelength spaced elements. The small distance between the antenna elements favors the use of beamforming, whereas the large inter-array distance is useful for spatial multiplexing. Therefore, such a structure may be used for the parallel transmission of two symbol streams $s1\&s2$, one from each beamforming antenna array. Thus, besides the increase in data transmission rate, there may be also SINR improvement due to the directive transmission of the symbols.

Figure 7 depicts a simple implementation that combines beam switching with SM. Two BM 2×2 provide the antenna arrays with the pre-defined necessary phase adjustments, in order to cover the desired sector with two beams instead of a typical sector beam from a single element. A BM 2×2 is practically an 90° hybrid coupler. An SPDT switch is used to select between the two ports of the coupler, which give a phase difference between the elements of either 90° or −90°. Figure 8 shows the resulting beams 1L and 1R. Depending on the channel condition and the DoA information the best beam should be chosen for transmission.

The beamforming concept described above can be extended to antenna configurations with more elements and Butler matrices with bigger dimensions. For example, a structure with a

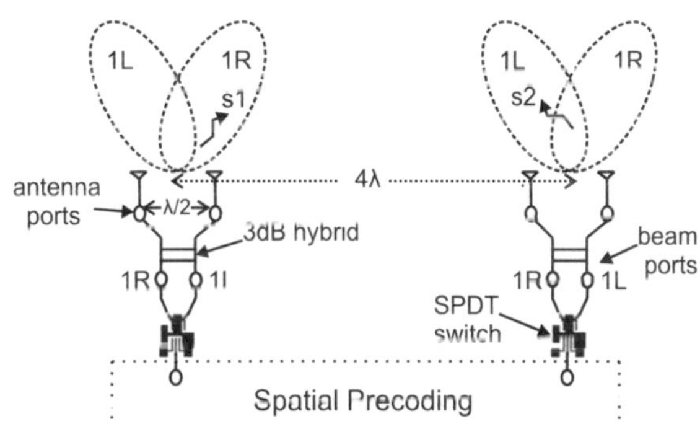

Fig. 7. Schematic of a transmitter combining spatial multiplexing and beam switching for a 2×2 MIMO scheme.

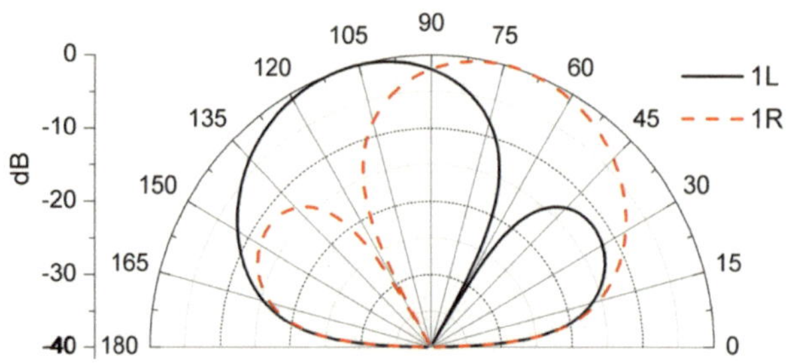

Fig. 8. Beams produced by an antenna array of two orthogonal microstrip patches fed by a BM 2×2.

BM 4×4 may be also used for a 2×2 transmission scheme. However, in this case there are more beam choices regarding the symbol transmission.

5. Conclusions

The research on smart beamforming systems has been started several decades before, when the first radars were developed for military purposes. The term beamforming refers to the function done by a group of co-operating antenna elements (called an antenna array), in order to form the desired radiation pattern and direct the radiated energy towards a specific target. Adaptive beamforming, namely the formation of a dynamically changing beam pattern that continuously directs a maximum towards the desired user and nulls towards the interfering signals, is theoretically an ideal operation. However, despite the great research done on this field, adaptive array systems have not been adopted by the wireless communications market, mainly due to their cost and complexity. Contrary to them, switched-beam systems (SBS) constitute probably the most easily implemented choice. Instead of adaptive array processors, a SBS uses a simple switching network (SN) to select the most appropriate beam from a set of predefined beams produced by a fixed beamforming network (BFN). The main scope of this chapter has been the investigation of the potential of the switched-beam approach in modern mobile communications.

The most popular fixed BFN for applications in mobile communications is the Butler Matrix (BM). The various beamforming possibilities of an improved BM based SBS have been discussed and extensive simulations took place, in order to compare their performance in terms of the required base station transmit power towards the desired mobile user. Besides the Typical Switching (TS) determined by the highest received power or SINR, a novel beamforming proposal has been also examined, which is the operation of a SBS that uses the DoA information of the desired user. The proposal is called DoA-based Switching (DoAS) and uses the neural network (NN) DoA estimation methodology developed by the authors in a previous work (Gotsis et al., 2009). The NN has low processing time compared to an adaptive array processor that runs a typical adaptive beamforming algorithm. This, together with the rapid response of the SN's switches provides fast decision and operation.

The simulations results lead to general conclusions concerning the operation of a SBS in a UMTS base station and also evaluate DoAS. The improved SBS besides supporting DoAS, it also provides a set of extra directive beams for more accurate transmission towards a target. The use of these fixed beams either in a typical way or in conjunction with DoAS gives improved performance results and shows very good potentiality. However, in order to further increase the overall power gain, the losses of the described BFN should be reduced. The design of a more power-effective BFN could be a next step of our work.

The chapter ends with a brief discussion regarding the evolution of beamforming, and especially the role of switched-beam techniques, from 3G towards 4G mobile communications systems. Modern implementations require that smart antenna configurations adapt to the varying conditions of a radiocommunications link and choose accordingly the most suitable smart antenna technique or even a combination of them. Within this context, a simple antenna structure has been described that applies beam switching (using a BM) in conjunction with spatial multiplexing (SM). Such a structure combines the increase in coverage and capacity due to beamforming, with the increase in the data transmission rate due to SM. SM is a Multiple Input-Multiple Output (MIMO) scheme that involves the parallel transmission of multiple data streams, using two or more antenna elements at both sides of the link. Contrary to that, beamforming is a Multiple Input-Single Output (MISO) technique, since it uses an array of antenna elements only at one side of the link (typically the base station).

As a future work in this field we intend to investigate the possibility of the simultaneous transmission of orthogonal beams from a single BM, in order to combine beam switching with SM into a single antenna array. The BM beam orthogonality may be useful in the context of MIMO processing, where low correlation between signals is needed.

6. References

3G-Americas (2009). MIMO transmission schemes for LTE and HSPA networks, *White Paper* .

3G-Americas (2010). MIMO and smart antennas for 3G and 4G wireless systems, *White Paper* .

3GPP (2004). Beamforming enhancements, TR 25.887 V6.0.0 (2004-03), *Release 6*, 3GPP.

3GPP (2007). Radio Frequency (RF) system scenarios, TR 25.942 V7.0.0 (2007-03), *Release 7*, 3GPP.

3GPP (2008). Radio Resource Control (RRC) Protocol Specification, TS 25.331 V7.11.0 (2008-12), *Release 7*, 3GPP.

Balanis, C. A. (2005). *Antenna Theory: Analysis and Design*, John Wiley and Sons, Hoboken, New Jersey.

Baumgartner, T (2003). *Smart Antenna Strategies for the UMTS FDD Downlink*, PhD thesis, Technischen Universität Wien, Wien.

Baumgartner, T. & Bonek, E. (2006). On the optimum number of beams for fixed beam smart antennas in UMTS FDD, *Wireless Communication, IEEE Transactions on* Vol. 5(No. 3): 560–567.

Butler, J. & Lowe, R. (1961). Beamforming matrix simplifies design of electronically scanned antennas, *Electronic Design* .

Chia-Chan, C., Ruey-Hsuan, L. & Ting-Yen, S. (2010). Design of a beam switching/steering Butler Matrix for phased array system, *Antennas and Propagation, IEEE Transactions on* Vol. 58(No. 2): 367 –374.

Christodoulou, C. & Georgiopoulos, M. (2001). *Applications of Neural Networks in Electromagnetics*, Artech House, Boston-London.

El Zooghby, A., Christodoulou, C. & Georgiopoulos, M. (2000). A neural network-based smart antenna for multiple source tracking, *Antennas and Propagation, IEEE Transactions on* Vol. 48(No. 5): 768–776.

Godara, L. C. (2004). *Smart Antennas*, CRC Press, Boca Raton.

Gotsis, K. A., Kaifas, T. N., Siakavara, K. & Sahalos, J. N. (2008). Direction of Arrival (DoA) estimation for a Switched-Beam DS-CDMA System using neural networks, *Journal Automatika* Vol. 49(No. 1-2): 27–33.

Gotsis, K. A., Kyriacou, G. A. & Sahalos, J. N. (2010). Improved Butler Matrix configuration for smart beamforming operations, *Antennas and Propagation (EuCAP), 2010 Proceedings of the Fourth European Conference on*, pp. 1 –4.

Gotsis, K. A., Siakavara, K. & Sahalos, J. N. (2009). On the Direction of Arrival (DoA) estimation for a switched-beam antenna system using neural networks, *Antennas and Propagation, IEEE Transactions on* Vol. 57(No. 5): 1399–1411.

Gotsis, K. A., Vaitsopoulos, E. G., Siakavara, K. & Sahalos, J. N. (2007). Multiple signal Direction of Arrival (DoA) estimation for a switched-beam system using neural networks, *Progress In Electromagn. Research Symp.*, Prague, Czech Republic, pp. 420–424.

Holma, H. & Toskala, A. (2007). *WCDMA for UMTS - HSPA Evolution and LTE*, Wiley.

Jensen, M. & Wallace, J. (2004). A review of antennas and propagation for MIMO wireless communications, *Antennas and Propagation, IEEE Transactions on* Vol. 52(No. 11): 2810 – 2824.

Kaifas, T. & Sahalos, J. (2006). On the design of a single-layer wideband Butler Matrix for switched-beam UMTS system applications [wireless corner], *Antennas and Propagation Magazine, IEEE* Vol. 48(No. 6): 193 –204.

Osseiran, A., Ericson, M., Barta, M., Goransson, B. & Hagerman, B. (2001). Downlink capacity comparison between different smart antenna concepts in a mixed service WCDMA system, *Proc. IEEE VTC*, pp. 1528–1532.

Pedersen, K. I., Mogensen, P. E. & Frederiksen, F. (1999). Joint directional properties of uplink and downlink channel in mobile communications, *Electronics Letters* Vol. 35: 1311–1312.

Pedersen, K. I., Mogensen, P. E. & Moreno, J. R. (2003). Application and performance of downlink beamforming techniques in UMTS, *IEEE Communications Magazine* pp. 134–143.

Peng, C., Wei, H., Zhenqi, K., Junfeng, X., Haiming, W., Jixin, C., Hongjun, T., Jianyi, Z. & Ke, W. (2009). A multibeam antenna based on substrate integrated waveguide technology for MIMO wireless communications, *Antennas and Propagation, IEEE Transactions on* Vol. 57(No. 6): 1813 –1821.

Pozar, D. M. (1990). *Microwave Engineering*, Addison-Wesley.

Ray, R. & Kailath, T. (1989). Esprit-estimation of signal parameters via rotational invariance techniques, *Acoustics, Speech and Signal Processing, IEEE Transactions on* Vol. 37: 984–995.

Sahalos, J. N. (2006). *Orthogonal Methods for Array Synthesis: Theory and the ORAMA Computer Tool*, John Wiley & Sons, New York.

Schmidt, R. O. (1986). Multiple emitter location and signal parameter estimation, *Antennas and Propagation, IEEE Transactions on* Vol. 34(No. 3): 276–280.

Ying, Z. & Boon, P.-N. (2010). Music-like DoA estimation without estimating the number of sources, *Signal Processing, IEEE Transactions on* Vol. 58(No. 3): 1668–1676.

12

Joint Cooperative Shared Relaying and Multipoint Coordination for Network MIMO in 3GPP LTE-Advanced Multihop Cellular Networks

Anthony Lo and Peng Guan
Delft University of Technology
The Netherlands

1. Introduction

LTE-Advanced (Parkvall et al., 2011; 3GPP TR36.814) is the successor of LTE (Long Term Evolution), which is specified by the Third Generation Partnership Project (3GPP). LTE-Advanced can provide downlink and uplink peak rates up to 1 Gb/s and 500 Mb/s, respectively, in 100 MHz of bandwidth. Similar to its predecessor, LTE-Advanced is an Orthogonal Frequency Division Multiplexing (OFDM)-based radio access technology, with conventional OFDM on the downlink and Discrete Fourier Transform Spread OFDM (DFTS-OFDM) in the uplink. In addition, LTE-Advanced includes several new key technological components, namely carrier aggregation, enhanced MIMO (Multiple-Input Multiple Output), Coordinated MultiPoint transmission and reception (CoMP), and relaying. In this chapter, we focus on the CoMP component and the relaying component.

CoMP is a means of coordinating the transmission and reception of data from/to a single mobile terminal using several geographically distributed base stations. Essentially, CoMP eliminates inter-cell interference by effectively having the multiple base stations act as a single transceiver. CoMP is especially effective for improving data rates of cell-edge mobile terminals where performance is degraded due to inter-cell interference since LTE-Advanced uses full-frequency reuse. Relaying is employed as a low-cost solution to enhance cell-coverage and -capacity. With relaying, the mobile terminal communicates with the base station via a relay node that is wirelessly connected to the base station using the same radio resources as for the mobile terminal that is directly connected to the base station. An LTE-Advanced relay node is divided into a transparent type and a non-transparent type. The main difference between the two types is in the amount of functionality and intelligence included in the relay node. A non-transparent relay node has more functionality and intelligence than a transparent one. This means it is also costlier. The simplest and the cheapest transparent relay is known as Amplify and Forward (AF).

The aim of this chapter is to leverage and combine the benefits of CoMP and relaying in order to improve the performance of cell-edge users, which is severely degraded due inter-cell interference. The joint relaying and CoMP technique can yield performance gains beyond what can be achieved using either one of the techniques alone. With the joint relaying and CoMP technique, an AF relay node is deployed at the intersection of two or more cells. The relay node amplifies and retransmits the received signals from multiple base stations to multiple mobile terminals in the downlink. In addition to the relayed signals, the mobile terminals also make use of the direct signals from the base stations to attain cooperative diversity. The coordinating base stations and the relay node form a network MIMO system. Our joint relaying and CoMP technique supports multi-user transmissions using precoding at the transmit side to precancel the co-channel interference. In order to evaluate the performance of the joint relaying and CoMP, we derive expressions for its achievable rates and compare them with the rates of CoMP.

2. LTE-Advanced multi-hop cellular network

Fig. 1 shows an overview of the network architecture of LTE-Advanced Multi-hop Cellular Networks (MCN), which is a flat all-IP network. The main architectural elements are the Mobility Management Entity and Gateway (MME/GW), the evolved Node B (eNB), the Relay Node (RN) and the User Equipment (UE). The eNB, which is a base station, connects to the MME/GW by the S1 interface through many-to-many relationship. Each eNB also connects to the neighbouring eNBs via the X2 interface, enabling direct communications. The RN is wirelessly connected to the eNB via the Un interface. A mobile terminal, which is the UE, connects to an RN via the Uu interface. In the case of direct communication, the UE connects to the eNB utilizing the same interface, Uu. In a cell, each eNB serves a number of RNs and UEs, which forms a tree structure with the eNB as the parent. The number of relays in a multi-hop chain is $n - 1$ for n number of hops. The complexity is related to the number of hops. Thus, 3GPP has limited n to two hops for LTE-Advanced MCN. In the rest of the chapter, we refer to the connection between an eNB and an RN as Relay Link (RL), the connection between an RN and a UE as Access Link (AL), and the connection between an eNB and a UE as Direct Link (DL).

2.1 Types of RNs

In LTE-Advanced MCN (3GPP TR36.814, 2010), depending on the number of protocol layers used to forward user data, an RN can be classified into a transparent type and a non-transparent type. A transparent RN is invisible to the UE. In other words, the RN just expands the cell coverage of the donor eNB. A transparent RN includes Amplify-and-Forward (AF) (Berger et al., 2009), Decode-and-Forward (DF) (Laneman et al., 2004), Compress-and-Forward (CF) (Kramer et al., 2005) and Estimate-and-Forward (EF) (Cover & El Gamal, 1979). AF amplifies before forwarding the received signal to the destination. DF decodes and re-encodes the received signal before it is forwarded. In CF, the RN compresses the source signal and forwards it to the destination without decoding it. An EF relay forwards an estimate of the received signal to the destination. A non-transparent RN appears as a mini-eNB to the UE. The non-transparent RN is also known as self-backhauling RN (Hoymann et al., 2008).

Irrespective of the relay class, an RN can operate in either full-duplex mode or half-duplex mode. A full-duplex RN can transmit and receive simultaneously, while a half-duplex RN alternate between transmitting and receiving states. The RL can operate on an orthogonal frequency band or share the same frequency band with the AL. The former is referred to as out-band RN while the latter as in-band RN. An in-band RN can have higher spectral efficiency than out-band RNs. However, it suffers from self-interference. Therefore, adequate isolation must be achieved between the transmitting part of one link and the receiving part of the other link. A half-duplex RN always transmits and receives on orthogonal channels typically in the time domain.

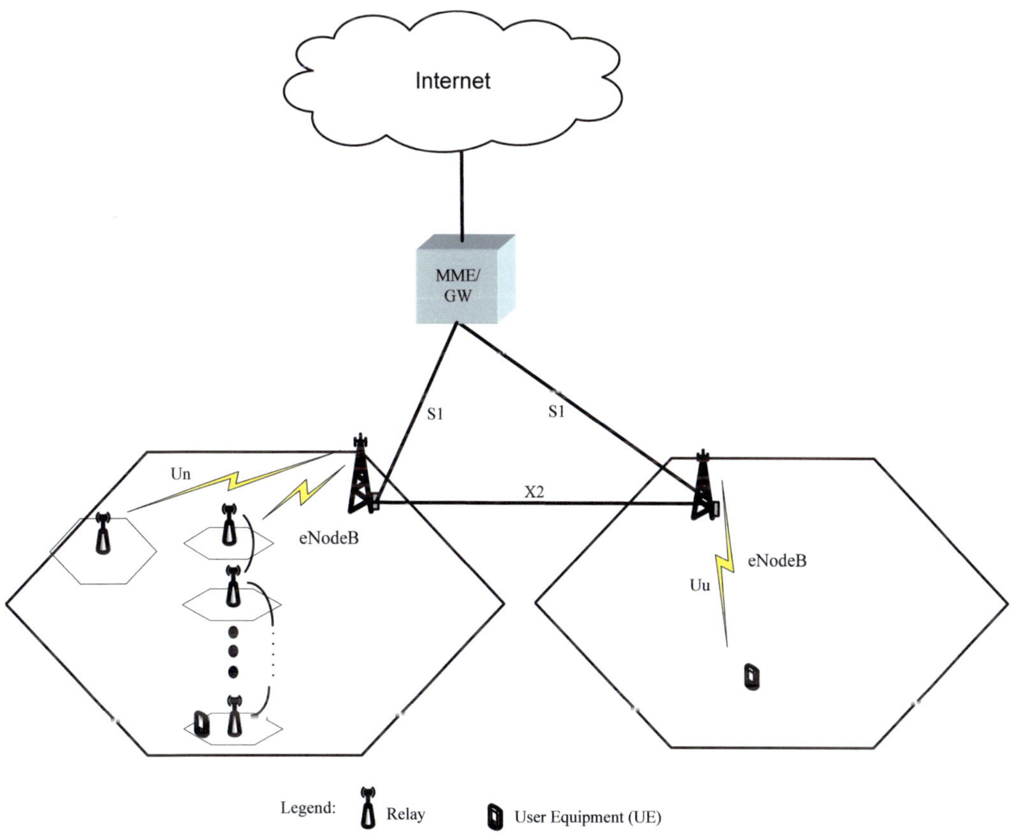

Fig. 1. LTE-Advanced Multi-hop Cellular Architecture

2.2 Coordinated Multipoint transmission and reception (CoMP)

LTE-Advanced uses full frequency reuse, which in turn leads to inter-cell interference. CoMP aims at mitigating the inter-cell interference and hence improves spectral efficiency of cell-edge users. Fig. 2 shows the CoMP architecture. The same spectrum resources are used in all cells, leading to interference for UEs at the edge between the cells, where signals from multiple eNBs are received with similar signal power in the downlink. Multiple eNBs can cooperate to mitigate the inter-cell interference. The eNBs are interconnected by the interface X2. Physically, this could be a direct fast fibre link. CoMP can be applied both in the uplink and the downlink. In the downlink, CoMP can be divided into two schemes:

- Joint Processing: user data to be transmitted to a single UE is available at each transmission, i.e., eNB.
- Coordinated scheduling and beamforming: user data is always transmitted from one eNB only, but user scheduling and beamforming decisions are made with coordination among cells.

In the CoMP uplink, multipoint reception implies coordination among multiple, geographically distributed eNBs. Uplink CoMP reception involves joint reception of the transmitted signal at multiple reception points and/or coordinated scheduling decisions among cells to control interference.

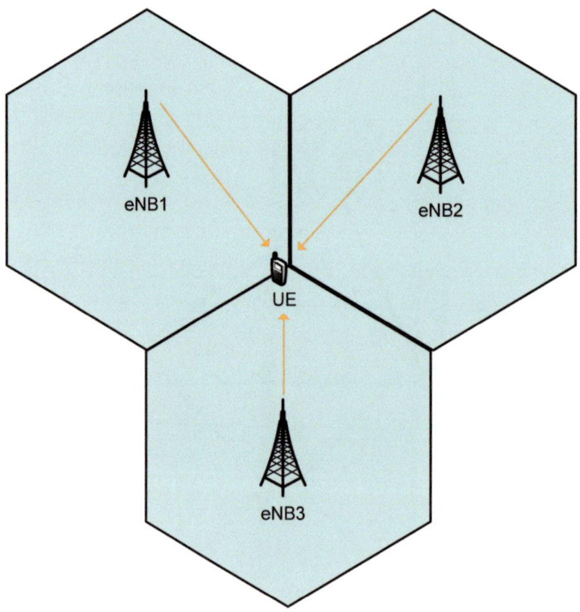

Fig. 2. CoMP Architecture

3. Joint cooperative shared relaying and Coordinated Multipoint transmission/reception

In this section, we leverage the benefits of the CoMP and the relaying techniques. The aim of the combined technique is to increase the cell-edge user data rates beyond what can be achieved by using either one of the techniques alone. Unlike users that are close to the eNB, cell-edge users experience lower signal strength because of the distance from the eNB and higher interference levels due to neighbouring eNBs. Furthermore, increasing transmission power does not necessarily lead to higher data rates due to an increase in inter-cell interference level. With the combined technique, an RN is employed to enhance the signal strength and CoMP to mitigate inter-cell interference due to neighbouring eNBs. The combined technique places an RN at the intersection of two or more cells. We propose a full-duplex AF RN for its high spectral efficiency. In the downlink, the RN amplifies and retransmits the received signals from the intersecting eNBs to multiple users. In the uplink, the RN amplifies and retransmits the received signals from multiple users to all intersecting eNBs. In either uplink or downlink, the destination (namely, UE in the downlink and eNB in the uplink) combines the relayed signal and the direct signals from all the sources (namely, eNB in the downlink and UE in the uplink). The RN and the sources form a network MIMO system. Such an RN is referred to as cooperative relay.

3.1 System model

Our system model considers an arbitrary hexagonal cellular network. The eNbs are located in the centre of each cell. The eNBs are grouped into L clusters. A cluster is composed of a single RN that is shared by M eNBs and K UEs. The eNB, the RN and the UE are equipped with one antenna element. In the downlink, each of the M eNBs transmits data streams of K UEs at the same time-frequency resources. In the uplink, the K UEs transmit to the M eNBs using the same time-frequency resources.

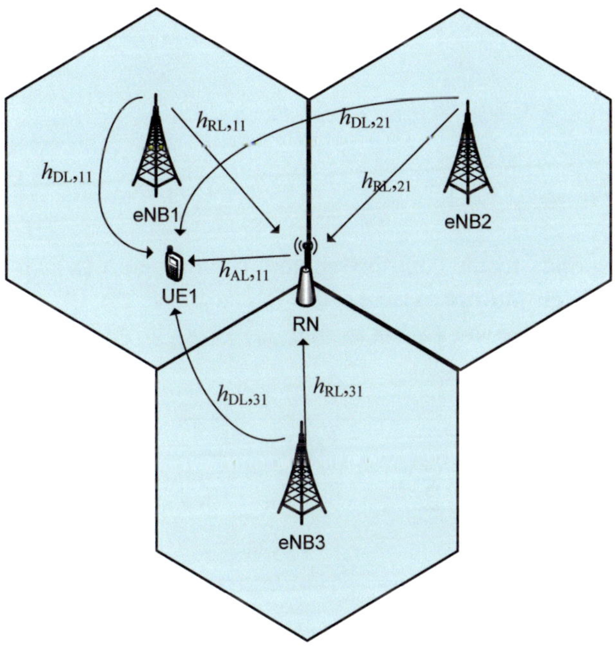

Fig. 3. A System Model for Joint Cooperative Shared Relaying and CoMP

Fig. 3 shows a typical cluster for joint cooperative shared relay and CoMP under the hexagonal cellular model. The relay RN which belongs to the lth cluster is placed at the corner of three adjacent cells. Thus, for this configuration, $M = 3$ and $K = 1$. In subsections 3.1.1 and 3.1.2, we derive the capacity equations for the downlink and uplink, respectively.
The rest of the chapter uses the following notation. Bold uppercase letters $\mathbf{A}$ denote matrices, bold lowercase letters $\mathbf{a}$ denote column vectors, and italic lowercase letters a denote scalars. $\mathbf{I}$ denotes the identity matrix, $\mathbf{A}^T$ is the transpose of matrix $\mathbf{A}$, $\mathbf{A}^*$ is the Hermitian transpose of matrix $\mathbf{A}$, ε is expectation operator, $\min\{x, y\}$ is the minimum of x and y, $|z|$ is the modulus of complex number z, and $||\mathbf{h}||$ is the Euclidean norm of vector $\mathbf{h}$.

3.1.1 Capacity of the joint CoMP and relaying

3.1.1.1 Downlink capacity

The signal received by RN is given by

$$\begin{aligned} y_{RN,l} &= \sum_{m=1}^{M} h_{RL,ml}\, x_{eNB,m} + n_{RN,l} \\ &= \mathbf{h}_{RL}\, \mathbf{x}_{eNB} + n_{RN,l} \end{aligned} \tag{1}$$

where $\mathbf{h}_{RL} = \left[h_{RL,1l}\ h_{RL,2l} \ldots h_{RL,Ml} \right]$ is the 1 × M RL channel vector (where $h_{RL,ml}$ corresponds to the complex-valued, channel gain between the mth eNb and the RN, which takes into account path loss), $\mathbf{x}_{eNB} = \left[x_{eNB,1}\ x_{eNB,2} \ldots x_{eNB,M} \right]^T$ is the M × 1 vector of the transmitted signal ($x_{eNB,m}$ is the signal transmitted by the mth eNB), and $n_{RN,l}$ is the additive white Gaussian noise observed at RN. The power of this noise term is $\varepsilon\{|n_{RN,l}|^2\} = \sigma^2_{RN,l}$. The signal received at the kth UE is a superposition of all the signals transmitted by eNBs and the RN, which is given by

$$\begin{aligned} y_{UE,k} &= \sum_{m=1}^{M} h_{DL,mk}\, x_{eNB,m} + h_{AL,lk}\, g\, y_{RN,l} + n_{UE,k} \\ &= \sum_{m=1}^{M} (h_{DL,mk} + g\, h_{AL,lk}\, h_{RL,ml}) x_m + (g h_{AL,lk}\, n_{RN,l} + n_{UE,k}), \qquad k = 1,2,\ldots,K \end{aligned} \tag{2}$$

where $h_{AL,lk}$ corresponds to the complex-valued, channel gain (which includes path loss, shadowing and Rayleigh fading) between the RN and the kth UE, $n_{UE,k}$ is the additive white Gaussian noise with power equals to $\varepsilon\{|n_{UE,k}|^2\} = \sigma^2_{UE,k}$, and g is the amplification gain of RN, which is defined as

$$g = \sqrt{\frac{P_{RN,l}}{\sum_{m=1}^{M} P_{eNB,m} |h_{RL,ml}|^2 + \sigma^2_{RN}}} > 1 \tag{3}$$

$P_{RN,l} = \varepsilon\{|g\, y_{RN,l}|^2\}$ and $P_{eNB,m} = \varepsilon\{|x_{eNB,m}|^2\}$ in Equation (3) are the transmit power of the RN and the mth eNB, respectively. Equation (2) can be expressed as

$$y_{UE,k} = (\mathbf{h}_{DL,k} + g\, h_{AL,lk}\, \mathbf{h}_{RL}) \mathbf{x}_{eNB} + \tilde{n} \tag{4}$$

where $\mathbf{x}_{eNB}$ is similar to the transmitted signal vector of Equation (1), $\mathbf{h}_{RL}$ is the RL channel vector as in Equation (1), $\mathbf{h}_{DL,k} = \left[h_{DL,1k}\ h_{DL,2k} \ldots h_{DL,Mk} \right]$ is the 1 × M DL channel vector ($h_{DL,mk}$ corresponds to the complex-valued, channel gain which includes path loss, shadowing and Rayleigh fading) between the mth eNb and the kth UE, and $\tilde{n} = g h_{AL,lk}\, n_{RN,l} + n_{UE,k}$ is effective noise term. The power of the effective noise term is $\varepsilon\{|\tilde{n}|^2\} = |g|^2 |h_{AL,lk}|^2 \sigma^2_{RN,l} + \sigma^2_{UE,k}$.

For K UEs, we can represent the network MIMO channels in the downlink by a system of linear equations as

$$\underbrace{\begin{bmatrix} y_{UE,1} \\ y_{UE,1} \\ \vdots \\ y_{UE,K} \end{bmatrix}}_{\mathbf{y}_{UE}} = \underbrace{\begin{bmatrix} \mathbf{h}_{DL,1} + gh_{AL,l1}\mathbf{h}_{RL} \\ \mathbf{h}_{DL,2} + gh_{AL,l2}\mathbf{h}_{RL} \\ \vdots \\ \mathbf{h}_{DL,K} + gh_{AL,lK}\mathbf{h}_{RL} \end{bmatrix}}_{\mathbf{H}} \underbrace{\begin{bmatrix} x_{eNB,1} \\ x_{eNB,2} \\ \vdots \\ x_{eNB,K} \end{bmatrix}}_{\mathbf{x}_{eNB}} + \underbrace{\begin{bmatrix} gh_{AL,l1}n_{RN,l} + n_{UE,1} \\ gh_{AL,l2}n_{RN,l} + n_{UE,2} \\ \vdots \\ gh_{AL,lK}n_{RN,l} + n_{UE,K} \end{bmatrix}}_{\tilde{\mathbf{n}}} \quad (5)$$

For multiuser transmission and reception, the transmitted signal vector $\mathbf{x}_{eNB}$ in Equation (5) is generated by a weighted linear combination of data symbols contained in a $K \times 1$ vector $\mathbf{d} = [d_1\ d_2 \cdots d_K]^T$, where d_k is the specific data symbol intended for the kth UE. Thus, the kth UE receives its own symbols as well as the other users' symbols. If the channel state information is perfectly known by each eNB then the zero-forcing method can be employed to nullify the undesired symbols. Using the zero-forcing method, the vector $\mathbf{x}_{eNB}$ can be generated at the eNB by precoding $\mathbf{d}$ with an $M \times K$ weight matrix as shown in Equation (6).

$$\mathbf{x}_{eNB} = \mathbf{W}\mathbf{d} \quad (6)$$

where

$$\mathbf{W} = \begin{bmatrix} \omega_{11} & \omega_{12} & \cdots & \omega_{1K} \\ \omega_{21} & \omega_{22} & \cdots & \omega_{2K} \\ \vdots & \vdots & \ddots & \vdots \\ \omega_{M1} & \omega_{M2} & \cdots & \omega_{MK} \end{bmatrix} \quad (7)$$

The weight ω_{mk} represents the precoding coefficient allocated by the mth eNB for the kth UE. The precoding matrix $\mathbf{W}$ can be simply obtained by inverting the channel matrix $\mathbf{H}$.
The capacity of the network MIMO system of K UEs is the sum of the capacity of each UE which can be obtained from Equation (5). Thus, the sum-rate capacity can be expressed as

$$C_{DN} = \sum_{k=1}^{K} \log_2\left(1 + \frac{\left|(\mathbf{h}_{DL,k} + gh_{AL,lk}\mathbf{h}_{RL})\mathbf{w}_k\right|^2}{|g|^2 \left|h_{AL,lk}\right|^2 \sigma_{RN,l}^2 + \sigma_{UE,k}^2}\right) \text{ b/s/Hz} \quad (8)$$

where $\mathbf{w}_k$ denotes the kth column of $\mathbf{W}$ for the kth UE. The zero forcing method ensures that interference due to multiuser is cancelled, i.e.,

$$(\mathbf{h}_{DL,k} + gh_{AL,lk}\mathbf{h}_{RL})\mathbf{w}_i = 0, \quad \forall k \neq i \quad (9)$$

3.1.1.2 Uplink capacity

The signal received by the RN is given by

$$\begin{aligned} y_{RN,l} &= \sum_{k=1}^{K} h_{AL,kl}\, x_{eNB,k} + n_{RN,l} \\ &= \mathbf{h}_{AL}\,\mathbf{x}_{UE} + n_{RN,l} \end{aligned} \quad (10)$$

where $\mathbf{h}_{AL} = \left[h_{AL,1l}\, h_{AL,2l} \ldots h_{AL,Kl} \right]$ is the 1 × K AL channel vector (where $h_{AL,kl}$ corresponds to the complex-valued, channel gain between the kth UE and the RN, which takes into account path loss, shadowing and Rayleigh fading), $\mathbf{x}_{UE} = \left[x_{UE,1}\, x_{UE,2} \ldots x_{UE,K} \right]^T$ is the K × 1 vector of the transmitted signal ($x_{UE,k}$ is the signal transmitted by the kth UE), and $n_{RN,l}$ is the additive white Gaussian noise observed at RN which is identical to that in Equation (1). The signal received at the mth eNB is a superposition of all the signals transmitted by all UEs and the RN, which is given by

$$\begin{aligned} y_{eNB,m} &= \sum_{k=1}^{K} h_{DL,km}\, x_{UE,k} + h_{RL,lm}\, g\, y_{RN,l} + n_{eNB,m} \\ &= \sum_{k=1}^{K} (h_{DL,km} + g\, h_{RL,lm}\, h_{AL,kl}) x_{UE,k} + (g h_{RL,lm}\, n_{RN,l} + n_{eNB,m}), \qquad m = 1,2,\ldots,M \end{aligned} \tag{11}$$

where $h_{DL,km}$ corresponds to the complex-valued, channel gain (which includes path loss, shadowing and Rayleigh fading) between the kth UE and the mth eNB, $n_{eNB,m}$ is the additive white Gaussian noise with power equals to $\varepsilon\{|n_{eNB,m}|^2\} = \sigma_{eNB,m}^2$, and g is the amplification gain of RN, which is given by

$$g = \sqrt{\frac{P_{RN,l}}{\sum_{k=1}^{K} P_{UE,k} |h_{AL,kl}|^2 + \sigma_{RN}^2}} > 1 \tag{12}$$

$P_{RN,l} = \varepsilon\{|g y_{RN,l}|^2\}$ and $P_{UE,k} = \varepsilon\{|y_{UE,k}|^2\}$ are the transmit power of the RN and the kth UE. Equation (11) can be expressed as

$$y_{eNB,m} = (\mathbf{h}_{DL,m} + g\, h_{RL,lm}\, \mathbf{h}_{AL}) \mathbf{x}_{UE} + \tilde{n} \tag{13}$$

where $\mathbf{x}_{UE}$ is similar to the transmitted signal vector of Equation (10), $\mathbf{h}_{AL}$ is the AL channel vector as in Equation (10), $\mathbf{h}_{DL,m} = \left[h_{DL,1m}\, h_{DL,2m} \ldots h_{DL,Km} \right]$ is the 1 × K DL channel vector, and $\tilde{n} = g h_{RL,lm} n_{RN,l} + n_{eNB,m}$ is effective noise term. The power of the effective noise term is $\varepsilon\{|\tilde{n}|^2\} = |g|^2 |h_{RL,lm}|^2 \sigma_{RN,l}^2 + \sigma_{eNB,m}^2$. Finally, we assume that the mth eNB normalizes $y_{eNB,m}$ by a factor $\zeta_m = \sqrt{|g|^2 |h_{RL,lm}|^2 \sigma_{RN,l}^2 + \sigma_{eNB,m}^2}$. This normalization does not alter the signal-to-noise ratio but simplifies the ensuing presentation.

For K UEs, we can represent the network MIMO channels in the uplink by a system of linear equations as

$$\underbrace{\begin{bmatrix} y_{eNB,1}/\zeta_1 \\ y_{eNB,2}/\zeta_2 \\ \vdots \\ y_{eNB,M}/\zeta_M \end{bmatrix}}_{\mathbf{y}_{eNB}} = \underbrace{\begin{bmatrix} \frac{1}{\zeta_1}(\mathbf{h}_{DL,1} + g h_{RL,l1} \mathbf{h}_{AL}) \\ \frac{1}{\zeta_2}(\mathbf{h}_{DL,2} + g h_{RL,l2} \mathbf{h}_{AL}) \\ \vdots \\ \frac{1}{\zeta_M}(\mathbf{h}_{DL,M} + g h_{RL,lM} \mathbf{h}_{AL}) \end{bmatrix}}_{\mathbf{H}} \underbrace{\begin{bmatrix} x_{UE,1} \\ x_{UE,2} \\ \vdots \\ x_{UE,K} \end{bmatrix}}_{\mathbf{x}_{UE}} + \underbrace{\begin{bmatrix} \frac{1}{\zeta_1}(g h_{RL,l1} n_{RN,l} + n_{eNB,1}) \\ \frac{1}{\zeta_2}(g h_{RL,l2} n_{RN,l} + n_{eNB,2}) \\ \vdots \\ \frac{1}{\zeta_M}(g h_{RL,lM} n_{RN,l} + n_{eNB,M}) \end{bmatrix}}_{\tilde{\mathbf{n}}} \tag{14}$$

The sum-rate capacity is

$$C_{UP} = \log_2 \det\left(\mathbf{I} + \sum_{k=1}^{K} \mathbf{g}_k \mathbf{g}_k^* P_{UE,k} \right) \quad \text{b/s/Hz} \tag{15}$$

where $\mathbf{g}_k$ is the kth column of $\mathbf{H}$, and $P_{UE,k}$ is the transmit power of the kth UE which is equal to $P_{UE,k} = \varepsilon\{|x_{UE,k}|^2\}$.

3.1.2 Capacity of CoMP

The capacity equations for CoMP can be derived in the same manner as in subsection 3.1.1. Unlike the joint technique, CoMP only involves direct transmissions between the UE and the eNB. In the downlink, the signals received by each of the K UEs can be summarized as

$$\underbrace{\begin{bmatrix} y_{UE,1} \\ y_{UE,1} \\ \vdots \\ y_{UE,K} \end{bmatrix}}_{\mathbf{y}_{UE}} = \underbrace{\begin{bmatrix} \mathbf{h}_{DL,1} \\ \mathbf{h}_{DL,2} \\ \vdots \\ \mathbf{h}_{DL,K} \end{bmatrix}}_{\mathbf{H}} \underbrace{\begin{bmatrix} x_{eNB,1} \\ x_{eNB,2} \\ \vdots \\ x_{eNB,K} \end{bmatrix}}_{\mathbf{x}_{eNB}} + \underbrace{\begin{bmatrix} n_{UE,1} \\ n_{UE,2} \\ \vdots \\ n_{UE,K} \end{bmatrix}}_{\tilde{\mathbf{n}}} \tag{16}$$

where $\mathbf{x}_{eNB}$ is given by Equation (6), $\mathbf{h}_{DL,k} = \left[h_{DL,1k}\ h_{DL,2k} \dots h_{DL,Mk} \right]$ is the 1 × M DL channel vector ($h_{DL,mk}$ corresponds to the complex-valued, channel gain which includes path loss, shadowing and Rayleigh fading) between the mth eNb and the kth UE, and $n_{UE,k}$ is the additive white Gaussian noise with power equals to $\varepsilon\{|n_{UE,k}|^2\} = \sigma_{UE,k}^2$. The sum-rate capacity is obtained from Equation (16) as

$$C_{DN} = \sum_{k=1}^{K} \log_2\left(1 + \frac{|\mathbf{h}_{DL,k}\ \mathbf{w}_k|^2}{\sigma_{UE,k}^2} \right) \quad \text{b/s/Hz} \tag{17}$$

In the uplink, the signals received by each of the eNB can be summarized as

$$\underbrace{\begin{bmatrix} y_{eNB,1}/\zeta_1 \\ y_{eNB,2}/\zeta_2 \\ \vdots \\ y_{eNB,M}/\zeta_M \end{bmatrix}}_{\mathbf{y}_{eNB}} = \underbrace{\begin{bmatrix} \frac{1}{\zeta_1}\mathbf{h}_{DL,1} \\ \frac{1}{\zeta_2}\mathbf{h}_{DL,2} \\ \vdots \\ \frac{1}{\zeta_M}\mathbf{h}_{DL,M} \end{bmatrix}}_{\mathbf{H}} \underbrace{\begin{bmatrix} x_{UE,1} \\ x_{UE,2} \\ \vdots \\ x_{UE,K} \end{bmatrix}}_{\mathbf{x}_{UE}} + \underbrace{\begin{bmatrix} \frac{1}{\zeta_1} n_{eNB,1} \\ \frac{1}{\zeta_2} n_{eNB,2} \\ \vdots \\ \frac{1}{\zeta_M} n_{eNB,M} \end{bmatrix}}_{\tilde{\mathbf{n}}} \tag{18}$$

where $x_{UE,k}$ is the signal transmitted by the kth UE, $\mathbf{h}_{DL,m} = \left[h_{DL,1m}\ h_{DL,2m} \dots h_{DL,Km} \right]$ is the 1 × K DL channel vector with each element $h_{DL,km}$ corresponds to the complex-valued, channel gain (which includes path loss, shadowing and Rayleigh fading) between the kth UE and the mth eNB, and $n_{eNB,m}$ is the additive white Gaussian noise with power equals to $\varepsilon\{|n_{eNB,m}|^2\} = \sigma_{eNB,m}^2$. Similar to Equation (14), the mth eNB normalizes $y_{eNB,m}$ by a factor

$\zeta_m = \sigma_{eNB,m}$. The sum-rate capacity is obtained from Equation (18) and it is identical to Equation (15) with the channel matrix $\mathbf{H}$ given in Equation (18).

4. Numerical results

In this section, the performance of the proposed joint scheme is evaluated by using Monte Carlo simulations. The parameters of the simulation are given in Table 1. The simulated cluster is composed of two cells. Only one UE was located in each cell. For the AL and DL links, we used the WINNER II C2 - Typical Urban Macro-cell Environment with Non Line-of-Sight (NLOS) channel model (Kyosti et al., 2007). The WINNER II B5a – LOS Stationary

Maximum total transmit power of eNBs	17 dBW
Maximum transmit power of the RN	14 dBW
Maximum total transmit power of UEs	5 dBW
RL (eNB-RN)channel model	WINNER B5a (Kyosti et al., 2007)
DL (eNB-UE) channel model	WINNER C2 NLOS (Kyosti et al., 2007)
AL (RN-UE) channel model	WINNER C2 NLOS (Kyosti et al., 2007)
Number of simulation runs	1000
Cell radius	876 m
Noise power ($\sigma^2_{UE,k}$, $\sigma^2_{RN,l}$ and $\sigma^2_{eNB,m}$)	-144 dBW
Number of eNBs (M)	2
Number of RN (L)	1
Number of UE (K)	2
Height of the eNB	25 m
Height of the RN	25 m
Height of the UE	1.5 m
Carrier frequency	2 GHz
Distance between eNB and UE in the same cell	700 m
Distance between eNB and UE in the different cell	1052 m
Distance between RN and UE	176 m

Table 1. Simulation Parameters

Feeder model (Kyosti et al., 2007) was used for the RL link. For the latter, a strong LOS signal is assumed. The assumption is valid because the position of the RN is fixed and it can be placed in such a way that a strong LOS signal is achieved. The RL channel is almost like in free space. Thus, the path loss does not depend on the antenna heights.

The height of the eNB is similar to the RN is similar, which was set to 25 m in the simulation. The height of the UEs was set to 1.5 m above the ground. The multi-path fading is a Rayleigh distribution. The carrier frequency was set to 2 GHz. The total maximum transmit power of eNB was set to 17 dBW. For simplicity, we assume the power is equally distributed between the two eNBs. The total maximum UE transmit power was set to 5 dBW. Similar to eNB, the power is equally distributed between the two UEs. The noise power ($\sigma^2_{UE,k}$, $\sigma^2_{RN,l}$ and $\sigma^2_{eNB,m}$) was set to -144 dBW. This noise power level corresponds to a 10-MHz channel.

Fig. 4 shows the downlink sum-rate capacity for the joint scheme and CoMP. The sum-rate capacity is the rate of the two UEs in the cells averaged over 1000 iterations. The joint scheme outperforms the CoMP by more than 4 b/s/Hz. The sum-rates of both schemes increase with higher eNB transmit power because of the higher Signal-to-Noise Ratio (SNR).

Uplink sum-rates are given in Fig. 5. As in the downlink case, the performance of the joint scheme is superior to CoMP. The achievable performance gain of the joint scheme is more than 7 b/s/Hz. The performance in the uplink is higher than in the downlink. This is because of the short-range connection between the UE and the RN. The signal received by the RN has very high SNR, which is then amplified and relayed to the eNB through a high quality channel.

Fig. 6 shows the downlink sum-rate for the joint scheme and the CoMP as a function of the UE distance from the RN. Both eNBs were set to transmit at the maximum power. Clearly, joint scheme outperforms CoMP. An interesting observation is made when both UEs are distanced from the RN, the joint scheme delivers a significant performance gain as compared with CoMP. The effect of the RN on the SNR at the receiver is clearly evidenced. The joint scheme benefits from the RN amplification gain and the low noise level of the RL link. Fig. 7 shows the uplink sum-rate for the same techniques. Unlike in the downlink case, the performance of the joint scheme in the uplink degrades as both UEs are distanced from the RN which leads to low SNR of the AL link. The performance degradation of the joint scheme is attributed to the amplified noise received at the RN through the AL link. This noise amplification offsets any gain resulting from using the RN. Thus, CoMP yields superior performance. However, the performance of the joint scheme is still better when the UEs are near the RN.

Figs. 8 and 9 show the downlink and uplink sum rate capacity as a function of relay transmit power, respectively. The sum-rates of the joint scheme are higher than CoMP. The plots of CoMP are constant because no relays are included in its model. Increasing the transmission power of the RN, the sum-rate is increased by an approximately 0.5 b/s/Hz. The joint scheme gives a roughly 60% and 200% capacity increase relative to CoMP in the downlink and uplink, respectively. The capacity gain in the uplink is higher than in the downlink because of the short-range connection between the UE and the RN which leads to high SNR. Thus, the noise amplification by the RN is minimal as compared with the downlink transmission which has longer range. The joint scheme benefits from both spatial diversity gain and amplification, which are provided by the RN.

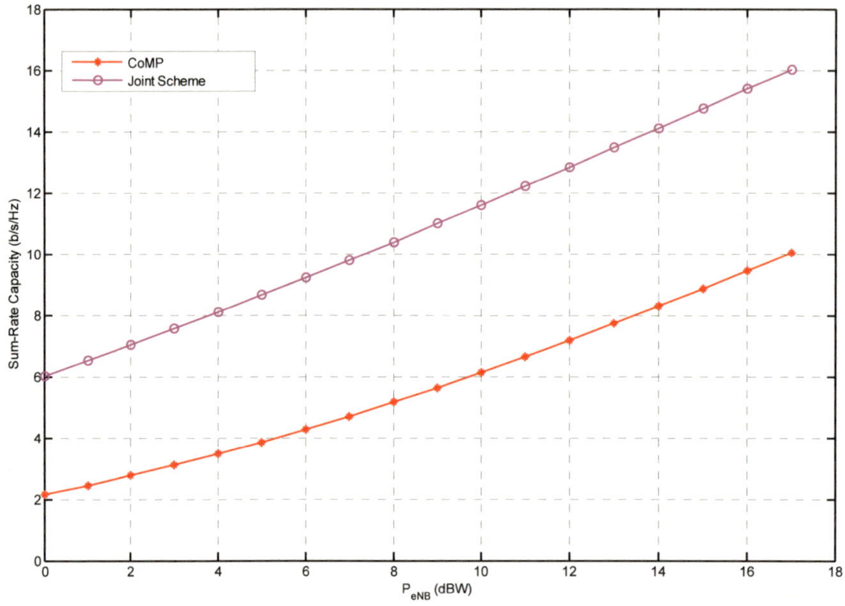

Fig. 4. Sum-Rate Capacity versus P_{eNB} in the downlink transmission

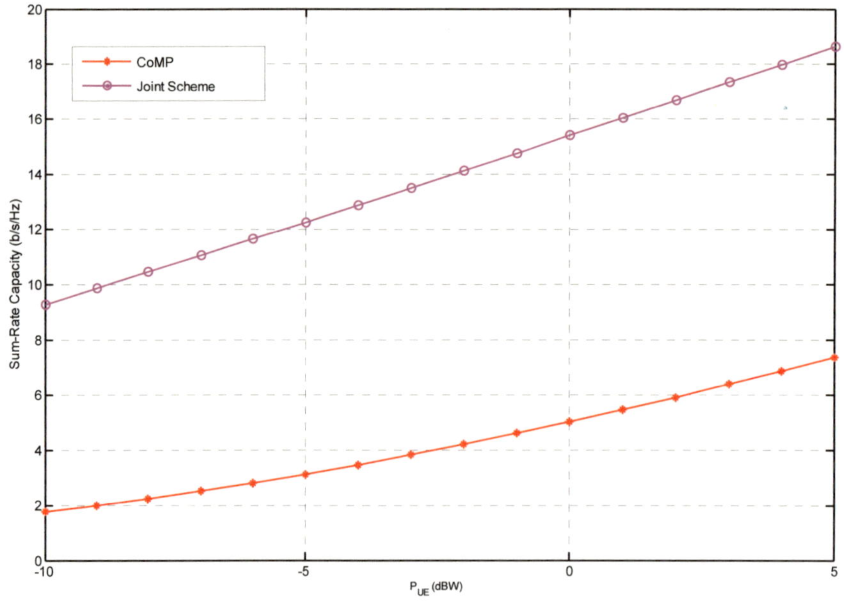

Fig. 5. Sum-Rate Capacity versus P_{UE} in the uplink transmission

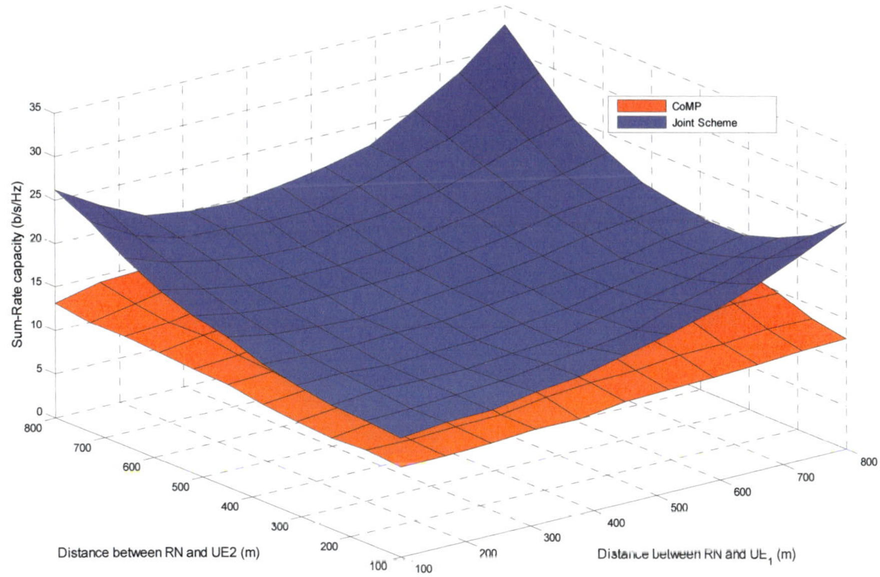

Fig. 6. Sum-Rate Capacity versus the UE distance relative to the RN in the downlink

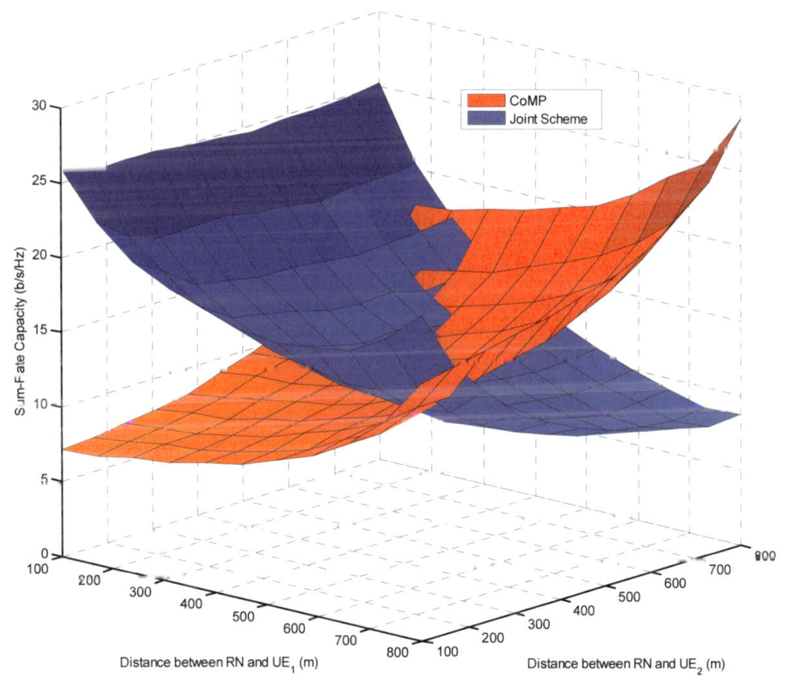

Fig. 7. Sum-Rate Capacity versus the UE distance relative to the RN in the uplink

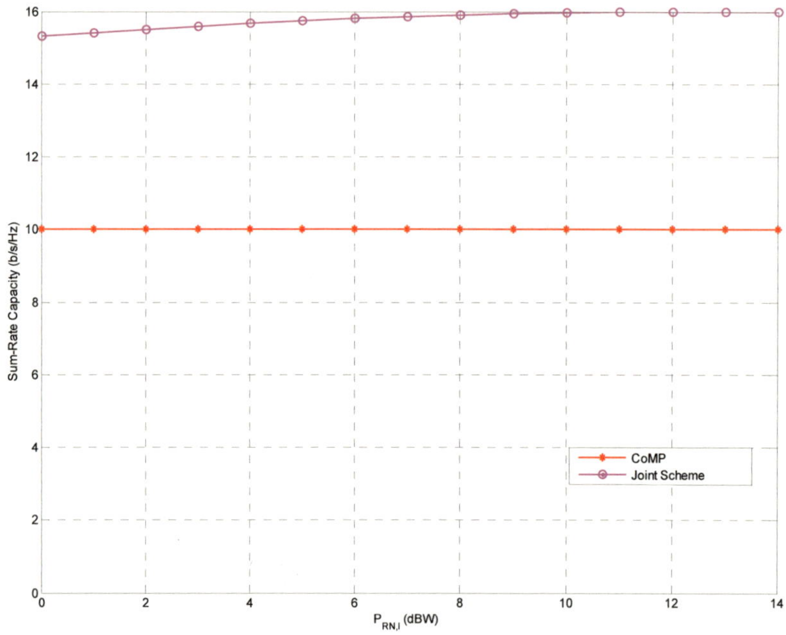

Fig. 8. Sum-Rate Capacity versus $P_{RN.l}$ in the downlink transmission

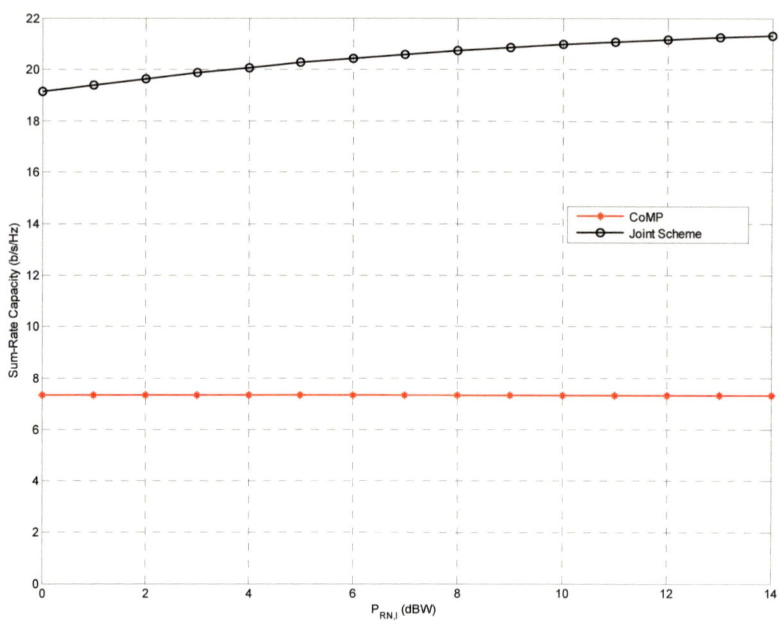

Fig. 9. Sum-Rate Capacity versus $P_{RN.l}$ in the uplink transmission

5. Conclusion

This chapter has proposed a joint cooperative shared relaying and CoMP technique to improve the performance of cell-edge users in LTE-Advanced multi-hop cellular networks. The performance of cell-edge users is severely degraded as a consequent of co-channel interference due to full-frequency reuse. The proposed joint technique considers a full-duplex amplify-and-forward relay due to its high spectral efficiency and low latency. The joint technique supports multi-user transmissions in order to increase system throughput. In the downlink, the relay node amplifies, combines and retransmits the received signals from the intersecting eNodeBs to multiple users. The zero-forcing method was used to nullify the interuser interference at the destination. In the uplink, the relay node amplifies combines and retransmits the received signals from multiple users to all intersecting eNodeBs. In both downlink and uplink, the destination combines the relayed signal and the direct signal from all the sources in order to attain cooperative diversity. We have derived the capacity equations for the joint technique and the CoMP as a baseline for comparison. Numerical results show that the performance of the joint technique is superior to CoMP in both uplink and downlink by at least a factor of three. The results also indicate that the channel quality of the relay link has a strong impact on the downlink performance than in the uplink. Therefore, it is crucial to deploy the relay node in a location that gives the best relay link quality.

6. References

Berger, S.; Kuhn, M.; Wittneben, A.; Unger, T. & Klein, A. (2009). Recent Advances in Amplify-and-Forward Tow-Hop Relaying. *IEEE Communications Magazine*, Vol.47, No.7, (2009)

Cover, T.M. & El Gamal, A.A. (1979). Capacity Theorems for the Relay Channel. *IEEE Transactions on Information Theory*. Vol.25, No.5, (1979)

Hoymann, C.; Racz, A.; Johansson, N.; & Lundsjo, J. (2008). A Self-backhauling Solution for LTE-Advanced. *WWRF*, 2008

Kyosti, P.; et al. (2007). WINNER II Channel Models. IST-4-027756 WINNER II Project, D1.1.2, 2007

Kramer, G.; Gastpar, M. & Gupta, P. (2005). Cooperative Strategies and Capacity Theorems for Relay Networks. *IEEE Transactions on Information Theory*. Vol.51, No.9, (2005)

Laneman, J.N.; Tse, D.N.C.; & Wornell, G.W. (2004). Cooperative Diversity in Wireless Networks: Efficient Protocols and Outage Behavior. *IEEE Transactions on Information Theory*, Vol.50, No.12, (2004)

Loa, K.; Wu, C.; Sheu, S.; Yuan, Y.; Chion, M.; Huo, D. & Xu, L. (2010). IMT-Advanced Relay Standards. *IEEE Communications Magazine*, Vol. 48, No.8, (August 2010)

Parkvall, S.; Furuskar, A. & Dahlman, E. (2011). Evolution of LTE toward IMT-Advanced. IEEE Communications Magazine, Vol.49, No.2 (February 2011)

Sawahashi, M.; Kishiyama, Y.; Morimoto, A.; Nishikawa, D. & Motohiro, T. (2010). Coordinated Multipoint Transmission/Reception Techniques for LTE-Advanced. *IEEE Wireless Communications*, Vol.17, No.3, (2010)

3GPP TR36.814 (2010). Further Advancements for E-UTRA Physical Layer Aspects (Rel 9). 3GPP TR 36.814 V9.0.0, 2010

13

Conventional and Unconventional Cellular Admission Control Mechanisms

Anna Izabel J. Tostes, Fátima de L. P. Duarte-Figueiredo and Luis E. Zárate
Pontifical Catholic University of Minas Gerais (PUCMG)
Brazil

1. Introduction

The wide range of services offered by third generation (3G) networks made them more popular around the world. Examples of these services are web browsing, video streaming, image transmission, downloads, videoconference, voice over IP (VoIP), voice calls and Short Message Service (SMS). When an user requests a particular service, a new call is requested. The greater the diversity of services required, the greater is the network resources management difficulty. Availability, reliability and performance are the major goals of its management.

Quality of Service (QoS) is related to the users' satisfaction (Steinmetz & Wolf, 1997). Quality is not measured only by the resources availability, but also by the performance. For example, when an user wants to start a videoconference, it is important, for him or her, low delay, low jitter and high throughput. If these requirements are not met, then the QoS may not be honored. As the number of calls in the network increases, higher will be the difficulty in deciding which requests should be accepted or not. The absence of an admission control mechanism does not guarantee that the network resources will be well distributed, leading to bad resources utilization and a consequent interference in the network availability and QoS assurance.

A Call Admission Control (CAC) is a QoS mechanism. This mechanism decides witch requests should be accepted according to the resource availability, to maintain QoS guarantee. A research challenge is the development of a CAC that solves the following problem: how to decide which application to accept in accordance with the network status. For instance, if a call is requested and the network is free, or a little bit congested, or even very congested, this new call can be accepted or not. This decision influences the QoS of the new call and the QoS of the already established calls. If the network can be guarantee the QoS, the CAC can accept the new call. Otherwise, the call must be blocked.

Another challenge is the CAC's precise knowledge in accept or reject a new call in accordance to its priority. Blocked calls can cause poor network resources utilization, which is unacceptable to cellular operators. Therefore, CAC needs to know which call should be accepted while it faces multiple calls requests in congestion times. Besides, CAC's decision should be taken in the shortest time with the lowest complexity. In summary, CAC's three main requirements are: (1) decides a new call's acceptance (or not), (2) selects which call to accept according to the pre-established services priorities, (3) the decision should be taken in the shortest time.

This chapter is organized as it follows. Section 2 presents how services are admitted through 3G networks, describing the services classes, the network architecture and the basic CAC concepts. Section 3 introduces references of conventional CACs. Section 4 presents a case study of two conventional CACs. Section 5 explains unconventional CAC's concepts. Section 6 presents a case study of two unconventional CACs. Section 7 describes how a CAC evaluation can be made while section 8 demonstrate some of this evaluations in a case study. Section 9 concludes this chapter and section 10 presents references.

2. How services are admitted through 3G networks

3rd Generation Partnership Project (3GPP) group has conceived the 3G cellular networks to meet three requirements: (1) at least 2 Mbps peak throughput, (2) multimedia transmission with QoS and (3) international roaming. To make this guarantee, the services were divided into four classes. These classes are used to conduct services prioritization in CACs.

2.1 Types of services

Applications and voice have specific QoS requirements. For example, the requirements are different for web browsing, video streaming, image transmission and SMS. The 3G network services are classified in four QoS classes for each traffic type (3GPP, 2010):

Conversational. This class of service represents real time (RT) applications, such as voice or telnet. The main QoS requirement is low delay. Low jitter and low packet loss are also important.

Streaming. This class of service represents multimedia and data transfer with continuous and stable stream processing. It's main QoS requirement is low jitter. Low delay and low packet loss should also be considered.

Interactive. This class of service is characterized by client/server requests with traffic bursts, such as web browsing, interactive games and email server access. Although this class is not sensible to jitter, the main QoS requirement is delay.

Background. This class of service represents not real time (NRT) applications, such as email (server to server), fax and transactions services, as SMS. This class is the most tolerant to delay. The main QoS requirement is a high throughput.

The difference between the classes of services is related to their sensibility to QoS parameters such as delay, jitter and throughput. Delay is the time interval in which the packet traverses the network until it reaches its destination. This interval corresponds to the time to send a package from a sender to a receiver. Jitter is the variation of consecutive delays. Mathematically, it consists in the difference between consecutive packets delay (Kurose & Ross, 2009). For example, multimedia and VoIP applications do not support jitter, so it should tends to zero. Throughput is the effective rate of transmission for an application, in bits per second. Some interactive applications like WEB browsing require high throughput. Some RT applications, like VoIP, require low delay.
In summary, table 1 shows the QoS parameters requirements for the higher-sensitive services of each class, specified by 3GPP (2010). For voice calls (1), the maximum of delay achieved must be less then 400ms. For video (2), jitter must be at most 2s. For web browsing (3), delay can be between 2 and 4s, or less. For SMS (4), the minimum throughput should be 2.8 kbps.

Class	Application	Demand	Delay		Jitter	Throughput
Conversational	Voice call	Higher	Best	< 150ms	< 1 ms	4–13 kbps
			Limit	< 400ms		
Streaming	One-way video	Low	Limit	< 10s	< 2s	32–384 kbps
Interactive	Web browsing	Higher	Limit	< 4s	–	> 20 kbps
Background	SMS	Medium	Limit	< 30s	–	2.8 kbps

Table 1. QoS requirements for 3G UMTS classes of service

2.2 3G infrastructure

Specified by 3GPP, the Universal Mobile Telecommunications System (UMTS) is the most popular 3G cellular networks. As it is shown by figure 1, UMTS architecture is composed by three interacting domains (3GPP, 2009; Kaaranen et al., 2005): (1) the call requested by a user equipment; (2) the air interface UTMS Terrestrial Radio Access Network (UTRAN); and (3) the Core Network (CN). When a call is requested, a solicitation is sent to the UTRAN. The responsible module for the communications with mobile users, over a coverage area (cell), is the Base Station (BS). The Radio Network Controller (RNC) controls the management of the BSs and it is also responsible for the soft handover (the user process of changing BS without loss of connection). Thereafter, the call request is sent to the CN, which is responsible for providing access to Internet and to other networks through the Public Switched Telephone Network (PSTN). Internet is accessed through the elements of Serving GPRS Support Node (SGSN) and Gateway GPRS Support Node (GGSN). The circuit switched domain through the Mobile Switching Centre (MSC), the Home Location Register (HLR), the Visitor Location Register (VLR) and the Gateway MSC (GMSC) achieves PSTN.

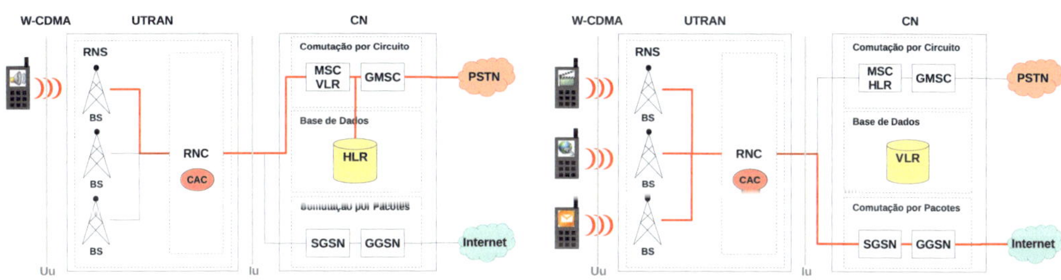

(a) Voice calls traffic (b) Video, web-browsing and e-mail traffic

Fig. 1. Tracing different service calls in 3G UMTS network architecture

Figures 1(a) and 1(b) present respectively the voice call traffic and the data traffic in 3G UMTS architecture. Independent of service type, the call request passes through UTRAN in the BS of its coverage area, managed by the RNC. In the CN, while the voice calls traffic passes through MSC and GMSC to access the PSTN, the data traffic passes through SGSN and GGSN to achieve Internet access.

2.3 Admission control

A Call Admission Control (CAC) is an algorithm of decision-making that provides QoS in the network by restricting access to the network resources (Ghaderi & Boutaba, 2006). Figure 2 explains the CAC's functionally. According to the requested call type, CAC decides to accept or block the new call according to the network resources availability. When there are not sufficient resources to ensure the call's quality or to keep the active calls' QoS (services already accepted – established), CAC blocks the new call. Otherwise, the call is accepted.

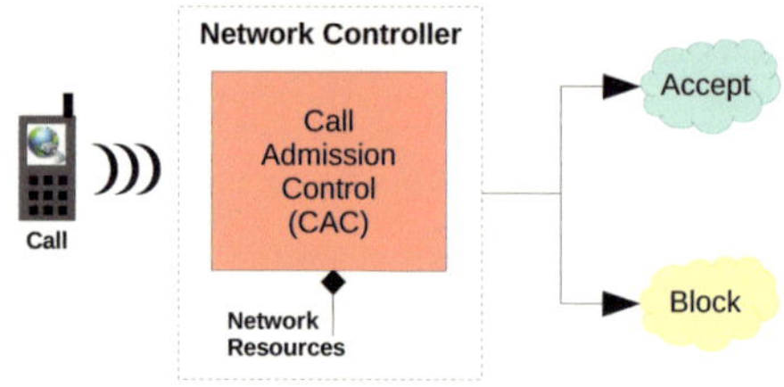

Fig. 2. Call admission control functionality

The literature presents several CAC techniques. Figure 3 shows a taxonomic tree of the CAC techniques used, adapted from (Ghaderi & Boutaba, 2006). The difference for Ghaderi & Boutaba (2006) tree is that they does not present a single tree. Ghaderi & Boutaba (2006) do not describe the thresholds technique, the homogeneous or the heterogeneous services and the application of computational intelligence in admission control as is presented in this work.

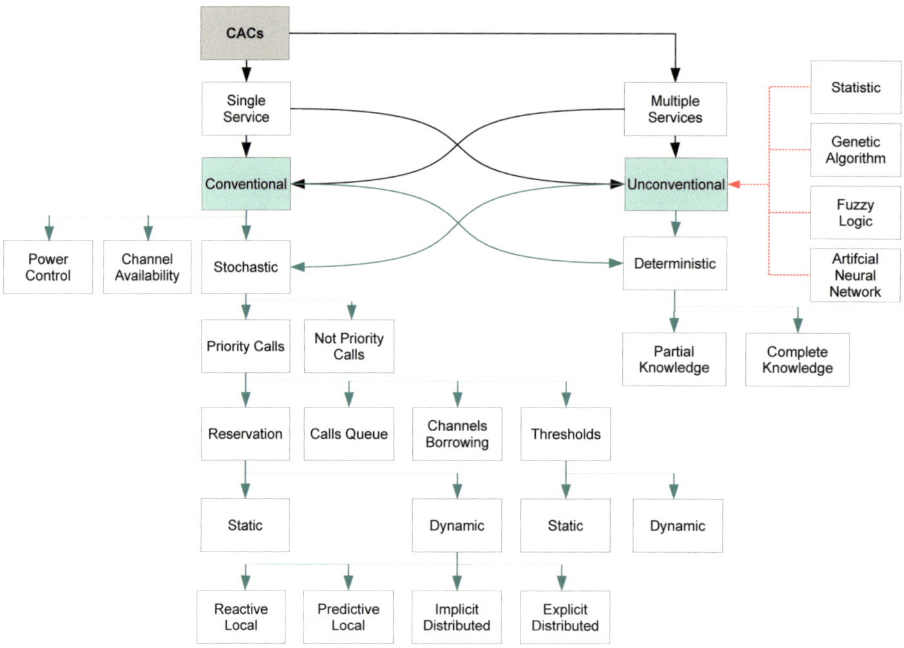

Fig. 3. Taxonomic tree of admission control techniques
Font. Adapted from Ghaderi & Boutaba (2006)

A CAC can work with a single service (voice, for example) or with several types of service (as 3G applications – multimedia, Internet access, email, voice or telnet). Admission control techniques are classified as conventional or unconventional. The conventional CACs use prioritization of calls, resources reservation and borrowing of channels. The CACs that apply computational intelligence methods associated with the traditional techniques are classified as unconventional CACs. Among the methods used, there are the artificial neural networks, fuzzy logic, genetic algorithms and statistical methods.

CACs can also be classified according to the confidence interval assured to QoS parameters. They can be deterministic, guaranteeing 100% of confidence, or stochastic, which are those that guarantee a probability of confidence (for example, 95% of confidence). The two main basic techniques are the power control, which analyze the uplink and downlink levels, and the channel availability, which tries to allocate an available channel for the call request.

The stochastic techniques can not use services, channels, resources or calls prioritization methods. CACs with services prioritization can use techniques of resource reservation, borrowing of channels, calls queue and acceptance thresholds. Acceptance thresholds technique set limits in some network parameter (for example, the network utilization) to decide, based on the call priority and the parameter level, if the request can be accepted or not. These limits may be static as defined by Service Level Agreement (SLA) or dynamic, updated according to the network status. In the calls queue technique, a queue for soft handover (SHO) is created. If a SHO attempt is blocked, the request can be kept in the priority queue. When there are channels available, the process allocates the channel to the first SHO waiting in the queue. On borrowing channels technique, a cell uses borrowed channels from their neighbors to SHO processes. A channel can only be borrowed if it does not interfere with existing calls. Importantly, when a channel is borrowed, other cells are prohibited from using it.

The last prioritization technique is the resource reservation, which can be static or dynamic. In the static reservation, there is a fixed amount of resources reserved for priority calls. In dynamic reservation, the amount of resource reserved varies according to external factors, such as network performance, availability and congestion level. The dynamic reservation can be a reactive or a predictive local approach. The reactive local approach consists in the use of channels thresholds for priority services. If the channels threshold is reached for a particular application type, no more channels can be allocated. In the predictive approach, the network global state is estimated through prediction models based on local information. The dynamic reservation can also be an implicit or a explicit distributed approach. When the reservation is implicitly distributed, the processing is local, but neighboring cells information are needed. The difference for explicit approach is that the processing is not local and the neighboring cells information are involved in the decision-making.

In addition to stochastic CACs, there are deterministic techniques. They can use partial or complete knowledge. The partial knowledge of CACs reserves resources in several cells to maintain a deterministic guarantee. The complete knowledge of CACs consists in an imaginary CAC of perfect knowledge that use all the user mobility information to make the best acceptance decision. In general, they are used for benchmarking and prediction purposes.

3. Conventional admission controls

This section describes some conventional CACs presented in the literature. The most basic CAC uses just the power control and channel availability. Stochastic techniques are the most common scheme used in admission controls. As we have said, it allows the ensuring of a probability of a confidence interval, generally of 95%–99%. The following techniques are generally used: services prioritization, borrowing and reservation of channels, queue calls and acceptance thresholds.

Deterministic schemes are only possible with partial knowledge. This is because the perfectionism of a complete knowledge is not tangible in admission controls, even in indoor environments (Lu & Bharghavan, 1996). Perfection is just theoretical, for benchmarks purposes. Talukdar et al. (1999) and Lu & Bharghavan (1996) propose deterministic CACs with a partial knowledge model. They are called worse case schemes because they must reserve resources in several cells to provide deterministic guarantees.

Katzela & Naghshineh (1996) presents a survey of channels borrowing schemes. In accordance with this study, the better borrowing technique for heavy loads is the hybrid one. The hybrid scheme combines two techniques: (1) a channels subset for nominally assigned in each cell, and (2) another channels subset to be borrowed to neighboring cells. In the hybrid technique, CAC can also reallocate calls from borrowed to nominal channels in order to minimize future calls from borrowing channels, which is the most common way in heavy loads.

In static reservation technique, we can reference (Hong & Rappaport, 1986). The authors make static reservation of permanent channels for SHO (called the guard channels), which are priority applications over new calls. They have shown that this reservation reduces the SHO blocking in comparison to CACs without priority calls. When the number of guard channels increase, the probability of dropping calls (forced termination of calls) decreases significantly, even when compared with the increase of new calls blocking.

In general, dynamic reservation technique is better than the static scheme, but it provide high overhead. Box & Jenkins (1990); Talukdar et al. (1999); Zhang et al. (2001) use dynamic predictive local reservation. In general, they assume that the control mechanism periodically measures the arrival rate and then compute the expected arrival rate from such online measurements through a simple exponentially weighted moving average. The channel reservation is calculated based on these parameters, which try to estimate the global network state.

Distributed dynamic reservation was introduced by Naghshineh & Schwartz (1996). The idea was that CACs should collaborate to make the acceptance decision in the combination of adjacent cells information with the local cell information. Although this paper technique was not stable and had violated the required dropping probability as the load increases, Levine et al. (1997) have evolved this technique including the shadow cluster concept. The idea is to use dynamic clusters for each user based on its mobility pattern instead of restricting itself to direct neighbors only. This technique is expensive and sometimes useless in practice (Ghaderi & Boutaba, 2006). Furthermore, it requires a precise knowledge of the mobile trajectory. In distributed dynamic reservation technique, it is also used analytical approaches, which involve huge matrix exponentiations (not acceptable for sequential computer architectures).

Chang et al. (1994) use a finite priority queue of waiting calls to gain access to available channels. In their model, Soft Handover (SHO) calls have priority over new calls. In contrast, Hong & Rappaport (1986) use an infinite queue of SHO attempts despite the channel reservation scheme. This queue can be used if, and only if, the SHO call is inside a handover area between cells. Results showed that using the queue improves the pure guard channel scheme performance, diminishing the dropping probability and maintaining the blocking probability.

Thereby, we perceive that CAC is a collection of techniques to improve the acceptance of new call in the network. The major goal is the better utilization of network resources, maintaining the commitment between performance and availability of the service's quality. Conventional CACs present techniques that can ensure a high efficiency, a low complexity and a low overhead. Nevertheless they are not adaptable, not stable and do not ensure 100% of confidence.

4. Case study: 3G conventional CACs

This section presents two conventional CACs: CAC-J (Antoniou et al., 2003) and CAC-RD (Tostes et al., 2010). CAC-J (Call Admission Control of Josephine Antoniou) is the most basic admission control, because it only does power control and channels availability. Differently, CAC-RD uses several CAC techniques.

In (Storck et al., 2008a;b; Tostes et al., 2010), the authors present CAC-RD: an UMTS call admission control based on resource reservation and a network diagnosis. It has the following modules: static blocking thresholds; channel reservation, network diagnosis and power control. CAC-RD uses a static threshold scheme for blocking lower-priority calls when the network utilization reaches certain limits. If the network utilization is up to 40%, all calls are accepted. If the utilization is between 40%–50%, non-real time (background class of service) is blocked. Between 50%–65% of utilization, background and interactive calls are blocked. Between 65%–75%, background, interactive and streaming calls are blocked. Above 75%, all new calls are blocked, including conversational calls. Results show that CAC-RD decrease 40% of SHO blocking and 11% of voice calls blocking.

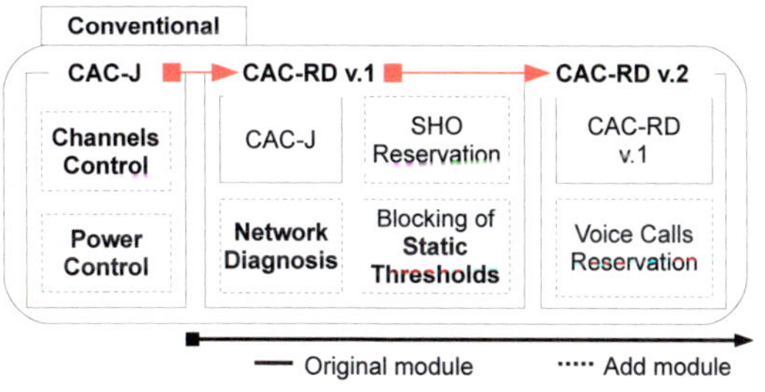

Fig. 4. Evolution line of CAC-RD

As figure 4 shows, two versions of CAC-RD were proposed. CAC-RD's first version (Storck et al., 2008a) reserves dynamically channels for SHO calls, which is the only priority call type. As the probability of conversational calls blocking attained unacceptable high levels, the second version of CAC-RD (Tostes & Duarte-Figueiredo, 2009) reserves dynamically channels for both SHO and conversational calls. Thereby, results presented a gain of 66.65% for conversational calls and of 82.70% for SHO processes.

5. Unconventional admission controls

In addition to the conventional CAC schemes, the academic community has proposed the unconventional CACs. In this category, computational intelligence is applied to conventional methods in order to attain better decision-making. As we have said, the most used methods are statistical approaches, genetic algorithms, Artificial Neural Networks (ANN) and fuzzy logic.

Deterministic algorithms are the most common traditional technique in unconventional CACs. Shen et al. (2000) developed a distributed deterministic CAC with intelligent techniques addressing the probabilistic estimation and prediction of mobility information.

Within genetic algorithms, Karabudak et al. (2004) propose the Genetic Admission Control (GAC) for next generation wireless systems. GAC aims high network utilization with the minimum cost, SHO latency and required QoS levels. Still, it incorporates the Markov decision model. Its main contribution is the provision of an efficient CAC with an intelligent approach to minimize the financial cost.

The self-learning ANN capacity is being applied in CACs to characterize the relationship between the traffic inputs and the system performance. Among ANN-based CACs, there are several approaches as (Hiramatsu, 1989), (S. A. Youssef, 1996) and (Cheng & Chang, 1997). Hiramatsu (1989) proposed a CAC with an ANN multilayer perceptron. The declared traffic parameters were used only for split connections into several services classes. The trained ANN has learned the relationship between the number of connections in each class and the required QoS, in accordance with the class' statistical characteristics.
S. A. Youssef (1996) developed a CAC with a trained ANN to compute the effective requested bandwidth rate. This is done so that CAC supports connections with different QoS requests. This CAC was proposed for Asynchronous Transfer Mode (ATM) networks. Results have shown that the CAC's adaptability to new traffic situations has been hit.
Cheng & Chang (1997) proposed the Neural Network-based Connection Admission Control (NNCAC) for ATM networks. The ANN was modeled with three pre-processed input parameters in order to simplify the training process and to increase the CAC's performance. According to Cheng & Chang (1997), ANN-based CACs provide learning and adaptability capacity, in general, to reduce the estimation error of conventional CACs and to achieve a similar performance of fuzzy logic-based CACs.
Several works with fuzzy logic are also presented in the literature. Ascia et al. (1997) present a fuzzy logic-based CAC for ATM networks. Fuzzy logic-based CAC was designed to know which decisions must be taken in case of traffic load level variations. Its fuzzy logic model consisted in three network parameters: (1) congestion, (2) quality and (3) network capacity. Ascia et al. (1997) analyzed the CAC's behavior in comparison with conventional CACs and the hardware implementation. Results have indicated that the fuzzy logic-based CAC is the most closer to the ideal behavior. Stability is one of the most important features in this CAC.
According to (Pedrycz & Vasilakos, 2000, p.67), fuzzy logic-based CACs have demonstrated the ability to make smart decisions for soft thresholds schemes, developing inaccurate quantity-based calculations and modeling linguistic rules. This technique emulates the decision-making of experts. Fuzzy logic-based CACs can model the acceptance decision as a linguistic variable, allowing a soft decision (accept, weakly accept, weakly blocked and blocked). This technique is particularly useful when precise mathematical models are impractical or unavailable.
Fuzzy Logic Connection Admission Control (FLCAC) was proposed in (Pedrycz & Vasilakos, 2000, chapter 3). Simulations compared FLCAC to conventional CACs. Results demonstrated an improvement in the network utilization and the better consumption of network resources while keeping the QoS contract. This has happened because FLCAC employs input variables with much more information than conventional models. Still, the fuzzy logic linguistic capabilities can handle the traffic complexity, providing a smooth control. For real applications, authors suggest the utilization of fuzzy-chips.
Another intelligent applied technique is neuro-fuzzy systems. In addition to the fuzzy logic-base CACs, Raad (2005)'s work has been extended with the neuro-fuzzy CAC proposal. Its distinguishing characteristic is the ANN adaptability. Cheng et al. (1999) had also used this approach, but in the context of high-speed multimedia networks.
Although there are different proposed unconventional CACs, the embodied knowledge in conventional methods is difficult to be incorporated into ANN or fuzzy logic design. To facilitate the representation of knowledge personified in conventional methods, two procedures for creating a ANN and fuzzy logic-based CACs were presented in Tostes (2010).

6. Case study: 3G unconventional CACs

In section 3, we have explained CAC-RD (Tostes & Duarte-Figueiredo, 2009), which is a conventional CAC. In order to improve CAC-RD, we have proposed two unconventional CACs as figure 5 shows: (1) Neural CAC-RD (CAC-RDN) (Tostes et al., 2008) and (2) Fuzzy CAC-RD (CAC-RDF) (Tostes et al., 2011). CAC-RDN represents CAC-RD's knowledge in an ANN while CAC-RDF makes a dynamic blocking of lower-priority calls.

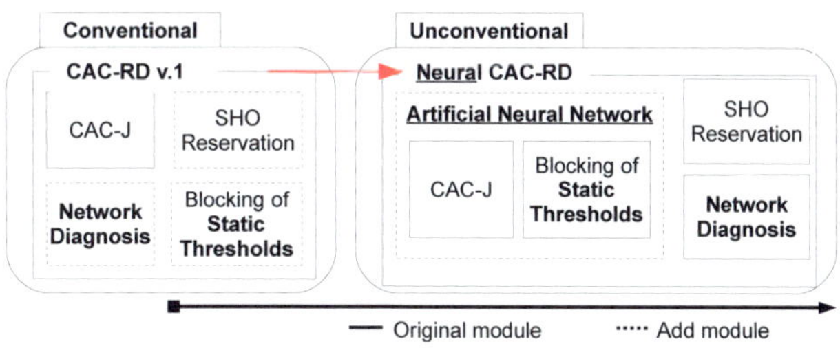

(a) Neural CAC-RD

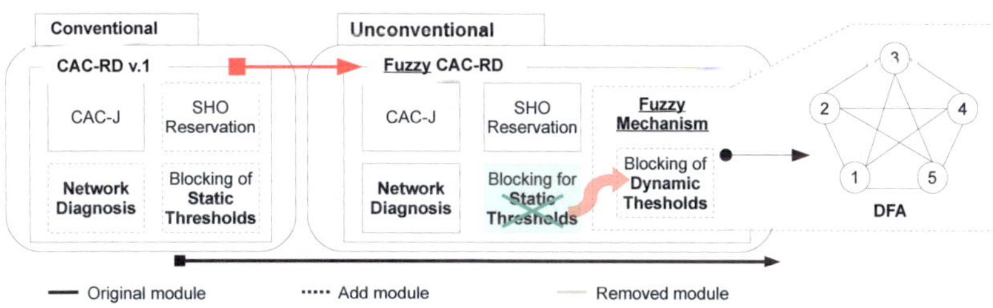

(b) Fuzzy CAC-RD

Fig. 5. Unconventional schemes of CAC-RD through neural network and fuzzy logic

The development of CAC-RDN has followed a CAC Neural Representation procedure. This procedure consisted in the creation of an ANN that replace the following CAC-RD modules, as figure 5(a) shows: static blocking thresholds, channel and power controls.

This neural procedure explains the ANN creation process in 10 steps. In step 1, ANN inputs (for instance, the network utilization or costs) and output (for instance, the decision of a new call acceptance) variables were chosen as well as the main CAC classifier attribute (in CAC-RD, it is the BS utilization). In step 2, a database of these variables was established. Each dataset sequence corresponded to the input values that result in a specific output. In step 3, a numerical coding was chosen for ANN variables. For example, each service class code – 1: conversational, 2: streaming, 3: interactive 4: background. Then, in step 4, we have calculated the size (x) of the training set. Step 5 consisted in balancing the database. In step 6, the database was reduced to the size of the training set (x) by removing outliers and selecting the best sequences. In step 7, we have selected which data sequence will go to the ANN training or to the ANN validation. Step 8 consisted in the ANN training, establishing a stopping criterion, and the ANN validation. The ninth step was the implementation of the trained ANN into CAC-RD. In step 10, we made the ANN-based CAC validation through simulations. Results have demonstrated that CAC-RDN reproduces CAC-RD with less complexity. CAC-RDN's major gain is the scalability: it is possible to simulate much more users and cells in less time.

In CAC-RDF, the dynamic blocking of calls occurs in accordance with the network congestion level. If the network is not congested, any new call is accepted. If the network is pre-congested, lower-priority calls are blocked in accordance with the QoS level met for established applications. In case of big network congestion, CAC-RDF does not accept any new call until some resource is released. In case of a soft congestion, some of the new calls are blocked. To represent these situations, we have used the fuzzy logic concepts, allowing the creation of a qualitative mechanism through natural rules based on linguistic descriptions.
CAC-RDF's architecture is divided into modules, as figure 5(b) presents. All CAC-RD modules were retained except the static blocking module. This module has been replaced by dynamic blocking module, which is a fuzzy logic mechanism. This fuzzy mechanism is represented in the figure by the deterministic finite automaton (DFA), which analyzes the network resources and decides which classes should be blocked. From the QoS level guaranteed, by class, the fuzzy engine defines the network state (output), which is one of the following five states of the DFA: (1) accept all calls (AA), (2) blocks background calls (BB), (3) blocks background and interactive incoming calls (BI), (4) blocks background, interactive and streaming calls (BS), (5) blocks all calls, including the conversational class (BC).
The fuzzy-based CAC development has followed seven steps. The first step is the CAC's thresholds definition: in CAC-RD, four thresholds. In step 2 and 3, the input (QoS parameters) and output (network status) variables and their inference functions are defined. Then the fuzzification process is done, transforming the input values (for instance, 100 ms of delay and 2 ms of jitter) in linguistic values (respectively low and high) from its inference function specified in step 3 ($f(x)$ with x variable). In step 4, a specialist defines the system linguistic rules, used by the inference mechanism to deduce the resulting output from the linguistic inputs. In the CAC-RDF, 27 rules were defined. The fifth step consists in selection of the defuzzification algorithm, which translates the output linguistic value to the corresponding discrete value from its inference function. The chosen algorithm was the gravity center, the most common in fuzzy literature. In steps 6 and 7, the fuzzy mechanism implementation and validation is made. Results have shown that CAC-RDF achieved two goals: greater acceptance of priority calls than CAC-RD and better distribution of available network resources. CAC-RDF also provides greater adaptability and stability in its decision-making.
Thereby, there are several CACs techniques reported in the literature: conventional and unconventional. We believe that unconventional CACs proposed for ATM networks may be adapted to 3G UMTS networks. Among the advantages of unconventional schemes are a high adaptability, a high stability, a high flexibility and a low complexity. The challenge of these new approaches is how to model and design an unconventional solution to ensure the availability and performance of networks. It is important to highlight that if the computational intelligence modeling is not well done, the CAC may not work correctly as it should. Notwithstanding, the problem of how to guarantee QoS and better network resources management is still opened.

7. Evaluation of admission controls

As it was previously explained, there are many parameters involved in the CAC's decision-making evaluation. Some parameters differ from CAC architecture. It is difficult to develop a CAC evaluation framework. Although there are several forms of CAC assessment, we present eight evaluation criteria (Ghaderi & Boutaba, 2006), as it follows:

Blocked calls. It refers to the number of services not accepted. This may happens due to the network congestion (scarce available resources or insufficient power).

Dropping calls. It measures the forced termination of calls in progress, after the CAC's acceptance decision. This evaluation has a more negative impact from the user's perspective, in general.

Soft handover failures. It measures the number of hard handover calls. A soft handover (SHO) failure occurs when the call path forces the network to discontinue the user call.

Efficiency. It refers to the achieved utilization level according to the network capacity requirement given a pre-defined QoS.

Complexity. It is related to the CAC's algorithm complexity in relation to its decision-making. A very complex CAC algorithm tends to be less efficient in time then a simple one.

Overhead. It refers to CAC's induced overhead. The CAC's processing time must be the minimum possible.

Adaptability. It measures the CAC's ability to react to the network conditions. For example, if the network has some performance problem, the CAC may reject more new requests.

Stability: It is the CAC's sensibility to traffic load fluctuations.

QoS performance parameters: The major network operator goal must be high availability with excellent performance. Though the CAC must look at the performance every time to avoid performance degradation when it accepts many new calls. Delay, jitter and throughput are the main important parameters to be seen.

8. Case study: Evaluation of the presented CACs

8.1 Blocked calls

Figure 6 shows a bar graph of CAC-RDN denied blocking calls in a 3G UMTS network with 10000 users and 100 cells. We can see that CAC-RDN can simulate more users than CAC-RD (1100 users in 3 cells), but it still blocks a lot of conversational calls. This happens because CAC-RDN was based on CAC-RD version 1, which has only resources reservation for SHO.

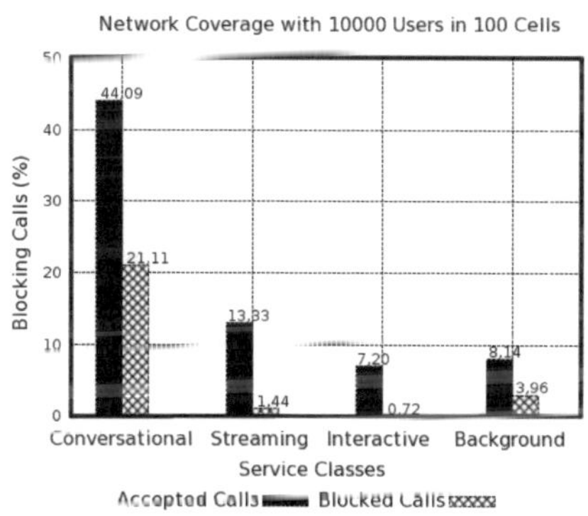

Fig. 6. Newly calls blocking evaluation in CAC-RDN

Figure 7 presents a different type of blocking calls graphics: the line point graphic. In this graphics, we can compare the performance of three different conventional CACs: CAC-J, CAC-RD version 1 and CAC-RD version 2. We can see that the high conversational blocking percentage, presented in CAC-J and CAC-RD, was reduced by CAC-RD version 2.

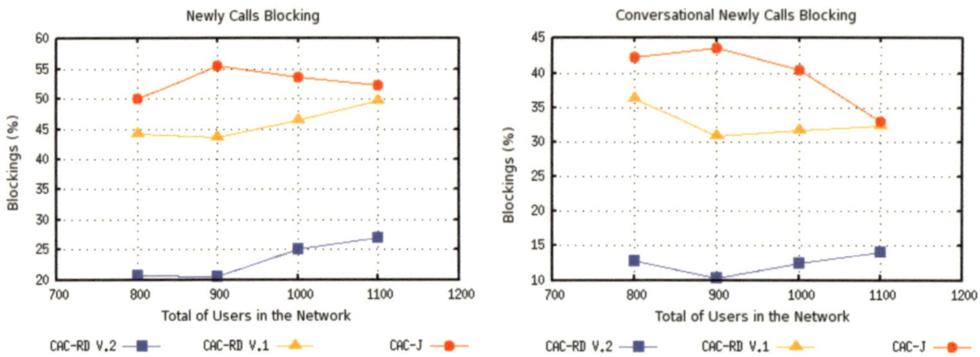

(a) Line points graphics

Fig. 7. Newly calls blocking evaluation by comparing different CAC schemes

8.2 Soft handover failures

Figure 8 presents SHO failures for three conventional CACs: CAC-J, CAC-RD version 1 and CAC-RD version 2. We can see that the new reservation scheme for conversational calls does not impact in the SHO failures. It only decreases the number of SHO failures to almost zero, in general, despite the total number of users in the network.

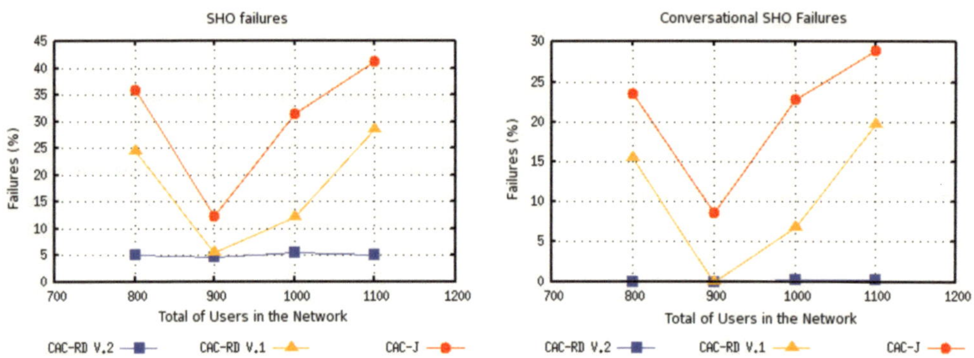

Fig. 8. SHO failures evaluation by comparing conventional CACs schemes

8.3 Efficiency

Figure 9 shows the network utilization comparison between CAC-RD and CAC-RDF. As the simulation time increases, the network moves toward a congestion state, but with a better utilization by CAC-RDF. The network utilization difference between CAC-RD and CAC-RDF is small in comparison to the network state. But as Tostes et al. (2011) analyzed, this difference is highly significant when evaluated the performance of each service class.

8.4 Complexity

Figure 10 shows the asymptotical analysis of CAC-RD and CAC-RDN. We perceive that CAC-RDN is better than CAC-RD in this aspect since it involves less computational operations.

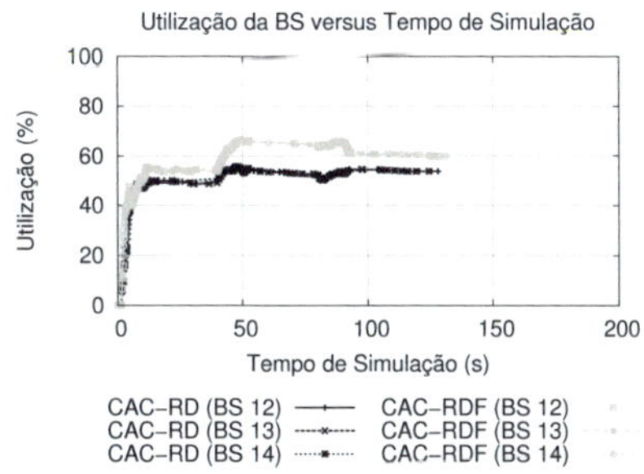

Fig. 9. Efficiency analysis of CAC-RD and CAC-RDF

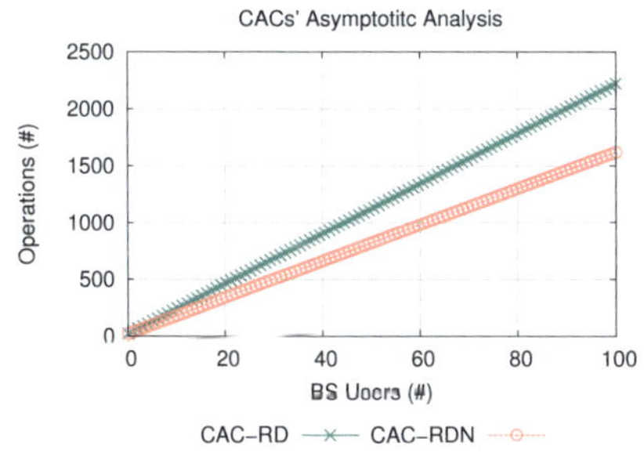

Fig. 10. Complexity analysis of CAC-RD and CAC-RDN

8.5 Adaptability

Figure 11 allows the evaluation of CAC's capacity to adapt while the network presents different load peaks. The graphics shows the load distribution as total number of calls increases.

Figure 11(a) presents conventional CACs assessment (CAC-J and CAC-RD) while figure 11(b) presents unconventional CACs assessment (CAC-RDN and CAC-RDF). When the total of requested calls increases, the amount of priority calls (conversational, streaming and interactive) also increases but until a certain extent. This happens due to the use of fixed thresholds and the scarce network resources availability. In the unconventional CACs, we can see that CAC-RDN is fairer due to the balanced calls load of priority classes. But CAC-RDF demonstrates adaptation and adjustment capabilities. It accepts more priority calls while the network tends to congestion (the greater number of required calls).

8.6 QoS performance parameters

Figure 12 shows two graphs, one that measures the jitter and another that measures the flow of the conversational class. The SHO process has been initiated in instance 120 s. In relation to jitter, before the SHO, the call failed to ensure QoS. However, after the SHO, jitter exceeded the limit of 1 ms for this class. This also happened with the throughput per user. After the SHO, the call could not get the minimum to have 4 kbps, which is specified by 3GPP. Therefore, it is clear that there is no QoS guarantee in the SHO process for conversational calls.

As it was shown in table 1, 3GPP has specified some QoS requirements for each class of service. For conversational class, 3GPP has specified a maximum 400 ms delay while for streaming class the delay limit was 10 s. As can be seen in figure 13, CAC-RD and CAC-RDF can guarantee both these requirements. For interactive class, 3GPP has specified a delay limit

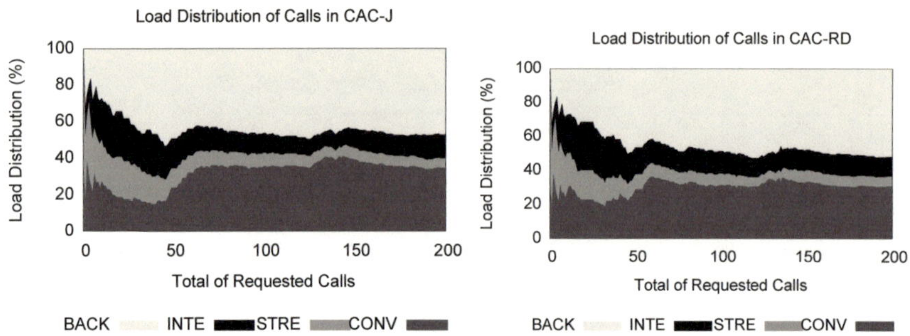

(a) Conventional CACs

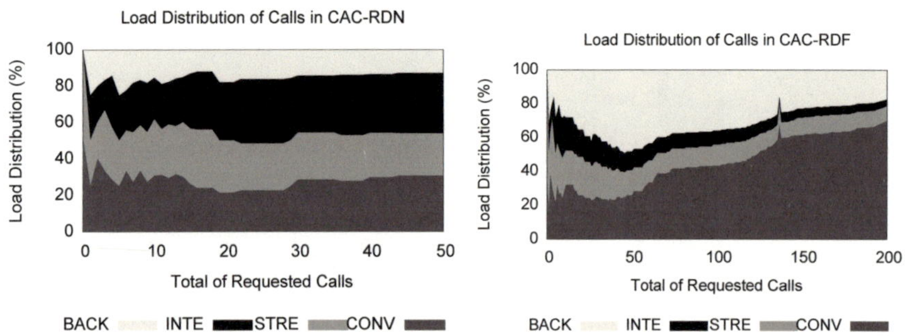

(b) Unconventional CACs

Fig. 11. Adaptability evaluation by comparing different CACs schemes

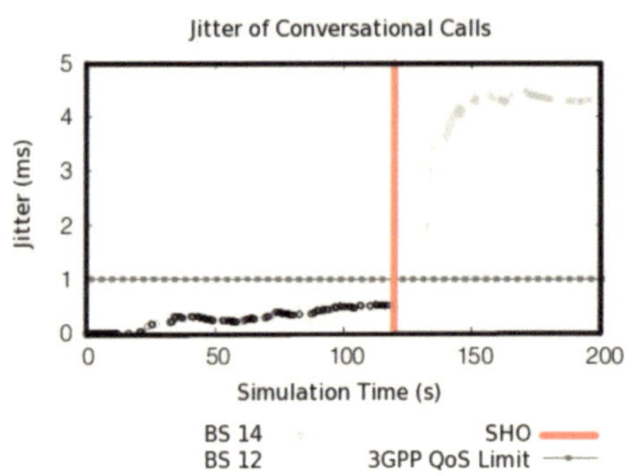

(a) Jitter

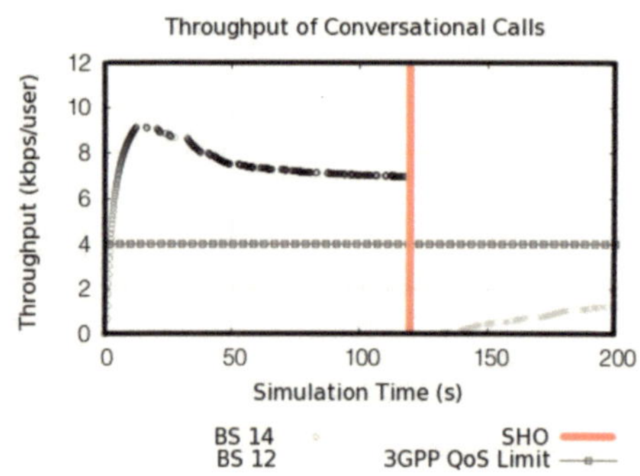

(b) Throughput

Fig. 12. QoS analysis of a call during soft handover process

of 4 s. Figure 13 shows that both CAC-RD and CAC-RDF do not guarantee this requirement all the simulation time. But CAC-RDF has a major advantage: it blocks new interactive calls when it perceives this network degradation. Although both CACs, CAC-RD and CAC-RDF, guarantee the minimum of 2.8 kbps throughput, CAC-RDF stops accepting this low priority class when it detects network degradation.

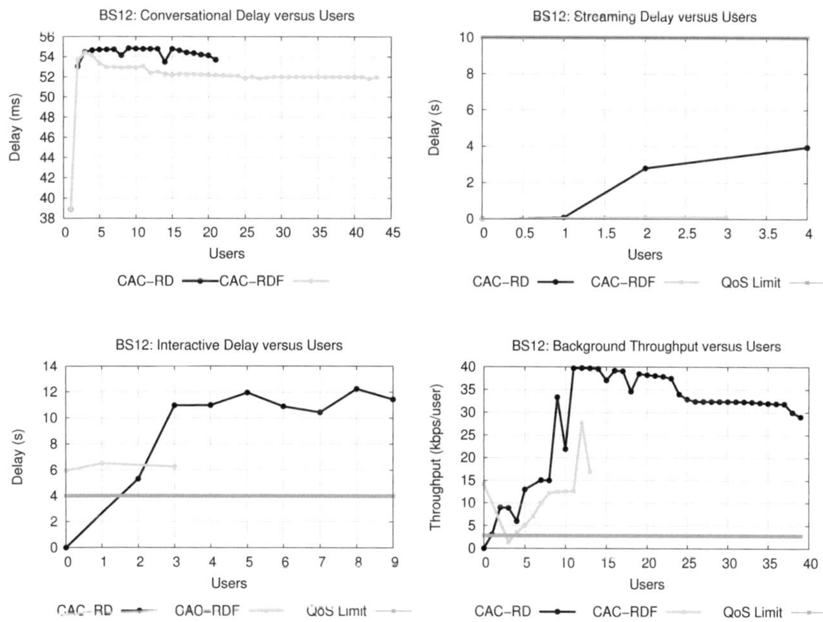

Fig. 13. QoS analysis of a call during soft handover process

9. Conclusion

Cellular networks of third generation (3G) allow us to be connected to the Internet anytime, anywhere, through smartphones, laptops and netbooks. Voice calls, video streaming, email, access to social networks and WEB browsing are some examples of 3G services. When an user accesses a specific service, a new call is requested. It is the responsibility of the cellular operators to ensure the quality of the requested services. As we have seen, the greater the diversity of the services required, the greater is the network resources management difficulty. Call Admission Control (CAC) is a very important system to a cellular network. It decides to accept or not of a new call depending on network conditions, regarding availability of channels, signal power and quality of service. Its greatest difficulty is the knowledge of which required calls have to be accepted and which have to be blocked, due to numerous parameters involved in the CAC's decision. When a new call is accepted, it increases the load over the network resources. The CAC's main challenge is to accept as much as it can, maintaining the network performance acceptable.

CAC mechanisms can be conventional or unconventional. The conventional ones use deterministic or stochastic techniques. The deterministic technique ensures a confidence interval of 100%, while the stochastic ones ensure a probability. The most common techniques are the stochastic ones, such as signal power control, borrowing and reservation of channels, queues and thresholds prioritization techniques. Although they can assure high efficiency, low complexity and low overhead, conventional CACs are not adaptable neither stable. To overcome these needs, the academic community has introduced the unconventional CACs. In this category, CACs use intelligent methods associated with the traditional techniques. Genetic algorithms, statistical methods, artificial neural networks and fuzzy logic are some of the intelligent methods used. The self-learning capacity of neural networks and the fuzzy

logic ability of modeling uncertainties are most often used in the unconventional literature, providing adaptive and stable CACs.
This paper presented some CAC's case studies in both categories: CAC-RD as a conventional control; Neural CAC-RD (CAC-RDN) and Fuzzy CAC-RD (CAC-RDF) as unconventional ones. CAC-RD is an admission control based on resource reservation, blocking of calls according to static network utilization thresholds, and a network diagnosis. But CAC-RD has the low scalability and the static blocking characteristics as a disadvantage. To overcome CAC-RD's scalability issue, we have presented CAC-RDN, which have learned CAC-RD's behavior through neural network training. On the other hand, CAC-RDF has been presented in response to CAC-RD's waste of resources. Its main contribution is CAC-RD's thresholds dynamization, with fuzzy logic (FL), in accordance with the quality of service (QoS) for the blocking of lower-priority calls.
As mentioned earlier, the assessment of CACs can be made through eight evaluation criteria: blocked calls, dropping calls, soft handover failures, efficiency, complexity, overhead, adaptability, stability and quality of service performance. We have presented the evaluation of CAC-RD, CAC-RDN and CAC-RDF. CAC-RD's main contribution is the commitment between performance and availability. With CAC-RDN, it was possible to simulate scenarios up to 10000 users and 100 cells. Furthermore, CAC-RDF had the following advantages over CAC-RD: (1) better utilization of network resources, (2) higher acceptance of priority calls, and (3) CACs' stability. Although it is not possible to say which CAC is the best, because it depends on the users and the cellular operator demands, we have seen that the intelligent techniques can improve the CAC scalability and stability, providing a better network utilization.
Some recommendations may be followed: (1) CAC must ensure not just high network availability, but also high performance; (2) the quality of service requirements must be guaranteed for each class of service; (3) one of the most important CAC evaluation parameter is the number of blocked calls; (4) computational intelligence techniques can offer high scalability and high efficiency characteristics for CACs; (5) the proposed CACs' performance must be analyzed in new scenarios and networks, such as Long Term Evolution (LTE) and next generation networks.

10. Acknowledgements

The authors acknowledge the financial support received from the Foundation for Research Support of Minas Gerais State, FAPEMIG, through Project CEX PPMIII 67/09, and the National Council for Scientific and Technological Development, CNPq, Brazil.

11. References

3GPP (2009). TS 25.211: Technical specification group: Physical channels and mapping of transport channels onto physical channels (fdd) (release 9).
URL: *http://www.3gpp.org*

3GPP (2010). TS 23.107: Technical specification group: Quality of service (QoS) concept and architecture (release 9).
URL: *http://www.3gpp.org*

Antoniou, J., Vassiliou, V., Pitsillides, A., Hadjipollas, G. & Jacovides, N. (2003). A simulation environment for enhanced UMTS performance evaluation.
URL: *http://citeseer.ist.psu.edu/642704.html*

Ascia, G., Catania, V., Ficili, G., Palazzo, S. & Panno, D. (1997). A VLSI fuzzy expert system for real-time traffic control in ATM networks, *Fuzzy Systems, IEEE Transactions on* 5(1): 20–31.

Box, G. E. P. & Jenkins, G. (1990). *Time Series Analysis, Forecasting and Control*, Holden-Day, Incorporated, San Francisco.

Chang, C.-J., Su, T.-T. & Chiang, Y.-Y. (1994). Analysis of a cutoff priority cellular radio system with finite queueing and reneging/dropping, *IEEE/ACM Trans. Netw.* 2(2): 166–175.

Cheng, R. G. & Chang, C. J. (1997). Neural network connection admission control for ATM networks, *IEE Proc. Commun.* 144(2): 93–98.

Cheng, R.-G., Chang, C.-J. & Lin, L.-F. (1999). A QoS-provisioning neural fuzzy connection admission controller for multimedia high-speed networks, *IEEE/ACM Trans. Netw.* 7(1): 111–121.

Ghaderi, M. & Boutaba, R. (2006). Call admission control in mobile cellular networks: a comprehensive survey, *Wireless Communications and Mobile Computing* 6(1): 69–93.
URL: *http://dx.doi.org/10.1002/wcm.246*

Hiramatsu, A. (1989). ATM communications network control by neural network, *International Joint Conference on Neural Networks*, Vol. 1, IEEE, USA, pp. 259–266.

Hong, D. & Rappaport, S. S. (1986). Traffic model and performance analysis for cellular mobile radio telephone systems with prioritized and nonprioritized handoff procedures, *Vehicular Technology, IEEE Transactions on* 35(3): 77–92.

Kaaranen, H., Ahtiainen, A., Laitinen, L., Naghian, S. & Niemi, V. (2005). *UMTS Networks: Architecture, Mobility and Services*, Wiley.
URL: *http://www.amazon.com/exec/obidos/redirect?tag=citeulike07-20&path=ASIN/0470011033*

Karabudak, D., Hung, C.-C. & Bing, B. (2004). A call admission control scheme using genetic algorithms, *SAC '04: Proceedings of the 2004 ACM symposium on Applied computing*, ACM, New York, NY, USA, pp. 1151–1158.
URL: *http://doi.acm.org/10.1145/967900.968135*

Katzela, I. & Naghshineh, M. (1996). Channel assignment schemes for cellular mobile telecommunication systems, *IEEE Personal Communications* 3: 10–31.

Kurose, J. F. & Ross, K. W. (2009). *Computer Networking: A Top-Down Approach*, 5th edn, Addison-Wesley Publishing Company, USA.

Levine, D. A., Akyildiz, I. F. & Naghshineh, M. (1997). A resource estimation and call admission algorithm for wireless multimedia networks using the shadow cluster concept, *IEEE/ACM Trans. Netw.* 5(1): 1–12.

Lu, S. & Bharghavan, V. (1996). Adaptive resource management algorithms for indoor mobile computing environments, *SIGCOMM '96: Conference proceedings on Applications, technologies, architectures, and protocols for computer communications*, ACM, New York, NY, USA, pp. 231–242.

Naghshineh, M. & Schwartz, M. (1996). Distributed call admission control in mobile/wireless networks, *IEEE Journal on Selected Areas in Communications* 14(4): 711–717.

Pedrycz, W. & Vasilakos, A. (2000). *Computational Intelligence in Telecommunications Networks*, CRC Press, Inc., Boca Raton, FL, USA.

Raad, R. (2005). *Neuro-fuzzy admission control in mobile communications systems*, PhD thesis, University of Wollongong.

S. A. Youssef, I. W. Habib, T. N. S. (1996). A neural network control for effective admission control in ATM networks, *IEEE ICC '96* pp. 434–438.

Shen, X., Mark, J. W. & Ye, J. (2000). User mobility profile prediction: an adaptive fuzzy inference approach, *Wirel. Netw.* 6(5): 363–374.

Steinmetz, R. & Wolf, L. C. (1997). Quality of service: Where are we, *Proc. 5th International Workshop on Quality of Service (IWQOS'97)*, IEEE Press, New York, USA, pp. 211–222.

Storck, C. R., Tostes, A. I. J. & Duarte-Figueiredo, F. L. P. (2008a). CAC-RD: A call admission control for UMTS networks, *International Workshop on Performance Modeling and Evaluation in Computer and Telecommunication Networks (PMECT-2008 in conjunction with IEEE ICCCN2008)*, PMECT-2008 - IEEE ICCCN2008, St. Thomas, U.S. Virgin Island.

Storck, C. R., Tostes, A. I. J. & Duarte-Figueiredo, F. L. P. (2008b). CAC-RD: Controle de admissão de chamadas para redes UMTS, *Simpósio Brasileiro de Redes de Computadores (SBRC)*, Rio de Janeiro, pp. 371–384.

Talukdar, A. K., Badrinath, B. R. & Acharya, A. (1999). Integrated services packet networks with mobile hosts: architecture and performance, *Wirel. Netw.* 5(2): 111–124.

Tostes, A. I. J. (2010). *Representações neural e fuzzy de controle de admissão de chamadas para redes UMTS*, Master's thesis, PUC Minas, Mestrado em Informática.

Tostes, A. I. J. & Duarte-Figueiredo, F. L. P. (2009). Simulação e análise de controles de admissão de chamadas para redes móveis de terceira geração 3G, *Concurso de Trabalhos de Iniciação Científica da SBC 2009*, Sociedade Brasileira de Computação, Bento Gonçalves.

Tostes, A. I. J., Duarte-Figueiredo, F. L. P. & Zárate, L. E. (2011). Controle de admissão fuzzy baseado em limites dinâmicos de congestionamento para redes de celulares, *Simpósio Brasileiro de Redes de Computadores (SBRC)*, Campo Grande, pp. 191–204.

Tostes, A. I. J., Duarte-Figueiredo, F., Novy, G., Storck, C., Dias, S. M. & ZÁrate, L. E. (2008). An artificial neural network approach for mechanisms of call admission control in UMTS 3G networks, *Hybrid Intelligent Systems, International Conference on* 0: 459–464.

Tostes, A. I. J., Storck, C. R. & de L. P. Duarte-Figueiredo, F. (2010). CAC-RD: an UMTS call admission control., *Telecommunication Systems* pp. 261–274.

Zhang, T., van den Berg, E., Chennikara, J., Agrawal, P., Chen, J.-C. & Kodama, T. (2001). Local predictive resource reservation for handoff in multimedia wireless IP networks, *IEEE Journal on Selected Areas in Communications* 19(10): 1931–1941.

14

Water Quality Dynamic Monitoring Technology and Application Based on Ion Selective Electrodes

Dongxian He, Weifen Du and Juanxiu Hu
China Agricultural University
East Campus in Xueyuan Rd.
Haidian Beijing,
China

1. Introduction

Water shortage and water pollution is become to a more serious problem in the world especially in developing country. Water pollution is mainly due to a large number of emissions from industrial wastewater, agricultural drainage, and urban wastewater to rivers, lakes and soil which is making surface water and groundwater quality deterioration. This water shortage is aggravated by the exacerbated water quality and thus to influence human health and society economy development [1]. In order to conserve water resources, prevent water pollution, and improve sustainable development of water environment and aquaculture production, water quality monitoring and control in dynamically and regularity is become more and more important. Water quality monitoring and control is not only able to understand the water quality changes and pollution migration patterns, but also provide scientific and reasonable technical support for water resources integrated planning, water environment assessment, water treatment and conservation technology [2]. Therefore, water quality monitoring is essential tool and basis foundation in water resource management and water pollution supervision and control for government administration.

Water quality monitoring of different water sources is difficult because of the intricate components and its mutual interference and thus depended mainly on chemical analysis and instrumental analysis. According to geochemistry, geographical and regional differences of water environment or water pollution, water quality monitoring is always conducted in instrument station and to construct a monitoring network for water quality management, assessment and planning via short-term monitoring, long-term monitoring, emergency monitoring, and dynamic monitoring [3, 4]. To supplement the instrument stations in laboratory level, water quality monitoring vehicle or shipping using flow sampling and online monitoring is used for water pollution tracking or iterative detection. Recent years, water quality dynamic monitoring using online sensors is applied for various applications and increasing with promoting commercialization of ions selective electrodes.

Since 1960, local, continents and federal agencies in the United State began to collect water quality data and gradually getting a unified specification data to establish STORET water quality system. 20 years later, the STORET water quality system is used to provide the water quality scientific data and water pollution management support for federal, state and local water authorities [5]. Water quality file analysis system WAP2 (Water Archive Phase 2) is developed in UK after STORET water quality system constructed. WAP2 system is focusing on relationship between the hydrological cycle and impact of human activities, and described in detail for hydrology, meteorology, water supply, drainage, wastewater treatment, industrial, and water quality [6, 7]. Compared to advanced countries, water quality monitoring technology in developing countries is developed and applied from 1980s and almost conducted by regional or local government. the national water quality monitoring network is still not established such as China and Indian, because of the different sampling methods and expensive instrument import form the United States and Germany [8].

Water quality dynamic monitoring is integrated from chemical analysis, computer, sensing, and communication technologies for simultaneous detection of several water quality parameters, mass data processing, and automatic water treatment in designated regions. The sampling method and monitored data will be basis foundation to represent water quality for in the regional water environment. On the other hand, suitable sample distribution and contrast handy chemical analysis are important in water quality dynamic monitoring especially for rivers with serious floating debris and soil erosion [9, 10]. Therefore, automatic cleaning, calibration, diagnostics, and alarm functions, data remote transmission, and networking construction will be required in water quality dynamic monitoring. This chapter is focus on water quality sensors based on ion specific electrodes, water quality dynamic monitoring system based on web-server-embedded technology and its application in rivers and freshwater detection, aquaculture production and hydroponic plant production.

2. Principle and application of ion selective electrodes

Ion selective electrode is a kind of electrochemical sensor which determine the ionic activity or concentration by membrane potential. The membrane potential is directly related to the ionic activity or concentration by interface of the sensitive membrane when ion selective electrode contact with the ion solution. Sensitive membrane of the ion selective electrode is a layer of special electrode film which is selectively responded to the specific ions. The relationship between electrode membrane potential and the ion content is conformed to Nernst formula [11]. The ion selective electrode is most widely used in electrochemical analysis for wild automatic and on-site continuous monitoring because of the good selectivity and short balance time. With a hundred years of technology development, more than 20 kinds of ion selective electrodes were commercially produced in the present now (Table 1). On-line monitoring based on ion selective electrode has become an independent scientific branch in electrochemical analysis, and is widely used for gas, water, soil, biological, chemical, marine, geology, biology, medicine, food, environment monitoring, and other fields [12,13].

Ion selective electrode consists of ion electrode probe, internal reference probe, internal reference solution, and sensitive membrane (Fig. 1). To a particular ion selective electrode, the electric potential mechanism caused in the sensitive membrane may be different, because the membrane material is responded sensitively to a specific ion. The electric potential arises of main ion selective electrode is based on the membrane materials and the interface exchange of ion solution [11]. According to the sensitive membrane material such as glass film, solid membrane, polymer membrane, gas infiltration membrane, and ion selective electrode can be classified into glass electrode, solid electrode, polymer electrode, and gas induction electrode.

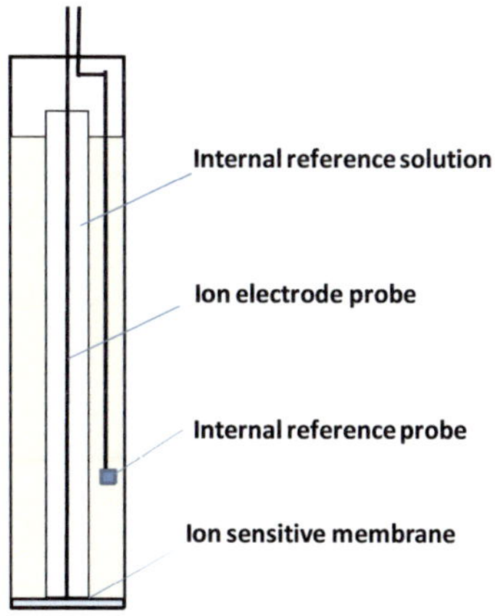

Fig. 1. Ion selective electrode

Glass film is made of silicon dioxide glass carrier which be added various chemicals, the common glass electrode includes pH electrode and Na+ selective electrode. Relatively unsolvable inorganic salts are used by solid electrodes membrane. The silver/sulfur ion, lead ion, copper ion, cyanide ion, thiocyanate ion, chloride ion, and fluorin ion electrodes are commonly commercialized. Polymer film is constituted by different ion exchange material which adding inert carrier such as PVC, PE, polyurethane or silicone rubber, potassium. The calcium ion, fluoride acid ion, nitrate ion, perchlorate ion, and water hardness are common produced by polymer electrodes. Gas sensing electrode including gas osmosis membrane and internal buffer liquid and without internal reference electrode for measuring ammonia, carbon dioxide, dissolved oxygen, nitrogen oxide and sulfur dioxide and free chlorine as dissolved gases.

Specific ion potential in solution is measured by ion selective electrode according to internal reference electrode with a constant potential. Electric potential difference of ion electrode and internal conference electrode depends on certain ions activity in the solution. However, the response of ion selective electrode will be affected by other ions in the sample solution. For certain ion selective electrode, impact of interfering ions can be quantitative corrected [14, 15]. If the sample solution is complex or contains large amounts of interfering ions, standard add method can be adopt [16]. Before the measuring, least 3 kinds of known concentrations of standard solution can be used to calibrate the ion selective electrode, and the sample concentration should be covered in the standard solution concentration. Sometimes, ionic strength adjustor (ISA) is used in order to ensure same ionic strength in the standard solution and sample solution. Usually, the ratio of sample solution and ISA is suitable in 50:1. Some ion selective electrode is used within certain pH range of solution, thus the pH of standard solution and sample solution can be adjusted by adding buffer to ionic strength adjustor. In addition, some chelating agent aiming to interference ion or preservatives aiming to antioxidant are added to ionic strength adjustor for dispelling interfering ions [17, 18].

Generally, ion selective electrode can be stabilized within 1 to 2 min, but more time will be needed when sensitive membrane is contaminated by solution with oil or particles. Electrode tip of glass electrode is washed by alcohol or a mild detergent such as laboratory cleaning detergent. Solid electrode is resumed by particles burnishing stick. Polymer electrode can be rinsed with freshwater [19]. The ion selective electrode needs to put in a low concentration of the standard solution for 2-4 h or more time. It is needed to calibrate after cleaning. The glass electrode can be stored in low concentration within the standard solution

Ion	Electrode type	Sensitive membrane material	Linear range	Interfering ions
Na^+	Rigid matrix electrode	NAS-11-18 glass	$1 \sim 10^{-6}$	H^+, Ag^+
Na^+	Rigid matrix electrode	ETH-157	$0.1 \sim 10^{-5}$	Li^+, Ca^+, H^+
K^+	Rigid matrix electrode	Valinomycin	$0.1 \sim 10^{-6}$	NH_4^+
K^+	Rigid matrix electrode	Di-tert-dibenz-30	$0.1 \sim 10^{-6}$	NH_4^+, Na_+
Ca^{2+}	Rigid matrix electrode	ETH-1001	$0.1 \sim 10^{-6}$	
Ca^{2+}	Mobile carrier electrode	Dioctyl-phenyl $Ca_3(PO_4)_2$	$0.1 \sim 10^{-5}$	Zn^{2+}, Mn^{2+}
Ag^+/S^{2-}	Homogeneous membrane electrode	Ag_2S	$1 \sim 10^{-7}$	Hg^{2+}
Pb^{2+}	Homogeneous membrane electrode	$PbS+Ag_2S$	$1 \sim 5*10^{-7}$	Ag^+, Hg^{2+}, Cu^{2+}
Cu^{2+}	Homogeneous membrane electrode	$CuS+Ag_2S$	$1 \sim 10^{-7}$	Ag^+, Hg^+
F^-	Crystalline membrane	LaF_3+EuF_2	$1 \sim 10^{-8}$	OH^-
Cl^-	Homogeneous membrane electrode	$AgCl+Ag_2S$	$1 \sim 5*10^{-5}$	S^{2-}, CN^-, I^-, SCN^-, Br^-
Cl^-	Homogeneous membrane electrode	Hg_2Cl_2+HgS	$0.1 \sim 5*10^{-5}$	S^{2-}, CN^-, I^-, SCN^-, Br^-
Br^-	Homogeneous membrane electrode	$AgBr+Ag_2S$	$1 \sim 5*10^{-6}$	S^{2-}, CN^-, I^-, SCN^-
I^-	Homogeneous membrane electrode	$AgI+Ag_2S$	$1 \sim 5*10^{-7}$	S^{2-}, CN^-, SCN^-
CN^-	Homogeneous membrane electrode	$AgI+Ag_2S$	$102 \sim 10^{-8}$	S^{2-}, I^-
NO_3^-	Mobile carrier electrodes	Trioctyl methyl ammonium chloride	$0.1 \sim 5*10^{-8}$	ClO_4^-, I^-, NO_3^-
BF_4^-	Mobile carrier electrodes	Triheptyl dodecyl amonium	$0.1 \sim 10^{-8}$	ClO_4^-, SCN^-
NH_3 molecule	Gas induction electrode		$1 \sim 10^{-6}$	Volatile amine
CO_2 molecule	Gas induction electrode		$10^2 \sim 10^{-4}$	

Table 1. General ion selective electrodes

without measuring. Solid electrode, polymer electrode, and gas induction electrode should be kept in a dry condition without measuring. Polymer electrode and gas induction electrode can store in a low concentration of the standard solution between the measuring.

3. Water quality dynamic monitoring parameters based on Ion selective electrodes

3.1 pH

pH as a basic parameter of water quality is to measure activity of hydrogen ion for indexing alkalinity or acidity of water. Component style and circulation of water are affected by pH value changed. The exorbitant or insufficient pH of water will cause serious consequence in the water cycling or aquatic biology. The exorbitant pH will interfere with carbon and iron utilization of phytoplankton and result in Illness or death of fish. The low pH value will restrict activity of nitrate reductase, and thus lead to nitrogen deficiency for hydrophyte [20, 21].

pH of water is usually measured by glass electrode with temperature compensation. pH glass electrode using doped glass membrane, glass H^+ ion sensitive electrode and additional reference electrode is sensitive to H^+ ion and active by the glass bubble tailored with varying amount of alumina (Al_2O_3) and other common constituent such as Na_2O, K_2O, B_2O_3. Inside of the pH electrode is usually filled with buffered internal filling solution of chlorides which is reference electrode usually made of silver wire covered with silver chloride (Ag/AgCl). Elective potential difference of the pH electrode is caused by H^+ activities of bubble on both thin glass sides and described by the Nernst formula. Since Nernst equation is a temperature-dependent function, temperature compensation is significantly required for measuring pH.

pH glass electrode is easy to be corroded because of fluoride in the water. Stibium (Sb) metal electrode could be adopted instead of glass electrode to measure pH in solution with fluoride. It is difficult to dissolve oxide membrane on the Sb electrode surface and issued in a reversible electron exchange between the oxide membrane and H^+ ion in the solution as follows:

$$2Sb + H_2O \Leftrightarrow Sb_2O_3 + 6H^+$$

$$Sb_2O_3 + 6H^+ \Leftrightarrow 2Sb_3+ +3H_2O$$

According to the Nernst formula, the elective potential difference between Sb metal and oxide membrane (Sb_2O_3) is measured for pH of the solution. In general, Sb metal electrode is widely used in solution containing cyanide, sulfide, fluoride, reduction sugar, alkaloids, etc. It shows measure accuracy in ±0.1 in the solution pH range from 2 to 7 and ±0.4 in the solution pH range from 7 to 12 [22].

3.2 DO

Dissolved Oxygen (DO) is amount of oxygen (O_2) dissolved in water. Since oxygen is needed for all forms of life including aquatic biology, DO measure is important in water quality monitoring of water environment. It is also produced by photosynthesis of plant byproduct and phytoplankton. DO is used in respiration and decomposition process of aquatic bacteria. A DO decrease is typically associated with an organic pollutant when organic matter such as animal waste or improperly treated wastewater entered to water, or algae growing increases [23]. DO of water is ranged from 0 to 18 mg/L, approximately 8 mg/L will be required for aquatic plant and animals, less than 5 mg/L will be taken biological stress. As DO levels constantly decrease, pollution intolerant organisms are replaced by pollution tolerant worms and fly larvae, and then result in water quality degenerating [24].

The dynamic monitoring of DO usually adopts membrane electrode which based on the membrane diffusion theory. This measure method is not affected by pH, salinity, oxidation-reduction substance, chromaticity, and turbidity, etc. Typical DO electrode contains a working electrode usually made of noble metal such as Pt or Au and a reference electrode of popular Ag/AgCl. Both electrodes are located in an electrolyte of KCl solution which is separated from the sample by a gas permeable membrane. The working electrode equaling to the cathode reduces the oxygen molecules to hydroxide ions occurred reaction following as $O_2 + 2H_2O + 4e \Leftrightarrow 4OH^-$. The anode occurred reaction following as $4Ag + 4Cl^- \Leftrightarrow 4AgCl + 4e$. In the above electrochemical reactions a current flows from the reference electrode to the working electrode. The current is generated directly proportional to the DO in the sample.

The DO measurement using ion selective electrode has the advantages of faster measurement speed, simple operation, and easy to automatically monitoring due to its gas permeable membrane easy to plug and the electrode aging. The maintenance of the DO electrode with periodical replacement of the membrane is needed. The solution flow with 0.2 to 0.3 m/s is required because of the oxygen contents decreased by cathode reaction during the measuring.

3.3 EC

Electrical Conductivity (EC) is used to indicate the purity of water and reflect the degree of the electrolyte existing in the water. Generally, the content of inorganic salt that is dissolved in the water can be estimated through measuring EC of the solution. EC means the ability to conduct electric current and equals the reciprocal of resistivity in physics. So according to the principle of electrolytic conduction, EC will be measured by the method of electric resistance measurement using metal electrode. Like the metal conductor, the electrolytic solution also obeys Ohm law, through the measurement of the electric current between the metal electrodes to get the EC of the solution. The resistance of the electrolyte is closely related with the concentration of solute in the solution [25]. In other words, the EC of the solution is closely related with the contents of inorganic acid, alkali, and slat in the solution.

The basic unit of the EC is Siemens (S). But S/cm is much more commonly used in the EC measuring because of the effect of electrode shape. At present, the 2-electrode based conductivity electrode also called as Kohlrausch electrode regularly is made of two metal plates or, metal cylinder fixed on the support holder with insulators. Its measuring range is usually 0 to 20 mS/cm, and the measuring range is different according to the different electrode constant, with conductivity reading accuracy of ±0.5%. In addition, another commonly used electrode is 4-electrode based conductivity electrode (Fig 2), which contains two coaxial current electrodes and two coaxial voltage electrodes. The current between the two current electrodes has a linear relationship with the conductivity of solution when measuring EC. The 4-electrode based conductivity electrode can avoid the influence by polarization effect usually occurred in 2-electrode based conductivity electrode. Its measuring range is usually 0 to 2000 mS/cm, and either the measuring range is different according to the different electrode constant, with conductivity reading accuracy of ±0.5%.

(a) (b)

Fig. 2. 2-electrodes (a) and 4-electrodes (b) conductivity sensors.

3.4 NH_4^+-N

The ammonium nitrogen (NH_4^+-N) in water body consists of molecular ammonium (NH_3) and ionic ammonium (NH_4^+). High content of NH_4^+-N in water body will cause the eutrophication phenomenon of the water body, and produce toxic hazard to the fish and other aquatic life, and then even result in a series of environmental problems [26]. Therefore, the content of NH_4^+-N is a significant symbol for evaluating the pollution degree of water body, as well as an important parameter for the water quality monitoring.

The ammonium ISE used to measure the NH_4^+-N in water is a diaphragm gas sensing electrode, in which there is a flat-plate pH glass electrode as the indicated electrode, and Ag/AgCl as a reference electrode which is immersed in the internal reference solution of 0.01 M NH_4Cl commix with inertia electrolyte, such as NaCl, KCl. At the end of electrode, the sensitive membrane of pH electrode is covered with a very thin layer of hydrophobic semi-permeable membrane usually made of PTFE that allows the molecular NH_3 only to pass through, but block off other ions in the solution. When measuring of NH_4^+-N in water with gas sensing electrode, the NH_4^+ in the solution will transform into molecular NH_3 by adjusting pH>11, and the NH_3 will pass through the membrane. There occurred a reversible chemical reaction as follow: $NH_3 + H_2O \Leftrightarrow NH_4^+ + OH^-$. Therefore, the NH_4^+-N content in the solution could be determined through measuring the change of electric potential utilizing the pH glass electrode in the gas sensing electrode. The ammonium ISE has the feature of high selectivity, and requires no titration and has no turbidity or color interferences, but is sensitive to the change of temperature [27]. At present, most commercial electrodes show the performance with the detection limit of 10^{-7}-10^{-6} M, measuring range of 10^{-7} - 1.0 M, and response time of 30 seconds.

3.5 ORP

Oxidation-Reduction Potential (ORP) is the measurement (in millivolt, or mV) of the oxidizing or reducing tendency of a solution. ORP is typically not a good method for measuring solute concentration, so it is not used to indicate the water quality independently, but used to integrate with other water quality parameters to reflect the ecological status of the water environment. When the ORP is positive, it means that the oxidation ability of water body is strong, that is the ability of the recursive organism which handles the biological metabolism is strong. If the pH, DO and temperature are suitable, the pollutants in the water will be decomposed fast and effectively [28].

The principle of ORP measurement is similar with the measuring of pH, which based on the selectivity of H^+ by glass electrode, while the principle of ORP measurement refers to the electrons' acceptance or release of noble metal electrode. The structure of a typical ORP electrode is almost same as that of pH electrode, adopting Pt or Au as the working electrode and the Ag/AgCl as the reference electrode inserted in the KCl internal reference solution. In the case of measuring ORP with a Pt metal electrode, a layer known as Helmholtze electric double layer which is equivalent to a capacitance will be formed. One end of capacitance is inked with the Pt metal, and another end is linked with reference electrode. Because the capacitance is undertook charging, this capacitance, therefore, the ORP of the testing solution could be obtained by measuring electric potential of the capacitance.

3.6 F^- Ion

In the case of the F- ISE, the selective membrane is a single crystal of Lanthanum Fluoride (LaF_3) doped with Europium Fluoride (EuF_2) which produces holes in the crystal lattice through which F^- ions can pass [29]. When immersed in a fluoride solution and connected via a voltmeter to an Ag/AgCl external reference electrode immersed in the same solution, the F^- ions in the solution pass through the crystal membrane by normal diffusion from high concentration to low concentration until achieves an balance between the force of diffusion and the reverse electrostatic force. On the other side of the membrane there is a corresponding build-up of positive ions. The potential between the analyte and internal reference solution faces of the crystal produced during the phase of F- ion exchange across LaF_3 is proportional to the logarithm of F- activities according to Nernst formula.

For the Fluoride ISE, the most important interfering species is OH^-. The OH^- complexes with the LaF3 crystal itself in much the same way as F^- does, with the chemical equilibrium of ($LaF_3 + OH^- \Leftrightarrow La(OH)_3 + 3F^-$) that produces additional F^-. Subsequently, the potential across the crystal will be a function of [OH^-] and will interfere with the fluoride determination. Another pH dependent effect is due to the basicity of F^-. HF is effectively a weak acid and at pH less than 5, acid-base equilibrium of ($2F^- + H^+ \Leftrightarrow HF + F^- \Leftrightarrow HF_2^-$) will reduce the concentration of F^- in solution. To remedy these problems, proton and hydroxide concentrations are controlled by buffering the pH of measurement solutions to between 5 and 7. In addition, sodium citrate can be used as masking agent for eliminating the interfering of polyvalent cations, such as Fe^{3+}, Al^{3+}, Be^{2+}, Si^{4+}, etc. which easy to form up complexes with F^- ion [30].

4. Water quality dynamic monitoring system and its application

4.1 Construction of the water quality dynamic monitoring system

Water quality dynamic monitoring system consists of water sampling module, water quality sensor module, data collection module, data processing module, communication module, and power supply module. Water sampling methods are usually used in pattern s of submerged style, flow circulating style, float style, and water logging style (Fig. 3). Submerged style sampling is combination of water sampling module and water quality sensor module which is fixed at a specific depth by mounting brackets for long-term dynamic monitoring of water quality. Water quality sensors are connected with water sampling module by using threaded bolt for sensors maintenance. Submerged style sampling is the most common pattern in water quality monitoring. Flow circulating sampling is widely used in poor or tanks in aquiculture and hydroponics. Flow circulating sampling module and water quality sensor module are fixed in main pipeline or bypass pipeline for water inlet or outlet. Float style sampling is widely used in water quality or pollutes monitoring of river, lake and reservoir [31-33]. Hydrological sensors, meteorological sensors, and biochemical sensors are always integrated with water quality sensors. Water logging style sampling is widely used in freshwater and sewage treatment and industrial wastewater treatment. The water logging size and the water exchange are depended with water treatment system. For water quality dynamic monitoring, ion selective electrodes as water quality sensor are widely used, other online sensors based on photoelectricity principle, fluorescence principle, scattering principle, fibre-optical principle, and bioinstrumentation principle are also used. For example, turbidity is measured dynamically by scattering principle; oil content of water is measured dynamically by ultraviolet-fluorescence analysis; nitrate content of water is measured dynamically by UV absorption method; chlorophyll and DO are measured dynamically by fluorescent method [34, 35]. Data collection module can be independent, or integrated with data processing module or communication module or both. Data collection and processing modules are to transfer analogue data measured by the water quality sensor into digital data. The monitoring data analysis and database store, uploads and download of data, remote operation functions are required in water quality dynamic monitoring system. Communication protocols are reported in RS-232, RS485, LAN, and CAN with wire or wireless [36, 37]. Security and efficient encryption protocols are used to improve the system security level. In the power supply module, 220V AC supplied for pumps and electromagnetic valves, 12V or 24V DC is supplied for water quality sensors.

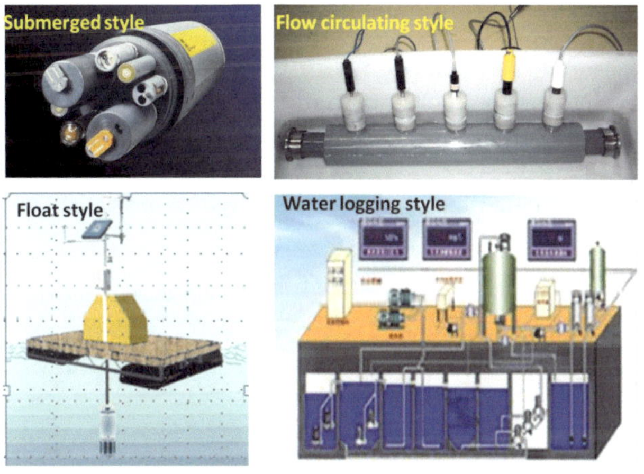

Fig. 3. Water sampling methods in water quality monitoring.

4.2 Water quality monitoring of Taihu Lake basin

4.2.1 Introduction of Taihu Lake basin

Taihu Lake is the third largest freshwater lake in China and covering Jiangsu, Zhejiang and Shanghai Province. The lake area is 2,338 square kilometers with 1.9 m mean depth of and 3.4 m greatest depth. Taihu lake basin including big cities of Wuxi, Suzhou, Changzhou, and Huzhou is about 3.65 millions square kilometers. It is also one of the most advanced industrial and agricultural areas in China. Water quality change of Taihu lake basin is related closely to surrounding water safe and Taihu regional economy development. Since the special geography heights decrease from west to east, excessive chemical fertilizer application caused soil nutrient excess, and a large number of industrial sewage, wastewater

urban drainage, water quality exacerbated and water pollution with nutrient-enrichment become to a serious problem in Taihu Lake basin.
In May to June, 2007, green algal blooms have broke out in Taihu Lake, thus source water quality deterioration directly influence freshwater quality to drink safety for surrounding threatened millions people. This emergency event was taken a great danger to human health and society economy. Water pollution in Taihu Lake is mainly due to nutrient enrichment with nitrogen and phosphorus. Past several years, water quality of Taihu Lake is always in the eutrophic condition [38]. Therefore, water quality monitoring and emergency warning in Taihu Lake basin has become an important problem for sustainable development of society and economy. In August 2009, water quality and disease monitoring platform around Taihu Lake basin funded by the China government was started in Wuxi, Suzhou, Changzhou and Huzhou city for water quality monitoring and disease warning [39-40].

4.2.2 Water quality and disease monitoring platform around Taihu Lake basin

Water quality monitoring in Taihu Lake basin is focus on source water and freshwater. In the water quality and disease monitoring platform around Taihu Lake basin, 15 instrumentation stations were deployed for source water quality monitoring and 14 instrumentation stations were deployed for freshwater quality monitoring (Fig. 4). Otherwise than water quality dynamic monitoring, standard measurement of water quality including algae amount, biomass of frustules, total nitrogen, total phosphorus, N/P ratio for source water, and chromaticity, turbidity, ammonia nitrogen, total phosphorus for freshwater were measured according same method at the first week of every month. On the other hand, shipping itineration detection with 15 sampling places is conducted once every month for source water quality monitoring by Wuxi Disease Control Center Cooperated with Wuxi Environmental Monitoring Station. Various water quality monitoring data will be update and issue to Jiangsu Province server for homepage publication. In addition, meteorological parameters such as wind direction, wind speed, air temperature, sunshine radiation rainfall are measured simultaneously.

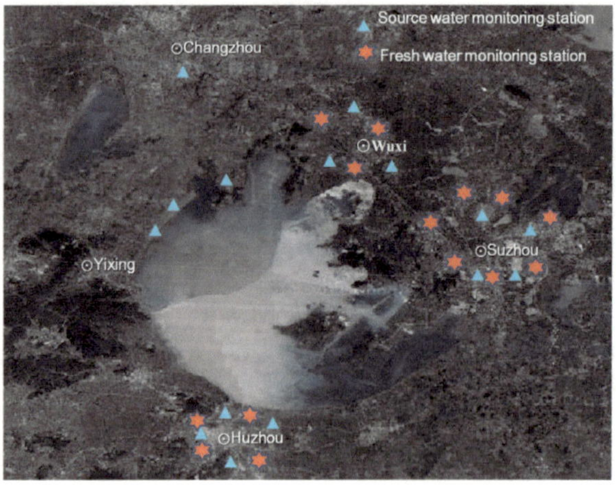

Fig. 4. Water quality monitoring in Taihu Lake basin for source water and fresh water.

Water quality and disease monitoring platform around Taihu Lake basin is a data-share-and-exchange platform developed under J2EE environment. The web interface with B/S construction is connected to Oracal database. All authenticated users can visit and manage the data. Using this platform, water quality change of source water and freshwater can be reported and alarmed automatically and clearly. Via the past two years operation, water quality parameters especially microcystin content were improved within national warning line

4.2.3 Water quality dynamic monitoring of Taihu Lake basin

Water quality dynamic monitoring in Taihu Lake basin based on multi-variables water quality monitoring instrument (YSI6820, YSI Inc., USA) were founded in 2004 for 13 invidious instrumentation station networked [41]. There are 13 stations of water quality/situation monitoring. It is succeed in monitoring and warning on water quality of Taihu Lake basin. Taihu Lake water situation monitoring determine water stage and rainfall with hydraulic remote-system stage gauge and rotary rain gauge every 5min. Water quality monitoring determine water temperature, pH, DO and ammonia nitrogen with electrode sensor every 5 min. There are 1 industrial personal computer, 1 I/O elevation model and 1 communication module in every station. Monitoring stations send data to the central station per hour, and examine data by central station or call phone at any time. The central station server can setup, dispose and synchronize monitoring station. There are 2 models to collect water sample: water quality sensor is set in water with PVC pipes surrounding; water quality sensor is set in water gauge that determining per 30 min. Water quality sensor applied with 12V D.C supply, sampling pump and other equipments applied with 220V A.C supply. Ammonia Nitrogen sensor has to be replaced every half a year, pH and DO sensor have to be replaced every year.

4.3 Water quality dynamic monitoring system based on web-server-embedded technology

4.3.1 Water quality monitoring problems in aquaculture production

Aquatic products play an important role in agricultural products. Aquaculture has becoming to an important industry in agriculture and the water quality monitoring is an important issue for improving aquatic production yield and its quality. Effective water quality dynamic monitoring are required for measuring and controlling water temperature, pH, DO, EC, and others in aquaculture production because of closer relationship between water quality parameters and aquatic organism (Fig. 5). In Europe, water quality variables were regulated by physical, chemical and biological methods to organize the aquaculture production by utilizing hazard analysis and critical control point operating rules. The computer-based process control technology proposed by United States for intensive aquaculture production was used in marine fish's growth environment control. Recently, water quality monitoring is trended to be smaller, more automatic and intelligent, and multi-functional automation. Although water quality monitoring technology has been researched in China at recent decade, there are still many problems such as the unstable sensor performance, less real-time monitoring parameters of water quality, low level of networked monitoring, and single system network monitoring. Many researches utilize new technology such as GPRS and neural networks to improve water quality monitoring level in China.

The aquaculture water pollution, aquatic production and quality issues, and the establishment of food safety traceability system make constructing effective dynamic monitoring network to obtain multi-parameters of water quality become an important issue in aquaculture. Distributed networking, real-time remote monitoring, the dynamic controlling of water quality using web-server-embedded technology and the sensor node technology are future trend development for intensive aquaculture production. In this chapter, a water quality dynamic monitoring system using ion selective electrodes and web-server-embedded technology was introduced for aquaculture production. The water quality information was transported safely using wireless communications technology based on CDMA services, WiFi and virtual private network (VPN) technology.

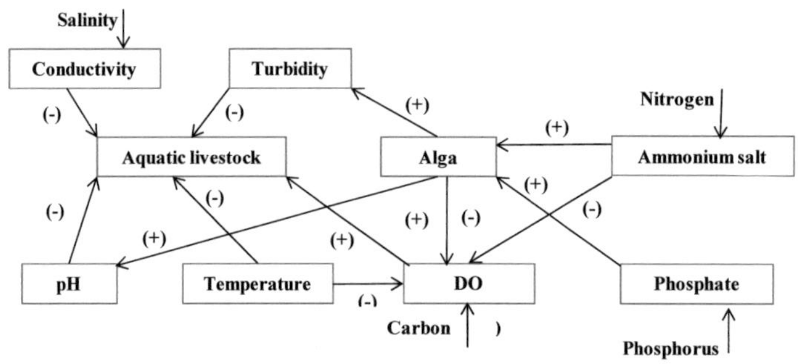

Fig. 5. Relationship between water quality parameters and aquatic organism.

4.3.2 Introduction of water quality dynamic monitoring system for aquaculture

The water quality dynamic monitoring system is composed of water quality dynamic monitoring devices and a remote information server. The water quality dynamic monitoring device consists of sensor module, data monitoring module, communication module, and power module (Fig. 6). The sensor module is composed of a water flow cell and water quality sensors including pH (ED201, Suzhou Han-star CO., China), EC (DJS-1T, Suzhou Han-star CO., China), DO (ED012, Suzhou Han-star CO., China), ORP (PC312, Suzhou Han-star CO., China), and temperature (STWB-1000, Beijing Saiyiling CO., China). The data monitoring module is a web-server-embedded chip (PICNIC2.0, TriState Co., Japan) to get water quality sensor signals. The communication module is composed of a Wi-Fi module (AirStation-G54, Buffalo Co., Japan), and a CDMA module with IPSec-based VPN function (InRouter210C, Beijing Inhand Co., China), for establishing wireless LAN network and achieving communication with local and remote information servers. The power module consists of 12V DC power supply and 220V AC power supply. The remote information server consists of an ADSL modem (DSL-300, D-Link Co., China), a VPN router (BV-601, Nesco Co., China), and personal PC as information server. The VPN router and information server connect to Internet via ADSL service. After the VPN router and CDMA device of the water quality dynamic monitoring device are connected to Internet, VPN connection is established via IPSec authentication.

The water quality dynamic monitoring system is an isolated local area network and can be used as a sensing network node. That means the system can be constructed to a large-scale wireless sensing network under CDMA signal covered areas. The system is easy to increase by 255 devices at most, and the monitoring device is also easy to increase sensor channels. Water quality data are transported by TCP/IP protocol and identified and stored in the remote information server. The CDMA module will be dynamically connected to the remote information servers via IPSec-based VPN security technology. In order to identify the specified remote information servers, the remote VPN router have to use a dynamic domain or a fixed global IP to support the remote VPN calling. Therefore, the water quality dynamic monitoring system deployed anywhere can be constructed to be a large monitoring network if the IPsec-based VPN tunnels were connected. Within the network, the information captured by all devices could be used as a local information network to conduct secure access. Therefore, all authorized users could visit or manage the remote sensing devices anywhere and anytime under Internet environment.

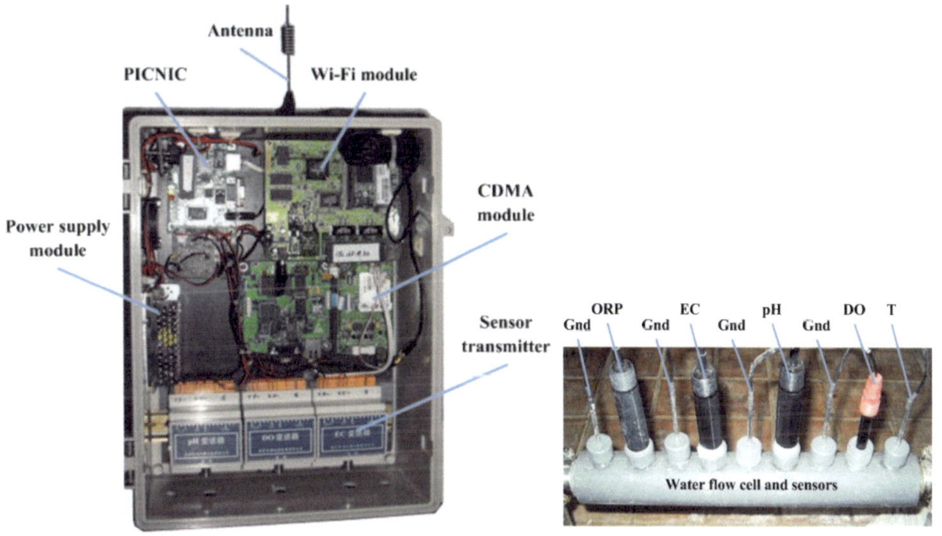

Fig. 6. Water quality dynamic monitoring system based on web-server-embedded technology.

4.3.3 Performance test of the water quality dynamic monitoring system

Two water quality dynamic monitoring systems were installed respectively in inlet and outlet in a seawater aquaculture farm (Fig. 7). The aquaculture production is using seawater and semi-circle mode for intensive aquiculture. In this testing, one remote information server with an IPsec-based VPN router is deployed in the China Agricultural University located in Beijing city. The testing experiments were conducted for three years.

The water quality data including pH, DO, EC, and water temperature throughout a week were dynamically storied or issued in webpage or Extensible Markup Language (XML) file by a special JAVA applet program in the remote information server (Fig. 8). In this testing, the 10-bit analog signals of the water quality sensors were obtained by the web-server-embedded chip without storage device. The water quality dynamic monitoring systems were communicated with the remote information server in Beijing by a 20-30 Kbps access speed with over 26-30 signal quality level of CDMA services. Under normal processing of Internet, 6 days data were randomly selected to analysis data loss via remote communication. As a result, high measuring accuracy of pH, DO, EC, and temperature and average packet loss rate is between 1.7 to 2.3% were reported. The water quality of seawater in inlet were relatively maintained in an appropriate level, but pH, DO and EC were changed greatly when rainfall. This is because the nature rainfall addition has decreased the EC and pH of seawater, thus led to low DO with drop in water temperature. In aquaculture production, input water has to be processed regularly via settling and filtered treatment in tanks, thus periodical water quality change is because of periodical water exchange of tank water. The growth and development of aquatic products will be affected by water quality change with rainfall and season change. Therefore, water quality dynamic monitoring is available for aquaculture production.

Fig. 7. Water quality dynamic monitoring systems installed in an aquaculture farm.

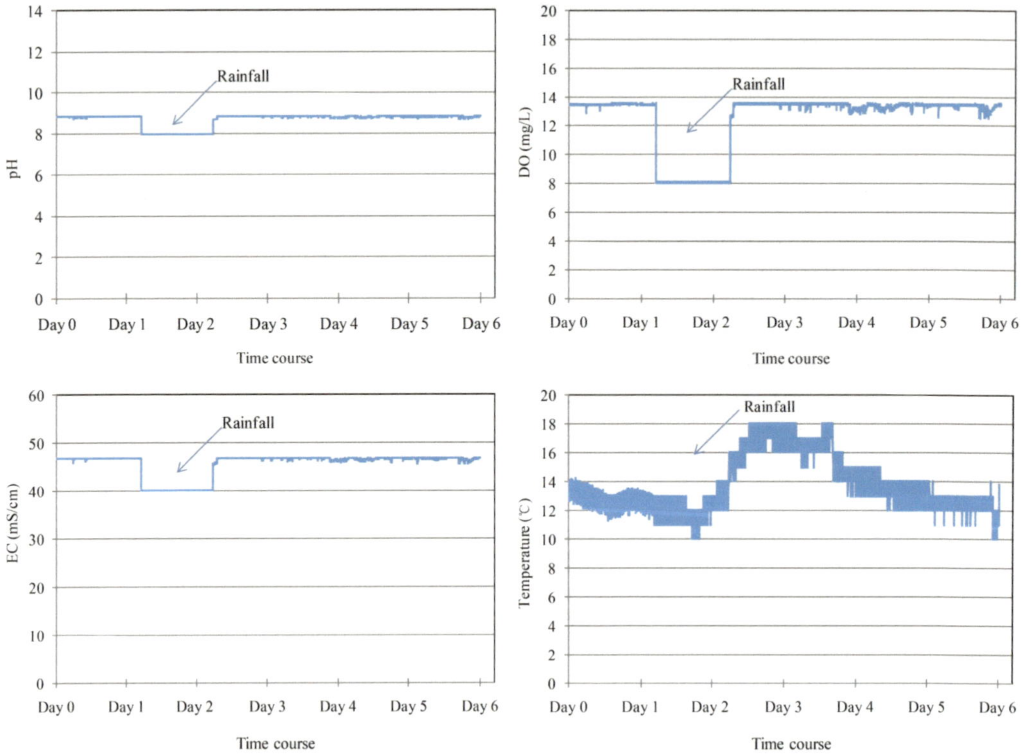

Fig. 8. Time course of water quality in the seawater aquaculture farm.

4.4 Water quality dynamic control system for hydroponic production

4.4.1 Water quality monitoring problems in hydroponic production

Cultivation of hydroponic plants as a clean and efficient plant production has been widely used in protected horticulture. Nutrient solution supply and control are important in hydroponic plant cultivation because of nutrient solution regulation is related directly to plant growth and development. Since nutrient solution is configured by mother solution periodly, nutrition solution is adjusted and replaced timely even popular used hydroponic production with deep flow technique (DFT) and nutrient film technique (NFT). The inorganic nutrition absorption of plants growth is dynamic and selective, so water quality of nutrient solution is considerably changed in the process of circulation or maintaining. In order to make stable water quality in nutrient solution, water quality of supply nutrient solution need be dynamically adjusted and controlled. In this chapter, a water quality dynamic control system using ion selective electrodes and web-server-embedded technology for hydroponic production is described.

Fig. 9. Hydroponic production with deep flow technique (DFT) and nutrient film technique (NFT).

4.4.2 Introduction of water quality dynamic monitoring system for hydroponics

Water quality dynamic control system for hydroponics consists of nutrient solution supply module and nutrient solution control module. The nutrient solution supply module includes supply solution box, return solution box, mother solution box, and cultivation groove. There are 4 pipes which are assembled 4 electromagnetic valves respectively between supply solution box and mother solution box for automatic adding 2 kinds of mother solution, acid solution, and alkaline solution. There is connect pipe and overflow pipe which are assembled a check valve between supply solution box and return solution box. The supply solution circulation between supply solution box and cultivation groove is forced by submersible pump. The solution circulation between cultivation groove and return solution box is operated by submersible pump or water level natural circulation. The water adding is automatic controlled by electromagnetic valve. Bypass is under the supply solution box contact with square circulation style sampling device to ssembl water quality sensors including temperature, pH, EC. There are supply solution pump, solution circulation pump, PTC heater and water level sensor in the supply solution box. Water level sensor is used for controlling high water level and alarming low water level. Nutrient solution control module consists of water quality transmitter module, data processing module, LAN hub module, and power supply module. Water quality transducer module make up of temperature transducer (PT1000, Hayashi Elect. Inc., Japan), pH tranducer (E201, Nanjing Chuandi Inc., China), and EC transducer (DJS-1, Nanjing Chuandi Inc., China). Data processing module is a web-server-embedded chip (PICNIC2.0, TriState Co., Japan) using TCP/IP protocol. The web browser can be used to communicate with the water quality dynamic control system in any local or remote area via Internet. LAN hub module is used connect more data processing modules. 220 V AC power supply and 12 V DC power supply were used in the power supply module. The water quality sensors and transducers are supplied by 12V DC, and submersible pump, electromagnetic valve, and other electrical equipment are supplied by 220V AC.

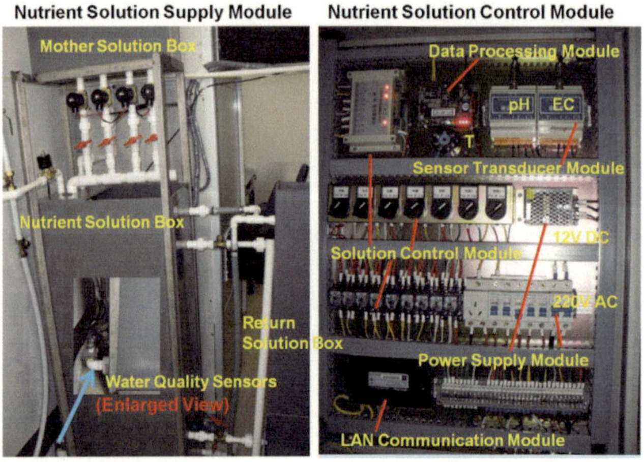

Fig. 10. Water quality dynamic control system for hydroponics.

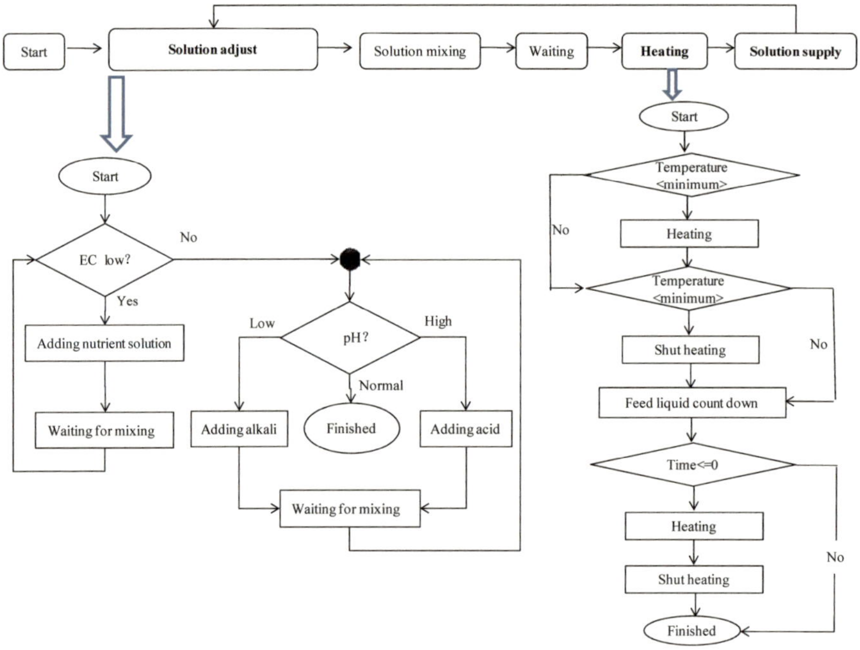

Fig. 11. Nutrient solution control process in the water quality dynamic control system for hydroponics.

The control process of nutrient solution in the water quality dynamic control system for hydroponics is always following the cycle as solution adjust, solution mixing, waiting, heating, solution supply (Fig. 11). Solution adjust process is following the cycle as water addition, mother solution addition, acid or alkaline solution addition. The mother solution includes A mother solution and B mother solution. The water addition and mother solution addition are used to adjust EC of the nutrient solution. The acid and alkaline solution additions are used to adjust the pH of the nutrient solution. Heating process is used after solution adjusts and before solution supply. Therefore, temperature, pH, EC, and water level of the nutrient solution is controlled automatically and dynamically by the water quality dynamic control system. Nutrient solution control algorithm can choose ON/OFF mode or prediction algorism. The control strategy can be specified by users. The water quality dynamic control system is easy to construct a water quality monitoring network because of the web-server-embedded technology. For networking convenience, the water quality dynamic control system is set to priority server control and the machine can also be run independently itself.

4.4.3 Performance test of water quality dynamic monitoring system for hydroponics

A small closed plant production system with artificial lighting is used to test the system performance. The closed plant production system is equipped with plant culture module, environmental control module, nutrient supply module, web-server-embedded environment/nutrition control module (Fig. 12). The hydroponic plant cultivation generally requires the preparation of concentrated stock solution (mother solution) and work nutrient solution. In order to prevent precipitation when the mother solution in the preparation, the compounds of the nutrient solution have to divide to two or three kinds because of precipitation formed when some ions mixed. Generally, the compound which reacts with calcium without precipitation is perpetrated as A mother solution, and the compound which react with phosphate without precipitation is perpetrated as B mother solution. In this test, the 200 times concentrated A and B mother solutions are used.

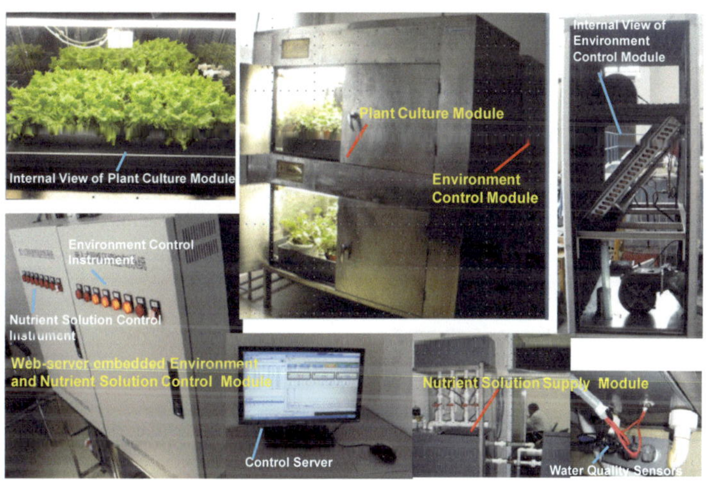

Fig. 12. Closed plant production system with artificial lighting.

When nutrient solution control target as: temperature 28°C, EC 4 mS/cm, pH 6.5, supply solution box begin dynamic control test while adding water, then nutrient solution control was cycle supplied after 1.5h from the start testing (Fig. 13). The total testing time was 2.5h. As a result, the regulation of nutrient levels would stable after 1h in dynamic controlling, and water quality parameters of nutrient solution can basically maintain at the target levels.

Because plant will selective absorb nutrient inorganic ions, the EC decreased significantly and there is little change in solution temperature and pH when the nutrient solution is started to supply and cycle for plants. The cycle supplied over a period of time, plants also tend to selectively absorb a dynamic equilibrium, then the EC levels of nutrient solution gradually dynamic control to the target level. A contract measurement is using a portable multi-parameter water quality instrument (HQ40, Hach Inc., USA) at 4:30 and 5:30 respectively, the solution temperature were 28.3℃ and 27.6℃, EC were 3.9 mS/cm and 4.3 mS / cm, pH were 6.7 and 6.8 respectively.

When the control target of nutrient solution as: temperature 28°C and pH 6.5 was unchanged and EC change from 4 mS/cm to 15 mS/cm, nutrient solution temperature and pH were basically stable after 30 min in the dynamic controlling (Fig. 14). EC stabled at the target levels after 1h after the target changed. After 20 min in the control target changed, solution temperature was 27.5°C±0.1°C; EC was 15 mS/cm±0.2 mS/cm; pH was 6.9±0.2. Therefore, the water quality dynamic control system can quickly correspond to the nutrient control change for different plant culture requirements.

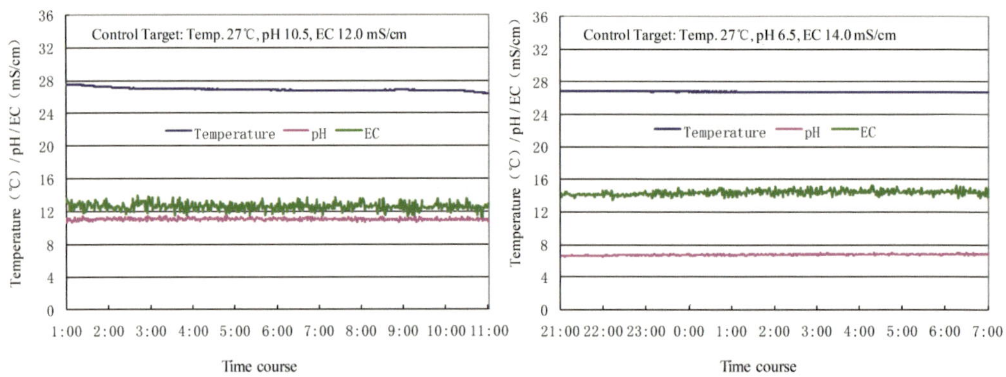

Fig. 13. Performance test of the nutrient solution dynamic control system for hydroponic plant production.

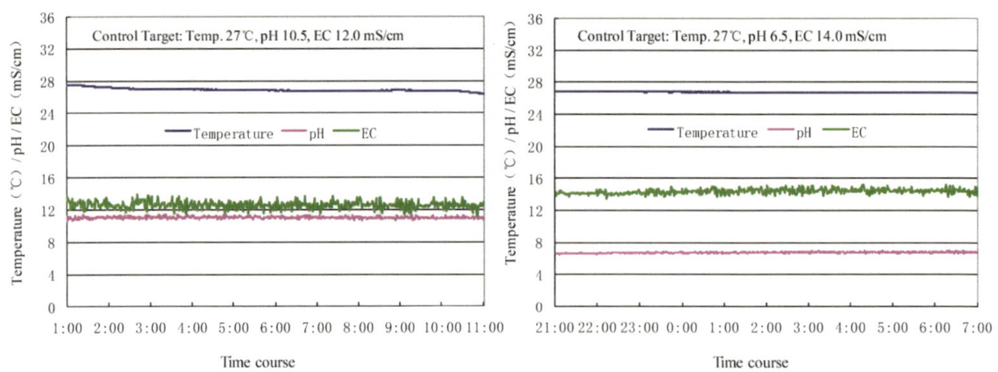

Fig. 14. Reliability test of the nutrient solution dynamic control system for hydroponic plant production.

In order for the 12h system stability testing, different level nutrient pH and EC were set up in two levels. The 10 h test data were resulted in temperature 26.9℃±0.2℃, pH 11.0±0.1, and EC 12.6±0.4 mS/cm for control target of temperature 27℃, pH 10.5, and EC 12 mS/cm; resulted in temperature 26.8℃±0.3℃, pH 6.7±0.1, EC 14.4±0.3 mS/cm for control target in temperature 27℃, pH 6.5, EC 14 mS/cm, respectively (Fig. 14).Therefore, the reliability of the water quality dynamic control system for hydroponics is available.

5. Conclusions

Water quality dynamic monitoring technology based on ion selective electrodes is an available toll for water quality monitoring in lake water and source water, in aquaculture and hydroponics production. Its system is used to achieve data collection in real time and data transmission in remote and wireless simultaneously. Except of the pH, EC, DO, ORP sensors, other water quality sensors in ion selective electrodes are not easy to apply in commercial applications. For the near future, development of water quality dynamic technology and application will be promoted by increasing commercialized ion selective electrodes.

6. Acknowledgments

Thanks Mr. Po Yang of Beijing Lighting valley Technology Company for his work and suggestions. These researches were supported by earmarked fund for Modern Agro-industry Technology Research System (CARS-25-D-04) and National advanced R & D Program of China (2006BAD10A02).

7. References

[1] Meng C. C., An Y. P. 2010. Recognition of water resources shortage risks. J. of North China Institute of Water Conservancy and Hydroelectric Power. 31(3): 5-7.

[2] Guo Z. Y., Yao H. L. 2011. A theoretical and empirical analysis of China's industrial waste water pollution. J. of Audit and Economics. 6(5): 97-103.

[3] Park S., Choi J. H., Wang S., Park S. S. 2006. Design of a water quality monitoring network in a large river system using the genetic algorithm. Ecological Modelling. 199: 289-297.

[4] Strobl R. O., Robillard P. D. 2008. Network design for water quality monitoring of surface freshwaters: A review. J. of Environ. Manage. 87: 639-648.

[5] Glasgow, H. B., Burkholder, J. M., Reed, R. E., Lewitus, A. J., Kleinman, J. E. 2004. Real-time remote monitoring of water quality: a review of current applications, and advancements in sensor, telemetry, and computing technologies. J. of Exp. Mar. Biol. Ecol. 300: 409-448.

[6] Halasz G., Szlepak E., Szilagyi E., Zagyva A., Fekete I. 2007. Application of EU Water Framework Directive for monitoring of small water catchment areas in Hungary, II Preliminary study for establishment of surveillance monitoring system for

moderately loaded (rural) and heavily loaded (urban) catchment areas. Microchemical J. 85: 72-79.

[7] Laszlo B., Szilagyi F., Szilagyi E., Heltai G., Licsko I. 2007. Implementation of the EU Water Framework Directive in monitoring of small water bodies in Hungary, I. Establishment of surveillance monitoring system for physical and chemical characteristics for small mountain watercourses. Microchemical Journal. 85: 65-71.

[8] Liu W. L., Dai J., Zhang J. D. 2011. Establishing drinking water monitoring warning system. Industrial Safety and Environmental Protection. 37(3):15-16.

[9] Champelya S., and Doledecb S. 1997. How to separate long-term trends from periodic variation in water quality monitoring. Water Research. 31(11): 2849-2857.

[10] Ouyang Y. 2005. Evaluation of river water quality monitoring stations by principal component analysis. Water Research. 39(12): 2621-2635.

[11] Huang D. P., Shen Z.S. 1982. Theory and application of ion-selective electrode. New times press, Beijing: 71-74.

[12] Zhu X. X., Wang D. J. 2007. Determination of ammonia-nitrogen in waste water by ion selective electrode-potentiometry. Physical Testing and Chemical Analysis Part B (Chemical Analysis). 43(4) : 283-284.

[13] Zou S. F., Men H., Wang P. 2004 .Current development of miniature electrochemical sensors. Chinese Journal of Sensors and Actuators. 2: 336-341.

[14] Gai P. P., Guo Z. Y. 2010. Progress in highly-sensitive polymeric membrane ion-selective electrodes. Chemistry. (12): 1080-1085.

[15] Zeng Y. H., Lin S.C. 2004. Analytical chemistry . Higher education press. Beijing: 247-279.

[16] Buek R. P., Lindner E. 1994. Recommendations for nomenclature of ion-selective electrodes. IUPAC Pure and Applied Chemistry. 66(12): 2527-2536.

[17] Solsky R. L. 1990. Ion-selective electrodes. Analytical Chemistry. 62(12): 21-33.

[18] Zhang Y. H. 1978. Ion-selective electrodes. Chinese Journal of Analytical Chemistry. 6(3): 213-223.

[19] Thomas J. D. R. 1986. Solvent polymeric membrane ion-selective electrodes. Analytica Chimica Acta. 180: 289-297.

[20] Mesner N. and Geiger J. June 2005. Understanding your watershed—pH. Utah State University Water Quality Extension.

[21] Magana N. and Laceya J.. 1984. Effect of temperature and pH on water relations of field and storage Fungi. Transactions of the British Mycological Society. 82(1): 71-81.

[22] [22] Tamai M. 1939. Studies on the antimony metal electrode for pH determinations. J. of Biochemistry. 29: 307-318.

[23] Dauer M. D., Rodi J. A., Ranasinghe J. A. 1992. Effects of low dissolved oxygen event on the macrobenthos of the lower Chesapeake Bay. Estuaries and Coasts. 15(3): 384-391.

[24] Cooper. C. M. 1992. Biological effects of agriculturally derived surface water pollutants on aquatic systems - A review. J. of Environmental Quality. 22(3): 402-408.

[25] Neugebauer C. A. and Webb M. B. 1962. Electrical conduction mechanism in ultrathin, evaporated metal films. J. of Applied Physics. 33(1): 74-82.

[26] Chapra S. C. 2008. Surface Water-quality modeling. Weveland Press, Illinois: 419-420.

[27] Magalhãesa M. C. S. J., Céspedesb F., Alegretc S., Machadoa A. A. S. C. 1997. Study of the temperature behaviour of all-solid-state nonactin ammonium electrodes with PVC membrane applied to graphite–epoxy supports of varied composition. Analytica Chimica Acta. 355(2-3): 241-247.

[28] Kishida N., Kim J., Chen M., Sasaki H., Sudo R. 2003. Effectiveness of oxidation-reduction potential and pH as monitoring and control parameters for nitrogen removal in Swine Wastewater treatment by sequencing batch reactors. Journal of Bioscience and Bioengineering. 96(3): 285-290.

[29] Sun B. Y., Ye Y. Z., Huanga H. W., Bai Y. 1993. Potentiometric determination of iron using a fluoride ion-selective electrode—the application of the apple II-ISE intelligent ion analyzer. Talanta. 40(6): 891-895.

[30] Konieczka P., Zygmunt B., Namiesnik J. 2000. Comparison of Fluoride Ion-Selective Electrode Based Potentiometric Methods of Fluoride Determination in Human Urine. Bulletin of Environmental Contamination and Toxicology. 64(6): 794-803.

[31] Wu C. H., Huang C. J., Zhang Z. Y. 2002. Application of water quality automatic monitoring system to aquiculture. Shanghai Environmental Sciences. 21(4) : 254-255, 261.

[32] Lou H. Q., Zhao G. Z. 2006. Design of regional water monitoring and measuring instrument based on embedded system. Mechanical and Electrical Engineering Magazine. 23(11) : 32-35.

[33] Charef A., Ghaucha A., Baussandb P., Martin-Bouyer M. 2000. Water quality monitoring using a smart sensing system. Measurement. 28(3): 219-224.

[34] Qin Z. L., Meng Q. H.. 1998. Sensor and their application in water quality monitoring. Environment Herald. 2 : 9-11.

[35] Zhu, M. R, Cao G. B., Jiang S. Y., Han S. C. 2006. Monitor and control the parameter of the industrial aquaculture. Journal of Fisheries of China. 19: 99-104.

[36] Telci, I. T., Nam, K., Guan, J., Aral, M. M. 2009. Optimal water quality monitoring network design for river systems. J. Environ. Manage. 90: 2987-2998.

[37] Yang W., Nan J., Sun D. Z. 2007. An online water quality monitoring and management system developed for the Liming River basin in Daqing, China. Journal of Environmental Management. 1-8.

[38] Gu S. L., Chen F., Sun J.L. 2011. Analysis of cyanobacteria monitoring and algal blooms in Taihu Lake. Water Resources Protection. 27(3): 28-32.

[39] Wang M. H., Shi W., Tang J. W. 2011. Water property monitoring and assessment for China's inland Lake Taihu from MODIS-Aqua measurements. Remote Sensing of Environment. 115(3): 841-854.

[40] Zhu Y., Wang W., Yu M. X., Zhang Z. H. 2009. The water quality for the bloom in Lake Taihu. Environmental Science and Technology. 22(6): 27-29.

[41] Hu G. D. 2004. R&D and implementation of automatic water quality monitoring system in Taihu Lake Basin. Automation in Water Resources and Hydrology. (4): 6-10.